# Environmentally Friendly Bio-Based Polymeric Materials

# Environmentally Friendly Bio-Based Polymeric Materials

Guest Editor

**Ana-Maria Manea-Saghin**

MDPI

Basel • Beijing • Wuhan • Barcelona • Belgrade • Novi Sad • Cluj • Manchester

*Guest Editor*
Ana-Maria Manea-Saghin
Research Center for
Environmental Protection and
Eco-Friendly Technologies
National University of
Science and Technology
Politehnica of Bucharest
Bucharest
Romania

*Editorial Office*
MDPI AG
Grosspeteranlage 5
4052 Basel, Switzerland

This is a reprint of the Special Issue, published open access by the journal *Polymers* (ISSN 2073-4360), freely accessible at: https://www.mdpi.com/journal/polymers/special_issues/TUP488136R.

For citation purposes, cite each article independently as indicated on the article page online and as indicated below:

| Lastname, A.A.; Lastname, B.B. Article Title. *Journal Name* **Year**, *Volume Number*, Page Range. |
| --- |

**ISBN 978-3-7258-7392-0 (Hbk)**
**ISBN 978-3-7258-7393-7 (PDF)**
**https://doi.org/10.3390/books978-3-7258-7393-7**

# Contents

# About the Editor

**Ana-Maria Manea-Saghin**

Ana-Maria Manea-Saghin is a Scientific Researcher at the National University of Science and Technology POLITEHNICA Bucharest, within the Research Center for Environmental Protection and Eco-Friendly Technologies (UPB-CPMTE), where she has been employed since 2011. She has been involved in research and managerial tasks for six national projects, two European Structural Funds, and eight European projects. She coordinated a postdoctoral project PD (2018–2020) as a Director and was responsible for a subsidiary contract POC-G (2019–2021). She obtained her Ph.D. in 2014 with a thesis on "*Nanostructured Materials Based on Natural Extracts with Potential Applications in Therapy and Biophotonics*". Her expertise includes the extraction and processing of plant extracts; instrumental techniques for complex natural mixtures; synthesis and characterization of nanostructured lipid carriers with vegetable biomolecules; thin films based on DNA and natural chromophores; linear spectroscopy (UV-VIS, IR, fluorescence); and nonlinear optical characterization of NLO materials by SHG and THG.

Her scientific output includes over 45 high-impact publications (more than 35 in ISI journals), two monographs, and a national patent on "*Procedure for obtaining agro-hydrogels based on collagen and starch*".

She has participated in many international conferences, including one keynote, 28 invited lectures, 18 oral presentations, and 20 posters. Two invited presentations received awards: First Prize (El Jadida, Morocco, 2012) and the Young Scientist Travel Grant Award (ICFPAM 2016, Daejeon, Korea) for outstanding research on DNA in polymer photonics.

# Preface

This Reprint brings together a selection of valuable papers that were published in the Special Issue "Environmentally Friendly Bio-Based Polymeric Materials" of the prestigious journal *Polymers*. It represents a curated collection of articles and reviews addressing key themes related to environmentally friendly polymeric materials.

This collection of papers falls within the scope of the utilization of bio-based and biodegradable raw materials. In this Reprint, it is once again demonstrated that sustainable development has become a demand in modern society due to environmental concerns and the concomitant decrease in fossil resources. Polymers constitute an inherent part of our daily lives, and renewable sources provide viable alternatives to conventional fossil-based raw materials, supporting sustainable development. Research shows that biopolymers can be manufactured by blending two polymers to develop new materials that exhibit properties that cannot be achieved using individual components alone. In addition, the nanostructure of that blend can be modified by adding an additive material, which enhances desirable properties.

Biodegradable polymers are a viable approach to addressing the issue of plastic pollution in marine ecosystems. In contrast to conventional plastics, these polymers are capable of rapidly decomposing over a brief span of a few months through natural mechanisms, such as enzymatic digestion, hydrolysis, oxidation, or mechanical degradation, thereby reducing their environmental impact.

This collection of thirteen scientific papers, united by the common topic of environmentally friendly materials, constitutes a significant contribution to scientific literature.

**Ana-Maria Manea-Saghin**
*Guest Editor*

polymers

MDPI

*Article*

# Innovative Collagen Based Biopolymers Tested as Fertilizers for Poor Soils Amendment

**Carolina Constantin [1], Daniela Simina Stefan [2], Ana-Maria Manea-Saghin [3,*] and Irina Meghea [4]**

1 Department of Inorganic Chemistry, Physical Chemistry and Electrochemistry, Faculty of Chemical Engineering and Biotechnologies, University Politehnica of Bucharest, 1-7 Polizu Str., 011061 Bucharest, Romania
2 Department of Analytical Chemistry and Environmental Engineering, Faculty of Chemical Engineering and Biotechnologies, University Politehnica of Bucharest, 1-7 Polizu Str., 011061 Bucharest, Romania
3 Research Center for Environmental Protection and Eco-Friendly Technologies, Faculty of Chemical Engineering and Biotechnologies, University Politehnica of Bucharest, 1-7 Polizu Str., 011061 Bucharest, Romania
4 Department of Mathematical Methods and Models, Faculty of Applied Sciences, University Politehnica of Bucharest, 313 Splaiul Independentei Str., 060042 Bucharest, Romania
* Correspondence: ana_maria.manea@upb.ro

**Abstract:** Improving soil quality is of growing interest and, among optimal solutions, the reuse and recycling of biopolymers of pelt waste from the tannery industry have been proposed, one of them being for collagen hydrolysate with micronutrients and polymers incorporated, to be used as fertilizers for poor soils rehabilitation. As functionalization agents, polyacrylamide, starch and dolomite were included into biopolymer matrixes in order to enhance their specific efficiency. These fertilizers were adequately characterized for their physical–chemical properties, including nutrient content, and tested on three poor soils, while a fourth sample of normal soil was chosen for comparative purposes. These soils were also characterized for their texture and physical–chemical properties in order to establish the fertility state of the soils as a function of nutrient content. In this respect, a series of agrochemical tests were developed at laboratory scale, simulating real agriculture environments in a vegetation room, where a significant plant growth in height was observed for all the agro-hydrogels with nutrients encapsulated, and multiplication of the nodosities number was observed in the case of the soybean culture. The most significant effect was obtained in the case of the fertilizer functionalized with starch. Finally, the application dose of the organic fertilizers for specific culture plants was estimated, such as field cultures (cereals, corn), field vegetables, vineyards or fruit-growing plantations. These agro-collagen fertilizers are particularly recommended for amendment of field cereals and vegetables. The novelty of this study mainly consists of the recovery and recycling of the pelt waste as efficient fertilizers after their adequate functionalization with synthetic or natural biopolymers.

**Keywords:** fertilizers; collagen hydrogels; agrochemical tests; poor soils rehabilitation; application dose on soil

## 1. Introduction

The continuous decline of soil fertility and crop productivity has resulted in an increasing interest to improve soil quality by adding various inorganic and organic amendments from different sources. A systematic study on using diverse biosolids in agriculture and their potential effects on soil and plant growth was reported by Sharma et al. [1]. Among the natural resources largely used for intensification of the bioproductivity of agricultural cultures, various natural zeolites, volcanic rock, gypsum and brown coal are mentioned [2–4], while biochar and straw have been applied in conjunction with chemical fertilizers [5]. Significant effects on growing soil fertility have been obtained by using organic composts with various types of waste, such as farmyard manure and crop residues [6], slaughterhouse waste [7], poultry litter [8], pig manure [9], spirulina from aquaculture waste [10] and micro-algal biomass [11]. Recently, an innovative poly-composite fertilizer was reported by recycling three organic wastes, sewage municipal waste, marine algae biomass and farmyard manure [12].

An alternative source to assure organic content is provided by the waste from the tannery industry, and significant research effort was devoted to produce protein composites with biochemical treatments using microorganism enzymes in order to obtain protein hydrolysates and protein binders with different uses. Organic biopolymers are a valuable source of raw materials for agriculture, as the protein waste matrix provides sufficient elements to improve the composition of poor and degraded soils, and many plants can benefit from the elements, such as carbon, nitrogen, calcium, magnesium, sodium and potassium [13,14].

Multicomponent, absorbent, hydrogel-type networks are next-generation materials with distinct three-dimensional structures and a high swelling capacity [15]. The applications of these materials have been diversifying in recent years, entering the fields of agriculture, food, pharmaceuticals, electrical devices and electronics [16], environmental protection and biomaterials.

Hydrolyzed collagen could be used in the food industry as a food supplement or in the cosmetic industry for skin care products [17,18]. By its functionalization with encapsulated nutrients, a product with applications as a biofertilizer in agriculture and horticulture has been obtained [19]. It is expected that such fertilizers will be cost-effective and available by recycling industrial units at prices at least five-times lower than similar marketable products, since there is no cost for the raw material. A quantitative estimation of the market cost of such fertilizers is in the range of 60–80 Euro/ton. Indeed, collagen can be recovered from pelt waste and used in various applications due to its high nitrogen content [20]. In this respect, the collagen obtained from wet blue leather waste after chromium extraction, for example, could be used as a nitrogen source and, after enriching with mineral potassium (K) and phosphorus (P), could be applied as NcollagenPK fertilizers for growing rice, soybean, sunflowers, maize, etc. [21–24], or as a potassium–phosphorus source from potato peel biochar [25].

Our previous research [26,27] developed new collagen composites produced by the incorporation of both mineral (phosphorus and potassium) nutrients and various synthetic and natural polymers into the collagen hydrogel matrix that was extracted with pelt waste processing in order to be proposed as fertilizers for the treatment of poor soils.

The novelty of the present work is focused on testing these innovative collagen composites for their fertilization qualities during application on three poor soil samples of Greek origin. For comparative purposes, a fourth Romanian normal soil sample is also tested. These soil samples are characterized with physical–chemical analyses, and the corresponding textural classes are assessed with specific granulometry measurements.

## 2. Materials and Methods

### 2.1. Materials

The synthesis of multi-polymeric, collagen-based agro-hydrogels was previously presented in our papers [26,27]. They were prepared using, as raw material, the limed hide waste (no haired) from the fleshing and trimming of bovine hides (lime fleshing) provided by SC PIELOREX tannery, Jilava, Ilfov county, Romania. Briefly, the gelatin hide obtained from collagen hydrolysate was further subjected to acid hydrolysis at 86–90 °C with 2.7–3.6% $H_2SO_4$ for 1.5–3.5 h in the presence of a $K_2HPO_4$ solution of 10–20% concentration, thus obtaining a reference fertilizer, denoted Ref-CH. In parallel, a polymeric solution of starch or polyacrylamide gel was prepared by continuous mixing, degasing and the addition of N,N′-methylene bis-acrylamide as the reticulant agent. Finally, the protein hydrolysate containing mineral nutrients encapsulated (phosphorus and potassium) was then functionalized with 5% of either a synthetic polymer—poly-acrylamide—resulting in the composite denoted POLY, or a natural polymer—starch—obtaining the composite STAR. In this way, we could compare the efficacy of a natural polymer, starch, with that of a synthetic polymer on the fertilization process. Another analysis was performed that compared the fertilization action of the two wastes, namely the organic biopolymer and starch, with a mineral additive, dolomite (prepared by adding 5% of dolomite suspension to Ref-CH), and thus obtaining the composite denoted DO. In this respect, while Ref-CH provided N-collagenPK nutrients only, the other additives incorporated into the collagen matrix provided supplemental contributions of nitrogen and organic carbon in the case of POLY, of organic carbon in the case of STAR, and of calcium and magnesium in the case of DO.

The plants selected for the tests on the poor soils were corn, sunflower and soybean for the following reasons.

Corn has good resistance to dry weather and many diseases and pests, good adaptability to various climate conditions and a good efficacy in the uptake of both organic and mineral fertilizers. The corn cob is a caryopsis containing starch, carbohydrates, protein substances, vitamins of the B group, E vitamin, iron, phosphorus, magnesium, zinc, potassium and oils.

Sunflower is one of the most widespread oleaginous plants; Romania occupies the fourth worldwide place after Ukraine, Russia and China. The sunflower seeds contain 50–54% fats, which are largely industrialized as oils, while the waste is used as animal fodder.

Soybeans contain over 30% protein substances and 17–25% oils. Its main role is to ameliorate the physical features of the soil as a result of installing symbiosis between the reticular system and nitrogen-fixing bacteria. In this way, many nodosities are developed to enhance the atmospheric nitrogen uptake and thus confer increased soil fertility.

### 2.2. Physical–Chemical Analyses of Soils and Harvested Crops

The determination of the particle size fractions of the soils was carried out with the pipette method for fractions < 0.002 mm, the wet sieving method for fractions 0.002 to 0.2 mm and the dried method for fractions > 0.2 mm [28].

The pH was determined with the potentiometric method in aqueous suspension of soil: water ratio 1:2.5 with a glass–calomel combined electrode; the potentiometer is calibrated with buffer solutions of known pH prior to the analysis of samples.

The organic matter (humus), OC%, was determined with the wet oxidation-modified Walkley–Black method.

The total nitrogen (TN%) was determined with the Kjeldahl method: digestion with $H_2SO_4$ at 350 °C, potassium sulfate and copper sulfate catalysts.

Exchangeable (accessible, mobile) phosphorus (EP%) was determined with extraction followed by spectrophotometry (CINTRA 404 UV–VIS spectrometer) with molybdenum blue after reduction with ascorbic acid [28].

Exchangeable (accessible, mobile) potassium (EK%) was determined with extraction followed by flame photometry (Flame Photometer S-935).

The alkaline soil carbonates of the air-dried soil fraction (particle size < 2 mm) were estimated with the calculation of measured carbon dioxide from carbonates decomposed with hydrochloric acid (1:3)—STAS 7184/16–80.

The total phosphorus was estimated with spectrophotometry (CINTRA 404 UV–VIS spectrometer), and the total potassium was estimated with flame photometry (Flame Photometer S-935) [29].

The four soil samples were thoroughly investigated according to ecopedologic indicators that were determined based on physical–chemical analyses: granulometry, pH, organic carbon (OC), total nitrogen (TN), exchangeable phosphorus (EP), exchangeable potassium (EK), carbonates ($CO_3^{2-}$) and heavy metals content.

The chemical analyses of the soils are detailed in Tables 1 and 2.

**Table 1.** Chemical analysis of soils.

| | pH | OC, % | TN, % | EP, mg/kg | EK, mg/kg | $(CO_3)^{2-}$, mg/kg |
|---|---|---|---|---|---|---|
| S1-L | 8.30 | 2.41 | 0.12 | Nd | 50 | 37.5 |
| S2-CL | 8.27 | 2.81 | 0.11 | Nd | 128 | - |
| S3-SiCL | 8.50 | 2.28 | 0.09 | Nd | 42 | 54.3 |
| S4-SCL | 8.30 | 2.96 | 0.19 | 17 | 144 | 1.26 |

OC—organic carbon, TN—total nitrogen, EP—exchangeable phosphorous, EK—exchangeable potassium, Nd—below detection limit.

**Table 2.** Metal content in tested soils, mg/kg.

| Soil Sample | Essential Nutrients | | Potential Toxic Metals | | | | | | | |
|---|---|---|---|---|---|---|---|---|---|---|
| | Cu | Zn | Cr | Co | Ni | Mn | Pb | Cd | As | Hg |
| S1-L | 47 | 82 | 53 | 15.4 | 107 | 940 | 16.4 | 0.233 | 5.54 | 0.032 |
| S2-CL | 63 | 108 | 63 | 24.8 | 139 | 1710 | 18.3 | 0.125 | 8.11 | 0.025 |
| S3-SiCL | 39 | 110 | 46 | 12.1 | 82 | 743 | 17.1 | 0.268 | 3.79 | 0.025 |
| S4-SCL | 32 | 64 | 48 | 10.8 | 22 | 978 | 8.67 | 0.087 | 0.18 | 0.010 |
| Alertthreshold * | 100/250 | 300/700 | 100/300 | 30/100 | 75/250 | 1500/2000 | 50/250 | 3/5 | 15/25 | 1/4 |

* Sensitive/less sensitive use of lands—types of land uses, which involve a certain quality of soils, characterized by a maximum accepted level of pollutants.

According to ecopedologic indicators [30], these four soil samples are classified as follows:

- All the four soils fall in the weak alkaline reaction class with low organic matter;
- S1-L and S2-CL are in the class with low total nitrogen contents;
- S3-SiCL and S4-SCL are in the class with very low total nitrogen contents;
- S4-SCL is in the class with low exchangeable phosphorus;
- S1-L and S3-SiCL are in the class with very low exchangeable potassium;
- S2-CL is in the class with low exchangeable potassium;
- S4-SCL is in the class with medium exchangeable potassium;
- S1-L, S3-SiCL and S4-SCL belong to the class with low content of calcium carbonate with values below 1%.

Analysis of the data from Table 1 reveals, for the three samples S1-L, S2-CL and S3-SiCL, a very low nutrient content, phosphorous being not detectable, while for the sample S4-SCL, the data are specific for a fertile soil having a higher NPK nutrient content.

In Table 2, the metal content in the tested soils is quantified with special reference to essential metal nutrients and potential toxic metals. For this last category, one may observe that the metal contents are below the limit of environmental requirements except for nickel and manganese, which are within the alert zone for the sensitive soils in the case of the clay loam sample S2-CL.

The definitions in accordance with Romanian regulations [31] are the following:

- The sensitive use of lands is represented by their use for residential and leisure areas, for agricultural purposes, as protected areas or sanitary areas with restricted regime, as well as the land areas provided for such uses in the future;
- The less sensitive land use includes all existing industrial and commercial uses as well as the land areas provided for such future uses.

The alert thresholds are the concentrations of pollutants in air, water, soil or in emission/discharges, that have the role of alerting the competent authorities to a potential impact on the environment and that trigger the additional monitoring and/or reduction in pollutant concentrations in emission/evacuations.

Total heavy metals (Cd, Co, Cr, Cu, Ni, Mn, Pb, Zn, As and Hg) were quantified with atomic absorption spectrometry (GBC 932 AA atomic absorption spectrometer) using a calibration standard solution for each determined metallic ion.

The texture of the soils has been examined and the type of them is obtained using the triangular diagram of the basic soil textural classes according to USDA particle sizes from [32], page 182, Figure 8.13 as the location inside a ternary system built with the granulometric fractions in the corners: amounts of clay, silt and sand. According to these data, the following textural classes are established: soil S1 belongs to textural class L—loam; soil S2 belongs to textural class CL—clay loam; soil S3 belongs to textural class SiCL—silty clay loam; and soil S4 belongs to textural class SCL—sandy clay loam. In line with these findings, our tested soils are further coded as follows: S1-L, S2-CL, S3-SiCL and S4-SCL. More information can be found in the Supplementary Materials, Tables S1 and S2.

The leaves harvested from the tested crops (sunflowers, corn and soybean) were first dried at ambient temperature, and then the dehydration was finished in the oven followed by grounding to a powder and, finally, sampling for chemical analyses.

### 2.3. Evaluation of Exchangeable NPK Nutrients of Agro-Hydrogels in Soils

Soil that was poor in nutritive elements sampled from the Western Greece region was placed in an aluminum vessel and mixed with the agro-hydrogel as liquid, dried or gel in various final proportions: 1%, 5%, 10%, 30%, 50%. Portions of these mixtures were sampled at different time intervals (e.g., 1, 3, 7, 14, 21 and 30 days) and analyzed for ammonium leaching. Changeable ammonium was evaluated by elution of the soil mixture with a solution 2.0 M KCl, and ammonium nitrogen was determined with a modified Kjeldal method.

### 2.4. Experiments in the Vegetation Room

The agrochemical experimentations were carried out in the vegetation room belonging to the National Research and Development Institute for Pedology—ICPA Bucharest. The seeds of green pea, sunflower, corn and soybean were sowed during the springtime (end of March) in plastic vegetation pots containing 1 kg of the tested soils sampled from Greece (two types of poor soils with very low nutrient contents characterized according to the data shown in Section 3.1). Both soils (S1-and S3-SiCL) were pretreated (amended) with 5% agro-hydrogels (Ref-CH, POLY and STAR) 4 months before the cultivation for a good homogenization of the agro-hydrogels added with soil and for adaptability of the existing microorganisms in the soils. The soil humidity in the vegetation pots was ensured at a permanent 70% level of the field capacity. According to the testing methodology, three replicates (three plants per single replication for the beginning and one month after rising and growth) were ensured for each experimental variant. The experiment was finished in September.

### 2.5. Estimation of Dose for Application of Organic Fertilizer

Agrochemical tests were conducted in the vegetation room on various plant species by monitoring the parameters related to the chemical composition of the soils, chemical composition of the plants and composition of microflora in the soil. Finally, the stage of plant growth was evaluated in order to assess which type of crops was most appropriate for the different soils.

The optimal dose of organic fertilizers in ton/ha of cultivated terrain was calculated based on the experiments in the vegetation room by applying a special formula and taking into account the humus content in the soil and the base saturation degree to evaluate the nitrogen index (IN), clay content and total nitrogen of the soil according to Equation (1) [29]:

$$\text{DOF, t/ha} = (a + b/IN) \times (c - d/Ag) \times (e/Ng) \quad (1)$$

where:
DOF = Dose of Organic Fertilizer, t/ha;
a, b, c, d empiric parameters with the following values:
a = 15 for field crops and fruit-growing plantations; 20 for field vegetables and vineyards;

b = 30 for field crops and field vegetables; 40 for vineyards; 50 for fruit-growing plantations;
c = 1.35 for all cultures;
d = 8 for all cultures;
IN = nitrogen index of soil (% humus × V%)/100;
Ag = clay content of soil (%);
e = medium standard content of total nitrogen (%Nt) of the organic fertilizer = 0.45% from wet mass;
Ng = total nitrogen content of the organic fertilizer, % from wet mass.

### 2.6. Field Experiments

Two of the fertilizers, S1-L and S3-SiCL, were incorporated into the soils, and the soil humidity was permanently maintained at 70% from the soil capacity. These tests were performed from the spring sowing until the autumn harvesting period.

### 2.7. Statistical Analyses

The agrochemical results represent the average (mean) of three replicates for soil and plant, respectively, and the standard deviation values {STDEV} have been calculated. The mean summarizes an entire dataset with a single number representing a typical value. It is one of several measures of central tendency. STDEV measures the amount of variability or dispersion, indicating how close the individual data values are from the mean value.

The factorial experimental design is also introduced, and the analysis of variance (ANOVA) is used as one of the primary tools for the statistical data analysis. There is a three-factor factorial experiment, the factors being the following: the first factor is for the considered four situations for the soils (no fertilizer soil, and the three soils treated with Ref-CH, POLY and STAR, respectively); the second factor is for two considered soils, i.e., loam soil S1-L and silty clay soil S3-SiCL; and the third factor is for the three cultivated plants taken into account in this discussion—corn, sunflower and soybean. Therefore, there is a total number of 4 × 2 × 3 × 3, the last factor representing the number of replicates.

## 3. Results and Discussion

### 3.1. Physical–Chemical Characteristics of the Fertilized Soils

The general evaluation of the fertility state of the soils was performed as a function of the soluble nutrient content available for plant growth [33], as shown in Table 3.

**Table 3.** General evaluation of the fertility state of the soil as a function of the soluble nutrient content available for plant growth.

| Fertility State | Nitrogen (mg/kg) $NO_3^- + NH_4^+$ | Phosphorus, Aqueous Extract (mg/kg), $P_2O_5$ | Potassium, Aqueous Extract (mg/kg), $K_2O$ |
|---|---|---|---|
| Low to reduced | <10 | <10 | <50 |
| Medium | 10–30 | 11–20 | 50–80 |
| High | >30 | >20 | >80 |
| S1-L | 12 | Nd | 50 |
| S2-CL | 11 | Nd | 128 |
| S3-SiCL | 9 | Nd | 42 |
| S4-SCL | 9 | 17 | 144 |

Nd—below detection limit.

When these values are compared with the nutrient content of our soils, one may conclude that, for nitrogen content, the three soils, S1-L—S3-SiCL, are at the limit of very low fertility, and phosphorus is not detectable, while for potassium, only sample 2 corresponds to a good fertility state. That is why the new NPK agro-hydrogels proposed in this work could provide a promising source of nutrients to enhance the fertility state of these poor soils.

Three fertilizers were tested, including potassium phosphate incorporated as P–K nutrients source and considered as the reference sample together with its poly-composites obtained by functionalization with polyacrylamide, dolomite and starch. Their action on the soil samples was demonstrated by a series of agrochemical tests simulating real environmental conditions in the vegetation room for four culture plants. Finally, the dose of organic fertilizer to be applied for various agriculture crops in the field was estimated.

In Table 4, the results of the physical–chemical analyses for the fertilized soils are presented.

**Table 4.** Physical–chemical analyses, mg/g of dried fertilized soils.

| Parameters | S1-L | S2-CL | S3-SiCL | S4-SCL |
|---|---|---|---|---|
| pH | 7.89 | 7.25 | 7.64 | 7.52 |
| Temperature at sampling (°C) | 2 | 4 | 10 | 14 |
| Total phosphorous | 0.012 | 0.011 | 0.015 | 0.014 |
| $PO_4^{3-}$ | 0.040 | 0.042 | 0.044 | 0.043 |
| $Cl^-$ | 7.5 | 7.5 | 7.3 | 7.4 |
| $Cl_2$ free | 1.0 | 0 | 0 | 0 |
| $NO_3^-$ | 10 | 9.0 | 11 | 10 |
| $NO_2^-$ | 0.02 | 0.02 | 0.03 | 0.05 |
| $S^{2-}$ | 0.021 | 0.017 | 0.028 | 0.029 |
| $SO_4^{2-}$ | 0.021 | 0.016 | 0.028 | 0.030 |
| $CO_2$ free | 5.17 | 2.53 | 1.54 | 2.2 |
| $NH_4^+$ | 0.001 | 0.001 | 0.01 | 0.05 |
| Total dissolved salts | 30.8 | 33.2 | 36.9 | 38.0 |

The average stem heights after one month from sowing of the three plant cultures are shown in Table 5.

**Table 5.** The average stem heights of the cultivated plants after one month from sowing (cm).

| Experimental Variant | Loam Soil, S1-L | | | Silty Clay Soil, S3-SiCL | | |
|---|---|---|---|---|---|---|
| | Corn | Sunflower | Soybean | Corn | Sunflower | Soybean |
| No fertilizer | 20 | 18 | 15 | 18 | 16 | 13 |
| Treated with Ref-CH | 42 | 34 | 26 | 41 | 35 | 23 |
| Treated with POLY | 44 | 38 | 26 | 40 | 37 | 25 |
| Treated with STAR | 45 | 42 | 28 | 42 | 38 | 26 |

From the analysis of these data, one may observe that, in all variants using agro-hydrogels with nutrients encapsulated, a significant plant growth in height was obtained, accompanied by a well-developed and profound reticular system, vigorous and healthy plantlets and multiplication of the nodosities number in the case of the soybean culture. The most important effect was produced by the agro-collagen hydrogel functionalized with starch (STAR).

These remarkable fertilizing effects are in good agreement with the monitoring results on the nutrients' inclusion into the hydrogel matrix and their release towards the cultivated plants for the two tested poor soils (Table 4).

The mean values considered in Table 5 can be considered in good agreement with our experiment, since we discovered interactions between the variables as revealed in Table 6, which contains the ANOVA results.

**Table 6.** Analysis of variance table for the three-factor, fixed-effects model.

| ANOVA | (Balanced Design) | | |
|---|---|---|---|
| Factor | Type | Levels | |
| Treatment | fixed | 4 | |
| Soil | fixed | 2 | |
| Plant | fixed | 3 | |
| Analysis of Variance for Stem Height | | | |
| Source | DF | F | P |
| Treatment | 3 | 18.69 | 0.003 |
| Soil | 1 | 4.33 | 0.049 |
| Plant | 2 | 1.26 | 0.045 |
| Treatment*Soil | 3 | 3.10 | 0.056 |
| Treatment*Plant | 6 | 0.03 | 0.067 |
| Soil*Plant | 2 | 0.64 | 0.071 |
| Treatment*Soil*Plant | 6 | 2.08 | 0.188 |
| Error | 24 | | |
| Total | 47 | | |

* Represents interaction of the analysis factors.

The *F*-ratios for all three main effects and the interactions are formed by dividing the mean square for the effect of interest by the error mean square. Since we selected $\alpha = 0.05$, the critical value for each of these *F*-ratios is $f_{0.05}(18,1) = 4.41$. We could use the *p*-value approach. The *p*-values for all the test statistics are shown in the last column of Table 6. An inspection of these *p*-values is revealing. We observe that the most important factor is the treatment applied. In addition, the type of plant and the type of soil are significant. However, there is some indication of an effect due to the interactions of the treatments and soils, treatment and plants, and soil and plants, since the *p*-values are 0.056, 0.067 and 0.071, respectively, which are not much greater than $\alpha = 0.05$.

### 3.2. Comparison of Functionalized Agro-Hydrogels Tested in Real Simulated Conditions

Agro-hydrogels containing starch and dolomite embedded into a collagen matrix with phosphorus and potassium nutrients incorporated were first tested on the normal soil S4-SCL for a green pea culture in order to ameliorate the field by soil enriching in biologically fixed nitrogen, which allows early clearing of the field to be further prepared for barley sowing. It is well established that a green pea culture needs nitrous fertilizers, mainly during the first development stage. Afterwards, this culture grows on account of the nitrogen uptake from the atmosphere through the bacteria formed in the root nodosities. The most appropriate soil reaction for a green pea culture is in the neutral pH domain, which requires amending the soils with lime, not directly, but by means of prior plant cultures.

In this respect, during the flourish phenological phase of the green pea culture, Diana variety, the number of nodosities on the roots were counted per plant on soil treated with 0.25–0.5 $kg/m^2$. For the untreated soil, 17 nodosities per plant were counted; for the soil treated with hydrogel functionalized with starch (STAR), this number was 39; for dolomite (DO), this number was 37 (refer to Figure 1).

One may conclude that these collagen hydrogels functionalized with natural compounds have a high agrochemical efficacy, the best effect being obtained for the soil treated with the fertilizer containing starch.

Other agrochemical tests performed by using leather industry solid waste have been previously carried out for growth of the common bean [20] or rice [22], but only the effect of collagen as an organic nitrogen source has been investigated. Instead, our research is more complex, as these original smart fertilizers are able to provide all three main nutrients, NPK, while starch brings a supplementary contribution of organic carbon.

**Figure 1.** Monitoring plant cultures tested on poor soils in simulated field experiments.

Another series of three agro-hydrogels with phosphorus and potassium nutrients incorporated, (Ref-CH)—as reference fertilizer, POLY—Ref-CH with polyacrylamide additive as the functionalization agent, and STAR—Ref-CH with starch incorporated, were extensively studied when applied on two Greek soils, S1-L—loam type and S3-SiCL—silty clay type soil, in order to increase their nutrient content.

As one can observe from the previous data of Table 3, the soils S1-L and S3-SiCL are characterized by a very low content of total nitrogen and extremely low content of exchangeable potassium, while the content in exchangeable phosphorus is not detectable. These experiments aimed at using the above-mentioned fertilizers on the two poor soils in order to stimulate the germination processes, accelerate the productive phase (plantlet growth), vigorously develop the plants, profoundly root the plants and strengthen the self-defense system of the plants.

The NPK content in the tested plants, sunflowers, corn and soybean, cultivated on the two Greek soils (S1-L and S3-SiCL) after one month from sowing, is quantified in Table 7.

**Table 7.** Nutrient content in plants cultivated on soils S1-L and S3-SiCL, mg/kg.

| Agro-Collagen | Soil Not Treated | | | Ref-CH | | | POLY | | | STAR | | |
|---|---|---|---|---|---|---|---|---|---|---|---|---|
| **Nutrient** | N | *p* | K | N | *p* | K | N | *p* | K | N | *p* | K |
| | | | | | **Loam Soil S1-L** | | | | | | | |
| **Sunflower** | 0.09 | Nd | 35 | 0.156 | 0.08 | 40 | 0.180 | 0.11 | 38 | 0.217 | 0.12 | 38 |
| **Corn** | 0.11 | Nd | 40 | 0.178 | 0.09 | 42 | 0.205 | 0.12 | 40 | 0.223 | 0.12 | 45 |
| **Soybean** | 0.12 | Nd | 42 | 0.216 | 0.11 | 44 | 0.245 | 0.14 | 41 | 0.246 | 0.13 | 48 |
| | | | | | **Silty clay Soil S3-SiCL** | | | | | | | |
| **Sunflower** | 0.05 | Nd | 28 | 0.142 | 0.07 | 35 | 0.168 | 0.09 | 35 | 0.203 | 0.10 | 37 |
| **Corn** | 0.08 | Nd | 35 | 0.152 | 0.08 | 37 | 0.189 | 0.10 | 37 | 0.214 | 0.11 | 42 |
| **Soybean** | 0.11 | Nd | 40 | 0.253 | 0.09 | 40 | 0.218 | 0.11 | 40 | 0.232 | 0.12 | 44 |

Nd—below detection limit.

A comparison of the monitoring results obtained on the treatment of these two poor soils showed that the nutrient leaching into the soil and plants was more efficient for the loam soil S1-L than for silty clay soil S3-SiCL, as expected [34], and the most significant effect was observed on soybean.

### 3.3. Estimation of Application Dose for Various Types of Culture Plants

This study demonstrated that the new agro-hydrogels belong to the category of organic GREEN fertilizers. In order to extend their applicative area, a complementary investigation was needed to select the most appropriate plant cultures with higher technical economic efficiency. Based on the experiments carried out in the vegetation room, the potential of application on various types of plant cultures, such as field cultures (cereals, corn), field vegetables, vineyards or fruit-growing plantations, was evaluated, and the optimal application dose by hectare was estimated using the Equation (1) defined in Section 2.5.

The results on organic fertilizer dose (DOF) in t/ha, estimated for the three agro-collagen hydrogels applied on the three poor soils, are listed in Tables 8–10.

**Table 8.** Estimated Dose of Organic Fertilizer, DOF, for Ref-CH agro-hydrogel in t/ha.

| No Crt | Soil Sample | | Field Cultures (Cereals, Corn) | | | Field Vegetables | | | Fruit-Growing Plantations | | | Vineyards Plantations | | |
|---|---|---|---|---|---|---|---|---|---|---|---|---|---|---|
| | | | N | $P_2O_5$ | $K_2O$ | N | $P_2O_5$ | $K_2O$ | N | $P_2O_5$ | $K_2O$ | N | $P_2O_5$ | $K_2O$ |
| 1 | Soil 1 loam | DOF | | 36.79 | | | 37.80 | | | 87.41 | | | 71.55 | |
| | | NPK | 0.82 | 0.50 | 0.79 | 0.84 | 0.52 | 0.81 | 1.95 | 1.20 | 1.88 | 1.6 | 0.98 | 1.54 |
| 2 | Soil 2 Clay loam | DOF | | 46.46 | | | 47.64 | | | 110.84 | | | 90.56 | |
| | | NPK | 1.03 | 0.64 | 1.0 | 1.06 | 0.65 | 1.03 | 2.47 | 1.52 | 2.39 | 2.02 | 1.24 | 1.95 |
| 3 | Soil 3 Silty clay loam | DOF | | 55.08 | | | 56.24 | | | 132.47 | | | 107.83 | |
| | | NPK | 1.23 | 0.75 | 1.19 | 1.25 | 0.77 | 1.21 | 2.9 | 1.81 | 2.85 | 2.4 | 1.48 | 2.32 |

**Table 9.** Estimated Dose of Organic Fertilizer, DOF, for POLY agro-hydrogel in t/ha.

| No Crt | Soil Sample | | Field Cultures (Cereals, Corn) | | | Field Vegetables | | | Fruit-Growing Plantations | | | Vineyards Plantations | | |
|---|---|---|---|---|---|---|---|---|---|---|---|---|---|---|
| | | | N | $P_2O_5$ | $K_2O$ | N | $P_2O_5$ | $K_2O$ | N | $P_2O_5$ | $K_2O$ | N | $P_2O_5$ | $K_2O$ |
| 1 | Soil 1 loam | DOF | | 46.03 | | | 47.29 | | | 109.37 | | | 89.52 | |
| | | NPK | 0.82 | 0.57 | 0.83 | 0.84 | 0.59 | 7.0 | 1.95 | 1.36 | 1.97 | 1.6 | 1.11 | 1.61 |
| 2 | Soil 2 Clay loam | DOF | | 58.13 | | | 59.60 | | | 138.67 | | | 113.30 | |
| | | NPK | 1.03 | 0.72 | 1.05 | 1.06 | 0.74 | 1.07 | 2.47 | 1.73 | 2.5 | 2.02 | 1.41 | 2.04 |
| 3 | Soil 3 Silty clay loam | DOF | | 68.91 | | | 70.36 | | | 165.73 | | | 134.91 | |
| | | NPK | 1.23 | 0.86 | 1.24 | 1.25 | 0.88 | 1.27 | 2.9 | 2.06 | 3.0 | 2.4 | 1.68 | 2.43 |

**Table 10.** Estimated Dose of Organic Fertilizer, DOF, for STAR agro-hydrogel in t/ha.

| No Crt | Soil Sample | | Field Cultures (Cereals, Corn) | | | Field Vegetables | | | Fruit-Growing Plantations | | | Vineyards Plantations | | |
|---|---|---|---|---|---|---|---|---|---|---|---|---|---|---|
| | | | N | $P_2O_5$ | $K_2O$ | N | $P_2O_5$ | $K_2O$ | N | $P_2O_5$ | $K_2O$ | N | $P_2O_5$ | $K_2O$ |
| 1 | Soil 1 loam | DOF | | 25.60 | | | 26.31 | | | 60.84 | | | 49.80 | |
| | | NPK | 0.82 | 0.37 | 0.68 | 0.84 | 0.38 | 0.70 | 1.95 | 0.89 | 1.62 | 1.6 | 0.73 | 1.33 |
| 2 | Soil 2 Clay loam | DOF | | 32.33 | | | 33.15 | | | 77.13 | | | 63.02 | |
| | | NPK | 1.03 | 0.47 | 0.86 | 1.06 | 0.48 | 0.88 | 2.47 | 1.13 | 1.67 | 2.02 | 0.92 | 1.67 |
| 3 | Soil 3 Silty clay loam | DOF | | 38.33 | | | 39.14 | | | 92.19 | | | 75.04 | |
| | | NPK | 1.23 | 0.56 | 1.02 | 1.25 | 0.57 | 1.04 | 2.95 | 1.35 | 2.45 | 2.4 | 1.1 | 1.99 |

Once again, we confirmed that the most efficient among the composite fertilizers tested was the collagen hydrogel functionalized with starch (STAR), and for the poor tested soils, the application dose varied between 25 t/ha on the loam soil cultivated with field cultures and over 92 t/ha for the silty clay soil cultivated with fruit-growing plantations. However, for economic reasons, these agro-collagen fertilizers are particularly recommended for amendment of field cereals and vegetables in the case of all poor soils tested.

## 4. Conclusions

This study focused on the systematic testing of the agrochemical features of three composite fertilizers based on collagen hydrolysate, incorporating both mineral nutrients (phosphorus and potassium) and natural or synthetic additives, denoted as follows: Ref-CH—collagen hydrogel matrix containing $KH_2PO_4$ encapsulated as the reference sample, POLY—Ref-CH functionalized with poly-acrylamide, and STAR—Ref-CH functionalized with starch.

In this way, we compared the efficacy of a natural polymer, starch, with that of a synthetic polymer on the fertilization process. Another analysis was performed by comparing the fertilization action of the two wastes, namely the organic biopolymer, starch, with a mineral additive, dolomite, denoted DO.

These composite fertilizers were tested in a vegetation room to follow the evolution of four plants from sowing to harvesting time: green pea, corn, sunflower and soybean. A comparison of the monitoring results obtained on the treatment of these soils showed that the nutrient leaching into the soil and plants was more efficient for the loam soil than for the silty clay soil, and the most significant effect was observed on soybean. In all variants using agro-hydrogels with nutrients encapsulated, a significant plant growth in height was obtained, and multiplication of the nodosities number was obtained in the cases of the green pea and soy cultures.

All the results were statistically analyzed based on the ANOVA test.

The most important effect was produced by the agro-collagen hydrogel functionalized with starch, which can be successfully applied in a dose of 25 t/ha for the amendment of all poor soils tested. Moreover, this biopolymer waste has the advantage of being cost-effective and biodegradable when applied on soil.

**Supplementary Materials:** The following supporting information can be downloaded at: https://www.mdpi.com/article/10.3390/polym15092085/s1, Table S1: Absolute locations of sampling points for Greek and Romanian soils; Table S2: Soil texture.

**Author Contributions:** Conceptualization, C.C. and A.-M.M.-S.; methodology, C.C., I.M. and A.-M.M.-S.; software, C.C. and D.S.S.; validation, I.M. and A.-M.M.-S.; formal analysis, C.C.; investigation, C.C., I.M., D.S.S. and A.-M.M.-S.; resources, A.-M.M.-S.; data curation, C.C. and D.S.S.; writing—original draft preparation, C.C., I.M., D.S.S. and A.-M.M.-S.; writing—review and editing, A.-M.M.-S.; visualization, C.C.; supervision, I.M. and A.-M.M.-S.; project administration, A.-M.M.-S.; funding acquisition, A.-M.M.-S. All authors have read and agreed to the published version of the manuscript.

**Funding:** This research was partially funded by University Politehnica of Bucharest.

**Institutional Review Board Statement:** Not applicable.

**Data Availability Statement:** Not applicable.

**Acknowledgments:** The Romanian authors acknowledge the support of the Romanian Ministry of Research through the UEFISCDI organism under Contract Number 11/2018, Code Project ERANET-INCOMERA-AGRO-SMARTGEL.

**Conflicts of Interest:** The authors declare no conflict of interest.

## References

1. Sharma, B.; Sarkar, A.; Singh, P.; Singh, R.P. Agricultural utilization of biosolids: A review on potential effects on soil and plant grown. *Waste Manag.* **2017**, *64*, 117–123. [CrossRef]
2. Eprikashvili, L.; Zautashvili, M.; Kordzakhia, T.; Pirtskhalava, N.; Dzagania, M.; Rubashvili, I.; Tsitsishvili, V. Intensification of bioproductivity of agricultural cultures by adding natural zeolites and brown coals into soils. *Ann. Agrar. Sci.* **2016**, *14*, 67–71. [CrossRef]
3. Ramos, C.G.; Querol, X.; Dalmora, A.C.; De Jesus Pires, K.C.; Schneider, I.A.H.; Oliviera, L.F.S.; Kautzmann, R.M. Evaluation of the potential of volcanic rock waste from Southern Brazil as a natural soil fertilizer. *J. Clin. Prod.* **2017**, *142*, 2700–2706. [CrossRef]
4. Walia, M.K.; Dick, W.A. Soil chemistry and nutrient concentrations in perennial rye grass as influenced by gypsum and carbon amendments. *J. Soil Sci. Plant Nutr.* **2016**, *16*, 832–845.
5. He, L.L.; Zhong, Z.K.; Yang, H.M. Effects on soil quality of biochar and straw amendment in conjunction with chemical fertilizers. *J. Integr. Agric.* **2017**, *16*, 704–712. [CrossRef]

6. Choi, W.J.; Kwak, J.H.; Lim, S.S.; Park, H.J.; Chang, S.X.; Lee, S.M.; Arshad, M.A.; Yun, S.I.; Kim, H.Y. Synthetic fertilizer and livestock manure differently affect δ15N in the agricultural landscape: A review. *Agric. Ecosys. Environ.* **2017**, *237*, 1–15. [CrossRef]
7. De Albuquerque Nunes, W.A.G.; Menezes, J.F.S.; De Melo Benites, V.; De Lima Junior, S.A.; Dos Santos Oliveir, A. Use of organic compost produced from slaughterhouse waste as fertilizer in soybean and corn crops. *Sci. Agric.* **2015**, *72*, 343–350. [CrossRef]
8. Watts, D.B.; Torbert, H.A.; Prior, S.A.; Huluka, G. Long-term tillage and poultry litter impacts soil carbon and nitrogen mineralization and fertility. *Soil Sci. Soc. Am. J.* **2010**, *74*, 1239–1247. [CrossRef]
9. Kang, Y.; Hao, Y.; Shen, M.; Zhao, Q.; Li, Q.; Hu, J. Impacts of supplementing chemical fertilizers with organic fertilizers manufactured using pig manure as a substrate on the spread of tetracycline resistance genes in soil. *Ecotoxicol. Environ. Saf.* **2016**, *130*, 279–288. [CrossRef]
10. Wuang, S.C.; Khin, M.C.; Chua, P.Q.D.; Luo, Y.D. Use of Spirulina biomass produced from treatment of aquaculture wastewater as agricultural fertilizers. *Algal Res.* **2016**, *15*, 59–64. [CrossRef]
11. Nastac, M.; Negreanu-Pirjol, B.S.; Negreanu-Pirjol, T.; Meghea, A.; Gheorghiu, K.A.; Resteanu, A.N. Multicomposite Biologic Fertilizer. Romanian Patent RO131272 A2, reg. a2015 00061, 29 July 2016.
12. Lacatusu, A.-R.; Meghea, A.; Rogozea, A.-E.; Constantin, C. Innovative polycomposite obtained by recycling and processing three organic wastes. *Agriculture* **2021**, *11*, 1021. [CrossRef]
13. Bajza, Z.; Vrucek, V. Thermal and enzymatic recovering of proteins from untanned leather waste. *Waste Manag.* **2001**, *21*, 79–84. [CrossRef] [PubMed]
14. Serrano, A.R.; Maldonado, V.M.; Kosters, K. Characterization of waste materials in tanners for better ecological uses. *J. Am. Leather Chem. Assoc.* **2003**, *98*, 43–48.
15. Abd El-Rehim, H.A.; Hegazy, E.-S.A.; Abd El-Mohdy, H.L. Effect of various environmental conditions on the swelling property of PAAm/PAAcK superabsorbent hydrogel prepared by ionizing radiation. *J. Appl. Polym. Sci.* **2006**, *101*, 3955–3962. [CrossRef]
16. Rong, Q.; Wenwei, L.; Liu, M. Conductive hydrogels as smart materials for flexible electronic devices. *Chem. A Eur. J.* **2018**, *24*, 16930–16943. [CrossRef]
17. Yazaki, M.; Ito, Y.; Yamada, M.; Goulas, S.; Teramoto, S.; Nakaya, M.-A.; Ohno, S.; Yamaguchi, K. Oral Ingestion of Collagen Hydrolysate Leads to the Transportation of Highly Concentrated Gly-Pro-Hyp and Its Hydrolyzed Form of Pro-Hyp into the Bloodstream and Skin. *J. Agric. Food Chem.* **2017**, *65*, 2315–2322. [CrossRef]
18. Yorgancioglu, A.; Bayramoglu, E.E. Production of cosmetic purpose collagen containing antimicrobial emulsion with certain essential oils. *Ind. Crop. Prod.* **2012**, *44*, 378–382. [CrossRef]
19. Zainescu, G.A.; Albu, L.; Constantinescu, R. Study of Collagen Hydrogel Biodegradability Over Time. *Rev. Chim.* **2018**, *69*, 101–104. [CrossRef]
20. Lima, D.Q.; Oliveira, L.C.A.; Bastos, A.R.R.; Carvalho, G.S.; Marques, J.G.S.M.; Carvalho, J.G.; de Souza, G.A. Leather Industry Solid Waste as Nitrogen Source for Growth of Common Bean Plants. *Appl. Environ. Soil Sci.* **2010**, *2010*, 703842. [CrossRef]
21. Nogueira, F.G.E.; Prado, N.T.; Oliveira, L.C.A.; Bastos, A.R.R.; Lopes, J.H.; Carvalho, J.G. Incorporation of mineral phosphorus and potassium on leather waste (collagen): A new NcollagenPK-fertilizer with slow liberation. *J. Hazard. Mater.* **2010**, *176*, 374–380. [CrossRef]
22. Nogueira, F.G.E.; Castro, I.A.; Bastos, A.R.R.; Souza, G.A.; Carvalho, J.G.; Oliveira, L.C.A. Recycling of solid waste rich in organic nitrogen from leather industry: Mineral nutrition of rice plants. *J. Hazard. Mater.* **2011**, *186*, 1064–1069. [CrossRef]
23. Zainescu, G.; Constantinescu, R.; Sirbu, C. Smart hydrogels with collagen structure made of pelt waste. *Rev. Chem.* **2017**, *68*, 393–395. [CrossRef]
24. Constantinescu, R.R.; Zainescu, G.; Stefan, D.S.; Sirbu, C.; Voicu, P. Protein biofertilizer development and application on soybean cultivated degraded soil. *J. Leather Footwear* **2015**, *15*, 169–178. [CrossRef]
25. Vaitkevičienė, N. A comparative study on proximate and mineral composition of coloured potato peel and flesh. *J. Sci. Food Agric.* **2019**, *99*, 6227–6233. [CrossRef]
26. Tzoumani, I.; Lainioti, G.C.; Aletras, A.J.; Zainescu, G.; Stefan, S.; Meghea, A.; Kallitsis, J.K. Modification of collagen derivatives with water-soluble polymers for the development of cross-linked hydrogels for controlled release. *Materials* **2019**, *12*, 4067. [CrossRef] [PubMed]
27. Stefan, D.S.; Zainescu, G.; Manea-Saghin, A.-M.; Triantaphyllidou, I.-E.; Tzoumani, I.; Tatoulis, T.I.; Syriopoulos, G.T.; Meghea, A. Collagen-based hydrogels composites from hide waste to produce smart fertilizers. *Materials* **2020**, *13*, 4396. [CrossRef]
28. Stoica, E.; Rauta, C.; Florea, N. *Methods for Chemical Analysis of Soil*; ICPA: Bucharest, Romania, 1988; 487p. (In Romanian)
29. Madjar, R.; Davidescu, V. *Agrochimie*; Faculty of Horticulture, University of Agronomic Sciences and Veterinary Medicine Bucharest: Bucharest, Romania, 2009.
30. Dumitru, M.; Manea, A.; Ciobanu, C.; Dumitru, S.; Vrinceanu, N.; Calciu, I.; Tanase, V.; Preda, M.; Risnoveanu, I.; Mocanu, V.; et al. *Soil Quality Monitoring in Romania 2011*; SITECH Publishing: Craiova, Romania, 2011.
31. Ministry of Waters, Forests and Environmental Protection. *Romanian Legislation, Order 756/1997 for the Approval of the Regulation Regarding the Assessment of Environmental Pollution*; Ministry of Waters, Forests and Environmental Protection: Bucharest, Romania, 1997.

32. Dondeyne, S. IUSS Working Group. WRB. 2022. World Reference Base for Soil Resources. International Soil Classification System for Naming Soils and Creating Legends for Soil Maps. 4. Available online: https://www.academia.edu/85684582/Citation_IUSS_Working_Group_WRB_2022_World_Reference_Base_for_Soil_Resources_International_soil_classification_system_for_naming_soils_and_creating_legends_for_soil_maps_4 (accessed on 19 January 2023).
33. Borlan, Z.; Rauta, C. *Agrochemical Soil Analysis Methodology in Order to Establish the Necessary Amendments and Fertilizers*; ICPA: Bucharest, Romania, 1981.
34. Rusu, M.; Marghitas, M.; Oroian, I.; Mihaiescu, T.C.; Dumitras, A. Treatease of Agrochemistry. *Ed. Ceres Bucuresti* **2005**.

polymers

MDPI

Article

# Influence of Chitosan and Grape Seed Extract on Thermal and Mechanical Properties of PLA Blends

Victoria Goetjes [1], Claudia L. von Boyneburgk [1], Hans-Peter Heim [1] and Marilia M. Horn [2,*]

[1] Institute of Material Engineering, Polymer Engineering, University of Kassel, Mönchebergstr. 3, 34125 Kassel, Germany
[2] Physical Chemistry of Nanomaterials, Institute of Chemistry and Center for Interdisciplinary Nanostructure Science and Technology (CINSaT), University of Kassel, Heinich-Plett Straße 40, 34109 Kassel, Germany
* Correspondence: mhorn@uni-kassel.de

**Abstract:** Blends based on polylactic acid (PLA), chitosan, and grape seed extract (GE) were prepared by extrusion and injection molding. The effect of chitosan (5% and 15% on PLA basis) and natural extract (1% on PLA basis) incorporated into the PLA host matrix was explored regarding the thermal and mechanical properties. GE showed antioxidant activity, as determined by the DPPH assay method. Chitosan and GE affect the degree of crystallinity up to 30% as the polysaccharide acts as a nucleating agent, while the extract reduces the mobility of PLA chains. The decomposition temperature was mainly affected by adding chitosan, with a reduction of up to 25 °C. The color of the blends was specially modified after the incorporation of both components, obtaining high values of b* and L* after the addition of chitosan, while GE switched to high values of a*. The elongation at break (EB) exhibited that the polysaccharide is mainly responsible for its reduction of around 50%. Slight differences were accessed in tensile strength and Young's modulus, which were not statistically significant. Blends showed increased irregularities in their surface appearance, as observed by SEM analysis, corresponding to the partial miscibility of both polymers.

**Keywords:** chitosan; PLA; grape seed extract; DPPH assay

## 1. Introduction

Sustainable development has become a demand in modern society due to environmental concerns and the concomitant decrease in fossil resources. Polymers are an inherent part of our daily life. Renewable sources provide polymers for the replacement of conventional fossil-based raw materials aiming at sustainable development [1]. Research shows that biopolymers can be manufactured by blending two polymers to develop new materials that exhibit properties that could not be achieved by using each individual polymer. In addition, the nanostructure of that blend can be modified by adding additive material, which will enhance desirable properties.

Thermoplastic polymers exhibit low process temperatures compared to thermosets, which is advantageous for the development of packaging materials. Polylactic acid (PLA) is one of the most promising thermoplastics, as it is considered a fully sustainable and biodegradable polymer [2]. Its production is based on the fermentation of dextrose, which is converted into lactic acid, followed through polymerization. After disposal, the PLA-based materials are degraded into the water, and carbon dioxide is consumed to grow more agricultural products [3]. Nevertheless, toughness and strong hydrophobic properties limit PLA for packaging and biomedical applications. Strategies to overcome this limitation are usually related to employing methods such as copolymerization [4] or the addition of fillers [5]. Nevertheless, the copolymerization process involves the use of chemicals or generates unsafe subproducts. Then, a sustainable approach in the development of ecofriendly PLA materials includes the use of reinforcement materials, such as flax fibers [6], cellulose nanofiber [7], silk fibroin [8], or montmorillonite clay [9]. Alternatively, simply blending with another polymer is one of the most effective solutions. Pradeep et al. [10] investigated the use of PLA/poly (butylene succinate-co-adipate) (PBSA) and observed an increment in the crystallinity of the composites. In a study of blend filaments of PCL/PLA, the authors showed the effect on tensile strength regarding the concentration of PCL [11]. In our study, the aim was to combine PLA with a biopolymer that should provide appropriate mechanical and morphological properties without losing its biodegradable behavior. In that concern, chitosan is a polymer that fits those conditions.

Chitosan is obtained from the chitin deacetylation process in the presence of a strong alkali [12]. Besides its biodegradability properties, chitosan is considered a renewable and green polymer, as it is derived from waste sources in the fish industry [13]. In this regard, this polysaccharide also has promising potential for the packaging industry, improving hydrophilic capacity and adding ductility to the blends.

The association of both renewable polymers has already been described in the literature. Suyatma et al. [14] described the preparation of PLA/chitosan films by the casting process and observed an increase in both tensile strength and elongation at break associated with the rise of chitosan concentration. Bonilla et al. [15] studied the PLA/chitosan mixture by extrusion, but only a low amount of the polysaccharide was considered in the research. Claro et al. [16] compared PLA/chitosan and PLA/cellulose acetate films and found that plasticizers are not necessary to produce blends with potential use applications.

Our study addresses the development of PLA and chitosan blends using the extrusion process and injection molding process, which is known as an efficient technique for incorporating chitosan into PLA host matrixes [15]. Additionally, the traditional casting method for film preparation could not be advantageous, as solvent residues can be left in the final product due to the difficulty of controlling the solvent evaporation process [17].

Grape seed extract mainly consists of flavonoids, tocopherol, and other phenolic compounds responsible for its antioxidant activity [18]. Then, the grape seed extract was chosen as the additive active material of the blends. Recently, Wang et al. [17] described the preparation of films of poly(ε-caprolactone)/chitosan loaded with grapefruit seed extract and found an application as a packaging material. Nevertheless, poly(ε-caprolactone) is not considered a renewable source polymer. At this point, our contribution is to describe a blend composed of two renewable and biodegradable polymers (PLA and chitosan) associated with an additive (grape seed extract) to improve the antioxidant features without compromising the mechanical and thermal characteristics of these blends.

## 2. Materials and Methods

### 2.1. Materials

Polylactic acid (PLA) (grade: 3052D Ingeo, L-lactic acid/D-lactic acid: 96/4 and MM = 116,000 g $mol^{-1}$) used in this study was supplied by NatureWorks LLC, Plymouth, MA, USA. Low-molecular-weight chitosan was purchased from Sigma Aldrich (DA 15% and MM 120,000 g $mol^{-1}$). The grape seed extract was obtained from Buxtrade GmbH (Germany), and it mainly contained oligomeric proanthocyanidins, as specified by the supplier.

### 2.2. Antioxidant Properties of Grape Seed Extract

The free radical scavenging activity of the grape seed extract was evaluated by 1,1-diphenyl-2-picryl-hydrazil (DPPH) following the method adapted from [18]. The DPPH solution was prepared in ethanol and mixed with a diluted extract solution. The volume was adjusted with ethanol to a final volume of 5 mL. After incubation in the dark for 30 min at room temperature, the absorbance was measured at 517 nm against ethanol as blank using a UV Perkin Elmer Lambda 900. The scavenging activity of the sample was compared with a control (DPPH solution and ethanol). The absorbance of the control and the extract sample was measured in triplicate, and the activity was calculated according to Equation (1).

$$DPPH\ scavening\ activity\ (\%) = \left[\frac{(A_c - A_s)}{A_c}\right] \times 100 \quad (1)$$

where $A_c$ is the absorbance of the control solution and $A_s$ is the absorbance of the extract sample solution.

### 2.3. Extrusion

PLA granules and chitosan powder were dried at 40 °C overnight to remove the absorbed moisture. Both biopolymers were mixed in a beaker in different ratios of PLA/CHI (95/05 and 85/15), namely P95C5 and P85C15, respectively. A control sample of PLA was prepared to allow for a comparison effect of chitosan addition in the binary blends. Blends containing the dye molecule, grape seed extract, were prepared following the same ratio of PLA and chitosan. A fixed amount of 10 mg was used in our study. Then, samples were labeled as PLAGE, P95C5GE, and P85C15GE, where the suffix "GE" means the addition of the grape seed extract.

Each mixture was extruded in a twin-screw extruder (Haake Rheomix, ThermoFischer, Dreieich, Germany), and the screw speed and processing temperature were maintained at 50 rpm and 160 °C, respectively. The recycle time was set up in 5 min. The extruded strips were collected and stored in a humidity control chamber (50% and 23 °C).

### 2.4. Differential Scanning Calorimetry (DSC)

The thermal properties of PLA and the extruded samples with or without grape seed extract were determined by differential scanning calorimetry (Perkin-Elmer DSC 7). The glass transition temperature ($T_g$) and the intrinsic degree of crystallinity ($X_c$) as a function of the composition changes (chitosan and grape seed extract addition). The samples were heated from 20 to 200 °C at a rate of 5 °C min$^{-1}$. All sample weights were between 6 and 8 mg.

The onset temperature of the glass transition was determined as $T_g$ in the software of the equipment. The $X_c$ (%) was calculated from Equation (2):

$$Xc(\%) = \frac{\Delta Hm}{w \times \Delta H^{\circ}m} \times 100 \quad (2)$$

where $\Delta H_m$ is the measured melting enthalpy, w is the mass fraction of PLA in the composite, and $\Delta H^{\circ}{}_m$ is the corresponding enthalpy of 100% PLA crystalline polymer (93 J g$^{-1}$).

### 2.5. Thermogravimetry (TGA)

TGA analyses were performed on a Perkin-Elmer, Pyris Diamond TG/DTA. Thermogravimetric curves were performed under a synthetic air atmosphere. Samples of approximately 8 mg were loaded into a platinum crucible and heated from 25 to 800 °C at a heating rate of 10 °C min$^{-1}$.

### 2.6. Tensile Tests

Using the extrudate produced as previously described, shoulder bars of type 5A were manufactured with the aid of a miniature injection molding machine (Haake MiniJet, ThermoFischer, Dreieich, Germany). Tensile tests were carried out using a Zwick Z010 from Zwick & Roell according to the standard DIN EN ISO 527 at a test speed of 2 mm/min. Tensile strength, maximum elongation at break, and Young's modulus were evaluated. At least three measurements with each blend (17.5 × 4 mm) were performed. Tensile strength (TS) was expressed in megapascals (MPa) and calculated by dividing the maximum load (N) by the specimen's initial cross-sectional area (m$^2$). EB was determined by the ratio of the final length (sample rupture) to the initial length of the specimen and expressed as a percentage. Young's modulus was evaluated by the slope of the linear portion of the stress–strain curve.

Tensile test results were statistically treated using analysis of variance (ANOVA) and Tukey's test with a significance level set at 5%. The test was performed with the Origin software and results were presented as the mean value and the standard deviation.

### 2.7. Color Measurements

The injection-molded test specimens were subjected to color spectroscopy. Color changes were examined by spectroscopy, based on the CIE L*, a*, and b* color space, in which L* represents the lightness from 0 (black) to 100 (white), a* express green/red, and b* depict the yellow–blue dimension. The measurement was performed on an Ultra Scan Pro (Hunterlab, Reston, VA, USA) in a replicate assessment. The results were plotted as a function of the color space diagram.

### 2.8. Scanning Electron Microscopy (SEM)

The morphology of the blends was examined using scanning electron microscopy (SEM). Approximately 0.5 $cm^2$ sample size was placed in the stubs and covered using a thin layer of palladium-gold to improve their conductivity properties. The coating was performed in a Sputter coater from Polaron, and a thickness of 10 nm was deposited. The SEM images were acquired in an S-4000 equipment from Hitachi.

## 3. Results and Discussion

### 3.1. DHPP Antioxidant Assay

Grape seeds are the main by-products of juice and wine production, considered a waste material. Nevertheless, they consist of approximately 60–70% of polyphenols in their composition, a valuable compound with potent antioxidant activity by effectively scavenging free radicals. The free radical scavenging activity of the grape seed extract was tested using the DPPH method through the change of absorbance due to the reduction in the DPPH radical.

The inhibition percentage of the DPPH radical, based on the concentration used in the blends, was 78.12 ± 5.03%. A previous study described values ranging from 83% and 99% [10], in which the values differ due to the method of extraction, the species of grape, and even the season of cultivation [19]. The antioxidant activity of the grape seed extract was shown to be effective, and its addition to the blends improves their functionality against free radicals.

### 3.2. Differential Scanning Calorimetry (DSC)

Extrusion is a convenient method to prepare environmentally friendly polymer blends without using an organic solvent. Additionally, it is an efficient way to improve the blending of the components and prevent the formation of phase separation.

As the PLA is the main compound of the blends, we want to analyze the effect of chitosan and grape seed extract addition on its crystallinity degree (Figure 1). Moreover, the glass transition temperature ($T_g$) is a practical criterion for analyzing the miscibility of the components. Indeed, the $T_g$ value should be affected by the partial miscibility of the components, and it is usually composition-dependent.

Semicrystalline polymers, such as PLA, under measurement conditions followed in this study, show endothermic peaks related to the glass transition temperature ($T_g$) and melting ($T_m$). The glass transition (PLA, $T_g$) is found to be 58.6 °C, which is a comparable value reported by [20]. The exothermic peak at 98.2 °C refers to the crystallization temperature ($T_{cc}$), a thermal event observed after the glass transition in semicrystalline polymers. In fact, after the Tg, there is an orientation and order in the PLA chains due to the increase in mobility during the heating process, which is reflected in the crystallization process. A flat heat flow in the cooling step of the curve is observed (not shown), which means the PLA does not crystallize during the cooling ramp at 10 °C $min^{-1}$. There are two main reasons to explain this behavior. First, a portion of the D isomer inhibits the organization and crystallization of PLA chains. Second, the high cooling rate does not give enough time for the PLA chains to rearrange and organize in a crystalline structure.

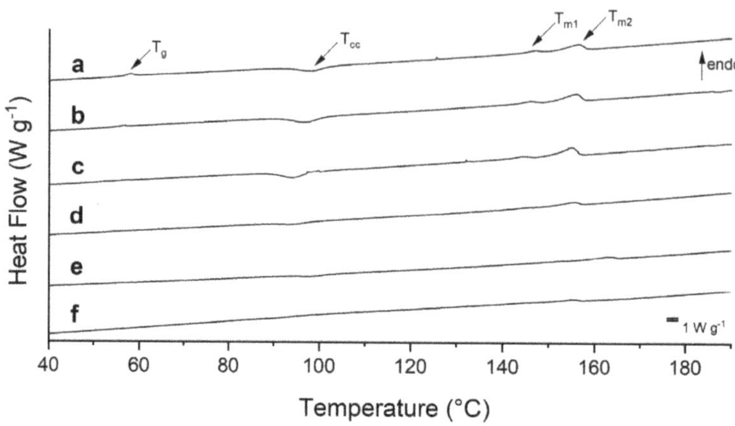


**Figure 1.** DSC curves for (a) PLA; (b) P95c5; (c) P85C15; (d) PLAGE; (e) P95C5GE; and (f) P85C15GE.

Additionally, two peaks observed at 147.3 °C and 156.3 °C are associated with double melting peaks for PLA [21], which is attributed to the melting of two kinds of PLA crystals with different lamellae thicknesses [22].

No significant modification of the glass transition is observed for the low-weight fraction of chitosan (Figure 1b), as a slight reduction of around 2 °C is noticed. However, sample P85C15 (Figure 1c) exhibits two peaks in the glass transition region (highlighted with * in Table 1), which confirms the phase separation of both components. Bonilla et al. [15] studied PLA/CHI ratios of 95:5 and 90:10 and did not observe any lack of miscibility of both polymers. Our study shows that increasing the chitosan ratio (P85C15) leads to phase separation. The cold crystallization temperature (Tc) decreases with the addition of chitosan, which means that chitosan acts as a nucleation agent that favors PLA crystallization. This observation agrees with the study by [23] on the thermal behavior of PLA and PLA/natural rubber. Similarly, the polymer at a low weight fraction allows a heterogeneous nucleation mechanism of the PLA, reducing the free energy barrier and fastening the crystallization process.

**Table 1.** Thermal properties of neat PLA and the blends obtained by DSC.

| Sample | $T_g$ (°C) | $T_{cc}$ (°C) | $\Delta H_{cc}$ (J g$^{-1}$) | $T_{m1}$ (°C) | $T_{m2}$ (°C) | $\Delta H_m$ (J g$^{-1}$) | $X_c$ (%) |
|---|---|---|---|---|---|---|---|
| PLA | 58.6 | 98.2 | 23.3 | 147.0 | 156.4 | 24.0 | 25.8 |
| P95C5 | 56.8 | 96.8 | 18.2 | 146.3 | 155.8 | 22.6 | 25.5 |
| P85C15 | 56.7/53.0 * | 94.2 | 17.7 | 144.3 | 155.0 | 23.2 | 29.3 |
| PLAGE | 55.9 | 94.0 | 16.7 | 145.3 | 155.5 | 16.9 | 18.2 |
| P95C5GE | 53.0 | 94.5 | 15.6 | 144.6 | 155.2 | 17.2 | 19.5 |
| P85C15GE | 54.1 | 93.0 | 12.9 | 144.9 | 155.4 | 18.9 | 23.9 |

* highlighted second glass transition peak for the sample P85C15

Comparably to the neat PLA sample, the blends show the two melting peaks ($T_{m1}$ and $T_{m2}$) due to the recrystallization process observed during heating. Then, the peak at a lower temperature is attributed to the reorganization of the less crystalline structure, while the $Tm_2$ refers to the more perfect crystalline structure of PLA [22]. The total melting enthalpy ($\Delta H_m$) of this thermal event was applied in Equation (1) to calculate the degree of crystallinity ($X_c$) of the samples (Table 1).

With the addition of the grape seed extract, the glass transition peak was shifted to lower temperatures compared to the transition temperature observed for neat PLA (Figure 1d) and blends without the extract (Figure 1e,f). Additionally, it affects the cold crystallization temperature and reduces the enthalpy involved in the thermal event. A reduction in the $\Delta H_{cc}$ might be related to a decrease in the recrystallization process after the glass transition event. Thus, chitosan and grape seed extract affect the degree of crystallization of the blends, as they act as a hindrance to disrupting the PLA crystalline sites.

A degree of crystallinity ($X_c$) of 25.8% was found for the pure PLA (Table 1), a comparable value found in the literature [20]. Interestingly, chitosan increases the $X_c$ mainly due to the phase separation observed in the sample P85C15, allowing the PLA to dense pack in a crystalline form, as the polysaccharide acts as a nucleating agent. The grape extract addition reduces the crystallinity of PLAGE compared to the control sample, which is related to the reduced mobility of PLA chains. In fact, even low weight contents of additives change the crystallization, and consequently the heat involved in the process, as observed by Yu et al., with the addition of Talc in PLA filaments [24].

### 3.3. Thermogravimetry (TGA)

The weight loss curves of pure PLA and blends are reported in Figure 2. A complete weight loss in a single step was observed for pure PLA, with an onset temperature of thermal decomposition at 318.7 °C (Table 2).

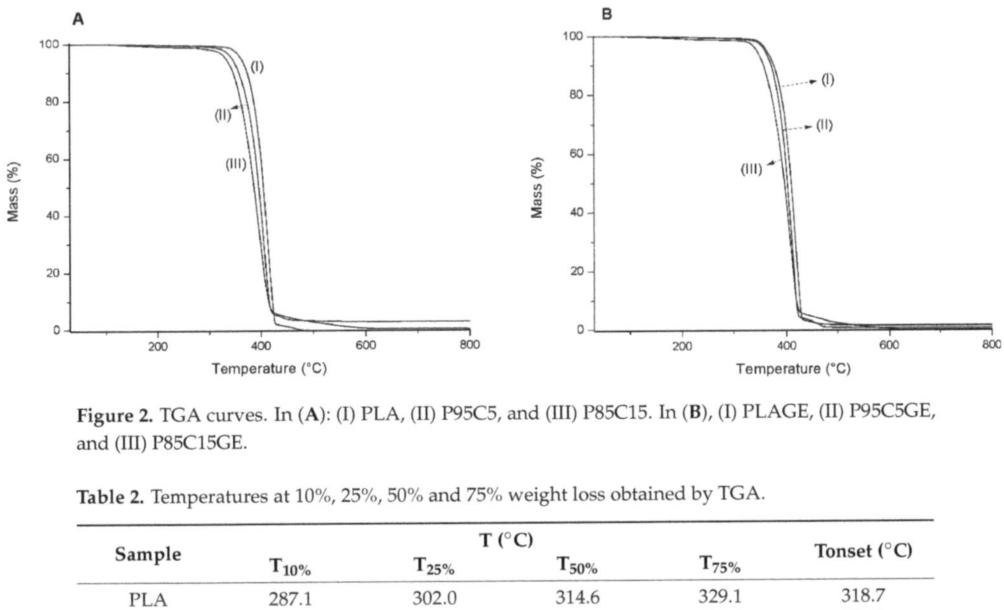


**Figure 2.** TGA curves. In (**A**): (I) PLA, (II) P95C5, and (III) P85C15. In (**B**), (I) PLAGE, (II) P95C5GE, and (III) P85C15GE.

**Table 2.** Temperatures at 10%, 25%, 50% and 75% weight loss obtained by TGA.

| Sample | T (°C) | | | | Tonset (°C) |
|---|---|---|---|---|---|
| | $T_{10\%}$ | $T_{25\%}$ | $T_{50\%}$ | $T_{75\%}$ | |
| PLA | 287.1 | 302.0 | 314.6 | 329.1 | 318.7 |
| P95C5 | 263.2 | 282.4 | 298.0 | 315.1 | 307.0 |
| P85C15 | 258.2 | 277.0 | 295.5 | 319.5 | 292.0 |
| PLAGE | 234.7 | 251.9 | 265.0 | 280.4 | 321.4 |
| P95C5GE | 282.5 | 298.2 | 311.7 | 327.4 | 309.3 |
| P85C15GE | 265.4 | 284.4 | 302.8 | 323.5 | 293.9 |

The onset temperature is the beginning of weight loss and reflects the thermal stability of the material. As PLA is the main compound in the polymeric blends, the inclusion of chitosan did not change the single-step profile; even then, chitosan is characterized by two significant degradation steps [25]. The single weight loss is associated with the relatively low chitosan ratio in the blends. Nevertheless, chitosan addition reduces the decomposition temperature up to 10 °C for the blend with a low amount of polysaccharide (5%) to around 25 °C for the blend with a high content (15%).

In general, the presence of grape seed extract does not change the thermal properties observed for PLA/chitosan ones, as comparable degradation temperatures were obtained. Lately, the low ratio of the extract could explain the behavior.

### 3.4. Tensile Tests

The average values and standard deviation of the mechanical properties of the specimens prepared in this study are reported in Table 3.

**Table 3.** Results of the tensile tests of neat PLA and the blends.

| Sample | Tensile Strength (MPa) | Maximum Elongation at Break (%) | Young's Modulus (MPa) |
|---|---|---|---|
| PLA | 55.4 ± 1.8 [a] | 4.1 ± 0.7 [a] | 3306 ± 217 [a] |
| P95C5 | 50.4 ± 4.4 [a,b] | 2.1 ± 0.5 [b] | 3415 ± 258 [a] |
| P85C15 | 49.4 ± 4.1 [a,b] | 2.2 ± 0.8 [b] | 3488 ± 431 [a] |
| PLAGE | 55.0 ± 0.4 [a] | 3.6 ± 0.5 [a] | 3319 ± 214 [a] |
| P95C5GE | 47.0 ± 6.0 [b] | 1.5 ± 0.2 [b] | 3582 ± 182 [a] |
| P85C15GE | 50.5 ± 1.2 [a,b] | 2.0 ± 0.3 [b] | 3808 ± 148 [a] |

The same letter in the column means no significative difference.

The effect on tensile strength (TS) of chitosan and grape seed extract incorporated in PLA-based injection-molded shoulder bars was observed as a function of the chitosan loading of 5% and 15% ($w/w$) and the addition of 1% of the GE regarding the dry weight of biopolymers (Figure 3A). The TS of the pure PLA specimen was found to be 55.4 ± 1.8 MPa, which means a very brittle behavior, a value in agreement with the literature [26]. Adding chitosan decreases the TS in the binary blends, displaying values up to 50.4 ± 4.4 MPa (P95C5) and 49.4 ± 4.1 MPa (P85C15). Even though a slight decrease in TS was noticed, the change is not statistically significant. In the same way, no change in tensile strength was observed after incorporating the grape seed extract into the neat PLA sample. Nevertheless, comparing this set of samples, the introduction of the extract reduces the TS values in P85C15GE (50.5 ± 1.2 MPa) and P95C5GE (47.0 ± 6.0 MPa).

The elongation at break (EB) exhibited that the inclusion of chitosan is the main responsible for its reduction (Figure 3B). Indeed, all the samples containing the polysaccharide showed a reduction of around 50% in the EB values. In fact, the simple fact of adding the grape seed extract does not change the EB by comparing PLA and PLAGE values. However, the inclusion of chitosan showed that the reduction was significant, as confirmed by the statistical analysis.

The behavior of tensile properties in composites is very sensitive due to interfacial adhesion in composite mixtures. In fact, strength and elongation at break decrease when interfacial affinity is not ensured [27]. Our results established that adding chitosan imparted mechanical properties due to the decrease in both analyzed features. Other authors observed the same characteristic when cellulose in different forms was combined with PLA. Sanchez-Garcia and Lagaron described the same reduction in mechanical properties by the mixing of PLA and cellulose nanowhiskers, while Petersson and Oksman showed a reduction in elongation at a break of about 16% when microcrystalline cellulose was used as reinforcement of PLA. Even though Zakaria et al. reported an increase in tensile strength when 5% chitosan was used in PLA–chitosan blends, a further increase in the polysaccharide concentration implied a reduction in the tensile strength [28]. Nevertheless, the reinforcement of PLA was intrinsically related to a better dispersion of the guest material and also the compatibility between the components, mainly defined by the interfacial affinity.

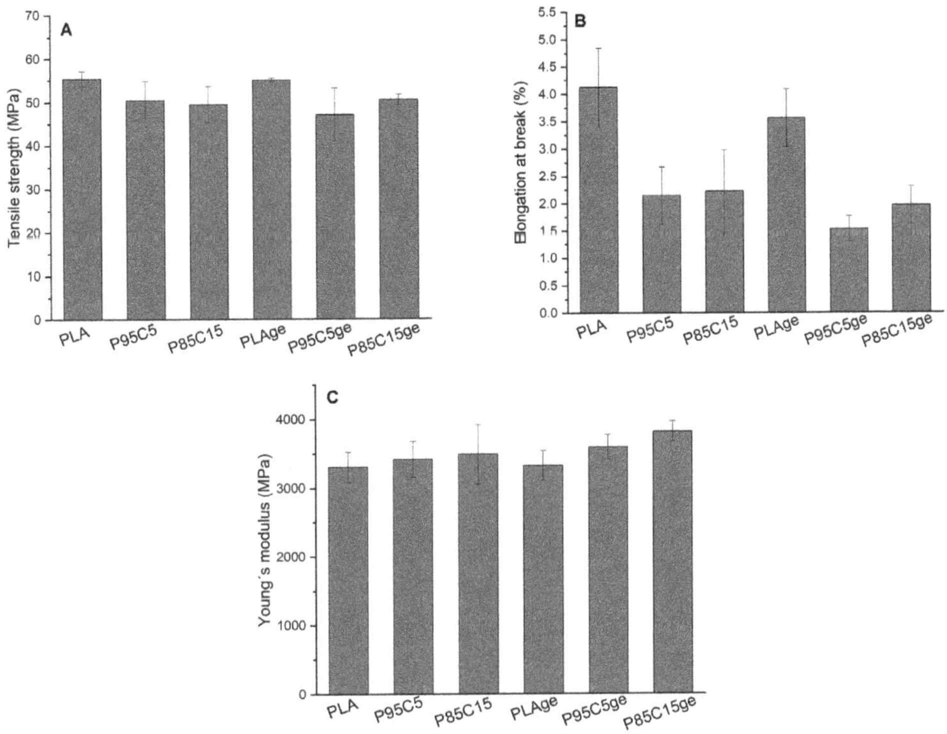


**Figure 3.** Mechanical properties of PLA and blends measured by tensile tests. In (**A**), tensile strength; in (**B**), elongation at break; in (**C**), Young´s modulus results.

Young's modulus is one of the most important structural parameters of materials, determined by the coefficient of proportionality as a linear relationship in stress–strain curves. Even though significant differences were not observed between Young's modulus values measured for the samples, a slight tendency to increase this property after the addition of chitosan was observed (Figure 3C). This indicates more rigidity of the samples, probably related to the rising of crystallinity [29] as calculated by DSC measurements. The extract showed no effect in the linear portion of the stress–strain curve.

### *3.5. Color Measurement*

Before carrying out the color spectroscopy, the injection-molded test specimens were first photographed (Figure 4). Even under nonstandardized light conditions, it can be clearly seen that adding chitosan at both 5% and 15% leads to a distinct reddish-yellowish tone. After adding the grape seed extract, the test specimens darkened to reddish brown.

In order to evaluate changes in coloration of the samples under standardized conditions, measurements were carried out by using the CIE coordinates values. To assess the effect of chitosan and grape seed extract incorporation on the color properties of the specimen, the parameters L (lightness), a (redness), b (yellowness), and total color difference ($\Delta E$) were measured. As shown in Figure 4, the PLA sample was visually the lightness one, which is consistent with the data obtained by color measurements (Table 4). The PLA control sample showed a higher $L^*$ value associated with the brightness characteristic (the white standard used as a background presents the L value of 100).

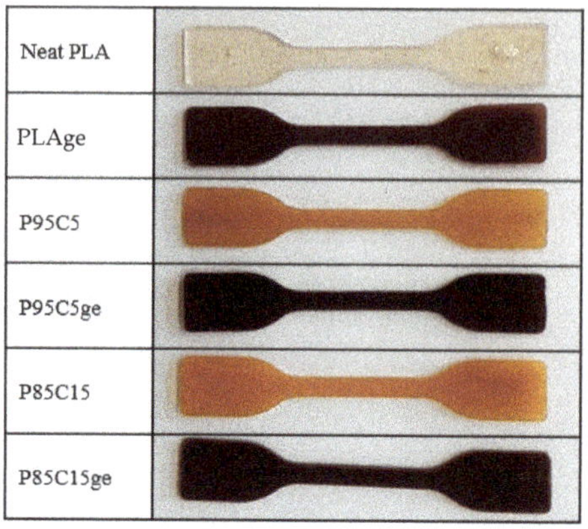


**Figure 4.** Photograph of the injection-molded test specimens prepared in this study.

**Table 4.** Color measurements values of neat PLA and the blends.

| Sample | Color Parameters | | | |
|---|---|---|---|---|
| | a* | b* | L* | ΔE |
| PLA | −0.40 | 9.43 | 78.09 | 0 |
| P95C5 | 5.33 | 26.49 | 46.55 | 36.32 |
| P85C15 | 8.38 | 20.59 | 40.44 | 40.24 |
| PLAGE | 5.40 | 3.34 | 26.53 | 52.25 |
| P95C5GE | 6.69 | 5.97 | 28.22 | 50.49 |
| P85C15GE | 5.51 | 5.05 | 27.42 | 51.20 |

The incorporation of chitosan decreased the L* parameter and is concentration-dependent, as—for P85C15—less brightness was observed. On the other hand, a considerable change in the b* value was noticed, as adding the polysaccharide shifts it to high values. In fact, the yellowish color is an intrinsic property of chitosan, related to its storage [30] and thermocompression molding [31].

All three samples containing the grape seed extract (GE) were in a similar color range, which is no longer assigned to yellow but to red due to the natural color of the extract. Additionally, the L* parameter decreased in PLAGE, P95C5GE, and P85C15GE samples. In practice, it means that the grape seed extract greatly impacts the color property of the blends, even if it was added in a low ratio.

The color difference (ΔE) of the samples was evaluated in comparison to pure PLA (Table 4). The ΔE was in the range of 40 after the addition of chitosan (P95C5 and P85C15), confirming the color changing behavior of the samples. As mentioned earlier, the high ΔE value can be associated with the thermocompression molding process used in sample preparation and was not significantly dependent on the polysaccharide concentration. As expected, the ΔE for the samples after the inclusion of the GE rose due to the intrinsic characteristic of natural extracts being colorful.

### *3.6. Scanning Electron Microscopy (SEM)*

Characteristic SEM images of pure PLA and blends with and without grape seed extract samples are shown in Figure 5. The surface of the PLA neat sample had a homogeneous structure, and no irregularities were observed (Figure 5A). Oppositely, P95C5 and P85C15 blends showed increased irregularities in their surface appearance (Figure 5B,C), corresponding to the partial miscibility of both polymers.

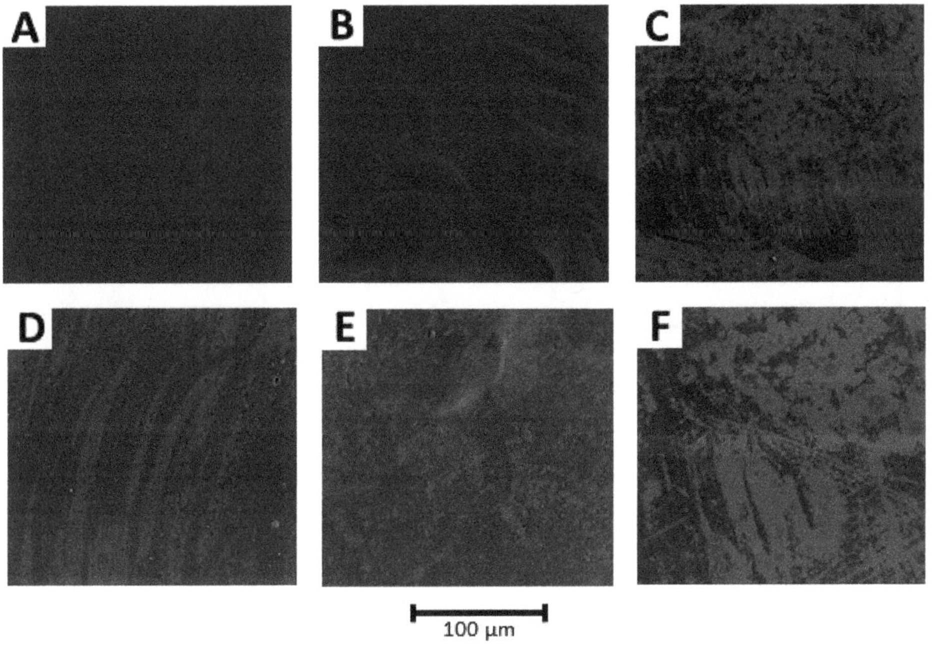


**Figure 5.** SEM images of the blends. PLA (**A**); P95C5 (**B**); P85C15 (**C**); PLAGE (**D**); P95C5GE (**E**); P85C15GE (**F**). Magnification of 200×.

When GE was incorporated into the blends (Figure 5D–F), a nonhomogeneous surface appearance with the presence of pits was noticed on their surface. Wang et al. [17] described that the addition of grape seed extract resulted in a poorer interfacial adhesion, with weak tensile strength and elongation at break in PCL/chitosan films. In our study, the GE did not affect the mechanical properties, probably due to the presence of chitosan, which acts as a main role in the interfacial adhesion properties, as visualized in elongation at break values, reducing this property to half of the value measured for the PLA neat sample.

## 4. Conclusions

The blends proposed in this study offer an alternative material, not only on the raw source side, but also on the disposal side, via promising end-of-life choices. The extrusion method was chosen to prevent extract aggregation, and homogeneous blends were obtained. Generally, the GE did not affect the mechanical and thermal properties of the blends but added antioxidant properties to the final materials. Chitosan is the main responsible for the reduction in elongation at break and thermal decomposition, mainly due to changes in interfacial adhesion properties. The addition of GSE leads to a nonhomogeneous surface appearance without compromising the mechanical properties of the material. Even though a reduction in elongation at break was observed after the inclusion of chitosan, the production of the proposed blend was achieved with comparable features found the pure PLA.

**Author Contributions:** Material preparation, data collection, and analysis were performed by C.L.v.B., V.G. and M.M.H. The first draft of the manuscript was written by M.M.H. and all authors commented on previous versions of the manuscript. H.-P.H. and M.M.H. were responsible for the funding resources. All authors have read and agreed to the published version of the manuscript.

**Funding:** This work was supported by the CINSaT Seed Money from Universität Kassel.

**Data Availability Statement:** The datasets generated during the current study are available from the corresponding author upon reasonable request.

**Conflicts of Interest:** The authors declare no conflict of interest.

## References


1. Pellis, A.; Malinconico, M.; Guarneri, A.; Gardossi, L. Renewable polymers and plastics: Performance beyond the green. *New Biotechnol.* **2021**, *60*, 146–158. [CrossRef] [PubMed]
2. Bajpai, P.K.; Singh, I.; Madaan, J. Development and characterization of PLA-based green composites. *J. Thermoplast. Compos. Mater.* **2014**, *27*, 52–81. [CrossRef]
3. McKeown, P.; Jones, M.D. The Chemical Recycling of PLA: A Review. *Sustain. Chem.* **2020**, *1*, 1–22. [CrossRef]
4. Li, X.; Lin, Y.; Liu, M.; Meng, L.; Li, C. A review of research and application of polylactic acid composites. *J. Appl. Polym. Sci.* **2023**, *140*, e53477. [CrossRef]
5. Ilyas, R.A.; Zuhri, M.Y.M.; Aisyah, H.A.; Asyraf, M.R.M.; Hassan, S.A.; Zainudin, E.S.; Sapuan, S.M.; Sharma, S.; Bangar, S.P.; Jumaidin, R.; et al. Natural Fiber-Reinforced Polylactic Acid, Polylactic Acid Blends and Their Composites for Advanced Applications. *Polymers* **2022**, *14*, 202. [CrossRef]
6. Oksman, K.; Skrifvars, M.; Selin, J.-F. Natural fibres as reinforcement in polylactic acid (PLA) composites. *Compos. Sci. Technol.* **2003**, *63*, 1317–1324. [CrossRef]
7. Ren, Q.; Wu, M.; Wang, L.; Zheng, W.; Hikima, Y.; Semba, T.; Ohshima, M. Cellulose nanofiber reinforced poly (lactic acid) with enhanced rheology, crystallization and foaming ability. *Carbohydr. Polym.* **2022**, *286*, 119320. [CrossRef]
8. Wang, F.; Wu, H.; Venkataraman, V.; Hu, X. Silk fibroin-poly(lactic acid) biocomposites: Effect of protein-synthetic polymer interactions and miscibility on material properties and biological responses. *Mater. Sci. Eng. C Mater. Biol. Appl.* **2019**, *104*, 109890. [CrossRef]
9. Martins, R.C.; Da Ribeiro, S.P.S.; Nascimento, R.S.V.; Nascimento, M.A.C.; Batistella, M.; Lopez-Cuesta, J.-M. The influence of montmorillonite on the flame-retarding properties of intumescent bio-based PLA composites. *J. Appl. Polym. Sci.* **2022**, *139*, 52243. [CrossRef]
10. Pradeep, S.A.; Kharbas, H.; Turng, L.-S.; Avalos, A.; Lawrence, J.G.; Pilla, S. Investigation of Thermal and Thermomechanical Properties of Biodegradable PLA/PBSA Composites Processed via Supercritical Fluid-Assisted Foam Injection Molding. *Polymers* **2017**, *9*, 22. [CrossRef]
11. Wang, Y.; Wang, Y.; Wei, Q.; Zhang, J.; Lei, M.; Li, M.; Li, D. Effects of the composition ratio on the properties of PCL/PLA blends: A kind of thermo-sensitive shape memory polymer composites. *J. Polym. Res.* **2021**, *28*, 451. [CrossRef]
12. Hahn, T.; Tafi, E.; Paul, A.; Salvia, R.; Falabella, P.; Zibek, S. Current state of chitin purification and chitosan production from insects. *J. Chem. Technol. Biotechnol.* **2020**, *95*, 2775–2795. [CrossRef]
13. Águila-Almanza, E.; Low, S.S.; Hernández-Cocoletzi, H.; Atonal-Sandoval, A.; Rubio-Rosas, E.; Violante-González, J.; Show, P.L. Facile and green approach in managing sand crab carapace biowaste for obtention of high deacetylation percentage chitosan. *J. Environ. Chem. Eng.* **2021**, *9*, 105229. [CrossRef]
14. Suyatma, N.E.; Copinet, A.; Tighzert, L.; Coma, V. Mechanical and Barrier Properties of Biodegradable Films Made from Chitosan and Poly (Lactic Acid) Blends. *J. Polym. Environ.* **2004**, *12*, 1–6. [CrossRef]
15. Bonilla, J.; Fortunati, E.; Vargas, M.; Chiralt, A.; Kenny, J.M. Effects of chitosan on the physicochemical and antimicrobial properties of PLA films. *J. Food Eng.* **2013**, *119*, 236–243. [CrossRef]
16. Claro, P.I.C.; Neto, A.R.S.; Bibbo, A.C.C.; Mattoso, L.H.C.; Bastos, M.S.R.; Marconcini, J.M. Biodegradable Blends with Potential Use in Packaging: A Comparison of PLA/Chitosan and PLA/Cellulose Acetate Films. *J. Polym. Environ.* **2016**, *24*, 363–371. [CrossRef]
17. Wang, K.; Lim, P.N.; Tong, S.Y.; Thian, E.S. Development of grapefruit seed extract-loaded poly(ε-caprolactone)/chitosan films for antimicrobial food packaging. *Food Packag. Shelf Life* **2019**, *22*, 100396. [CrossRef]
18. Habib, H.M.; El-Fakharany, E.M.; Kheadr, E.; Ibrahim, W.H. Grape seed proanthocyanidin extract inhibits DNA and protein damage and labile iron, enzyme, and cancer cell activities. *Sci. Rep.* **2022**, *12*, 12393. [CrossRef]
19. Unusan, N. Proanthocyanidins in grape seeds: An updated review of their health benefits and potential uses in the food industry. *J. Funct. Foods* **2020**, *67*, 103861. [CrossRef]
20. Ludwiczak, J.; Kozlowski, M. Foaming of Polylactide in the Presence of Chain Extender. *J. Polym. Environ.* **2015**, *23*, 137–142. [CrossRef]
21. Di Lorenzo, M.L.; Androsch, R. Influence of α'-/α-crystal polymorphism on properties of poly (l-lactic acid). *Polym. Int.* **2019**, *68*, 320–334. [CrossRef]
22. Frone, A.N.; Berlioz, S.; Chailan, J.-F.; Panaitescu, D.M.; Donescu, D. Cellulose fiber-reinforced polylactic acid. *Polym. Compos.* **2011**, *32*, 976–985. [CrossRef]
23. Suksut, B.; Deeprasertkul, C. Effect of Nucleating Agents on Physical Properties of Poly(lactic acid) and Its Blend with Natural Rubber. *J. Polym. Environ.* **2011**, *19*, 288–296. [CrossRef]
24. Yu, W.; Wang, X.; Ferraris, E.; Zhang, J. Melt crystallization of PLA/Talc in fused filament fabrication. *Mater. Des.* **2019**, *182*, 108013. [CrossRef]

25. Timur, M.; Paşa, A. Synthesis, Characterization, Swelling, and Metal Uptake Studies of Aryl Cross-Linked Chitosan Hydrogels. *ACS Omega* **2018**, *3*, 17416–17424. [CrossRef]
26. Espino-Pérez, E.; Bras, J.; Ducruet, V.; Guinault, A.; Dufresne, A.; Domenek, S. Influence of chemical surface modification of cellulose nanowhiskers on thermal, mechanical, and barrier properties of poly(lactide) based bionanocomposites. *Eur. Polym. J.* **2013**, *49*, 3144–3154. [CrossRef]
27. Petersson, L.; Oksman, K. Biopolymer based nanocomposites: Comparing layered silicates and microcrystalline cellulose as nanoreinforcement. *Compos. Sci. Technol.* **2006**, *66*, 2187–2196. [CrossRef]
28. Zakaria, Z.; Islam, M.S.; Hassan, A.; Mohamad Haafiz, M.K.; Arjmandi, R.; Inuwa, I.M.; Hasan, M. Mechanical Properties and Morphological Characterization of PLA/Chitosan/Epoxidized Natural Rubber Composites. *Adv. Mater. Sci. Eng.* **2013**, *2013*, 629092. [CrossRef]
29. Ren, L.; Yan, X.; Zhou, J.; Tong, J.; Su, X. Influence of chitosan concentration on mechanical and barrier properties of corn starch/chitosan films. *Int. J. Biol. Macromol.* **2017**, *105*, 1636–1643. [CrossRef]
30. Leceta, I.; Guerrero, P.; De La Caba, K. Functional properties of chitosan-based films. *Carbohydr. Polym.* **2013**, *93*, 339–346. [CrossRef]
31. Jakubowska, E.; Gierszewska, M.; Nowaczyk, J.; Olewnik-Kruszkowska, E. Physicochemical and storage properties of chitosan-based films plasticized with deep eutectic solvent. *Food Hydrocoll.* **2020**, *108*, 106007. [CrossRef]

polymers

MDPI

*Article*

# Neodymium Recovery from the Aqueous Phase Using a Residual Material from Saccharified Banana-Rachis/Polyethylene-Glycol

**Byron Lapo [1,2,*], Sandra Pavón [3,4], Martin Bertau [3,4], Hary Demey [1], Miguel Meneses [5] and Ana María Sastre [1]**

1 Department of Chemical Engineering, Universitat Politècnica de Catalunya, ETSEIB, Diagonal 647, 08028 Barcelona, Spain; ana.maria.sastre@upc.edu (A.M.S.)
2 School of Chemical Engineering, Universidad Técnica de Machala, UACQS, BIOeng, Machala 070151, Ecuador
3 Institute of Chemical Technology, TU Bergakademie Freiberg, Leipziger Straße 29, 09599 Freiberg, Germany; sandra.pavon.regana@ikts.fraunhofer.de (S.P.); martin.bertau@chemie.tu-freiberg.de (M.B.)
4 Fraunhofer Institute for Ceramic Technologies and Systems IKTS, Fraunhofer Technology Center for High-Performance Materials THM, Am St.-Nicklas-Schacht 13, 09599 Freiberg, Germany
5 Department of Chemistry, Universidad Técnica Particular de Loja, San Cayetano Alto, Loja 110150, Ecuador; mameneses@utpl.edu.ec
* Correspondence: byron.lapo@upc.edu; Tel.: +593-990778741

**Abstract:** Neodymium (Nd) is a key rare earth element (REE) needed for the future of incoming technologies including road transport and power generation. Hereby, a sustainable adsorbent material for recovering Nd from the aqueous phase using a residue from the saccharification process is presented. Banana rachis (BR) was treated with cellulases and polyethylene glycol (PEG) to produce fermentable sugars prior to applying the final residue (BR–PEG) as an adsorbent material. BR–PEG was characterized by scanning electron microscopy (SEM), compositional analysis, pH of zero charge ($pH_{pzc}$), Fourier transform infrared analysis (FTIR) and thermogravimetric analysis (TGA). A surface response experimental design was used for obtaining the optimized adsorption conditions in terms of the pH of the aqueous phase and the particle size. With the optimal conditions, equilibrium isotherms, kinetics and adsorption–desorption cycles were performed. The optimal pH and particle size were 4.5 and 209.19 µm, respectively. BR–PEG presented equilibrium kinetics after 20 min and maximum adsorption capacities of 44.11 mg/g. In terms of reusage, BR–PEG can be efficiently reused for five adsorption–desorption cycles. BR–PEG was demonstrated to be a low-cost bioresourced alternative for recovering Nd by adsorption.

**Keywords:** rare earth recovery; sorption; lignocellulosic waste; PEG

## 1. Introduction

The recovery of REE, and particularly the Nd element, has gained attention due to the growing demand for the production of technological Nd-containing products. For example, the Nd element is part of permanent NdFeB magnets used in wind turbines, vehicles, mobile phones, CoSm magnets, hard disk drives, computers and peripherals [1]. The Nd element is considered a key element in the green energy technology market [2], particularly in road mobility and energy generation [3]. Over the last 30 years, a total of 880 kt of Nd was extracted from mines, 64% of which has not been recycled [4]. Accordingly, Nd was classified as a critical REE by the U.S. Department of Energy [5], which involves several actions, including promoting the development of technology in recycling areas to minimize the future provisioning issues. On the other hand, Nd has been recognized as a new contaminant in water resources [6]. Thus, the recovery of Nd is needed to ensure the future global demand and to minimize the environmental affectation caused by the discarding of technological devices.

Several methods have been developed for the recovery of $Nd^{3+}$ ions from the aqueous phase such as solvent extraction, ion-exchange, co-precipitation and adsorption [7,8]. Among these technologies, adsorption and, particularly, biosorption are sustainable methods for recovering this element due to the use of bio-based materials. Lignocellulosic resources have gained attention in recent years because these biosorbent materials have demonstrated the capability for removing metals, dyes, REE and other contaminants [9–11]. Banana rachis (BR) is a lignocellulosic waste material which is able to recover various REE ions by the absorption of what is owed to the affinity of its carboxylic groups to REE [12]. Banana cultivation is of considerable economic importance; about 130 countries produce and export this fruit. The production of one metric ton of fruit generates four metric tons of wastes, which means that the production of approximately 115 million metric tons of fruit per year produces around 460 million metric tons of waste, including rachis, pseudostem and peel [13,14]. This means BR is not only an interesting and promising source for bio-sorbent material fabrication, but it also holds great potential for alleviating both waste problems and rare earth supply problems.

A now-established approach to valorizing lignocellulosic wastes is producing second-generation ethanol [15], but the solid residue after saccharification could be used as adsorbent material. The solid remnants mainly consist of residues of cellulose, hemicelluloses and lignin [16], which conserves some functional surface groups that are able to attach metals by the adsorption process.

Ethanol production from banana waste would start with converting cellulose into glucose as a fermentable sugar. Several approaches have been tested, including biological, physical-mechanical, chemical approaches or their combinations [17,18]. Polyethylene glycol (PEG) is a non-toxic, non-ionic surfactant for which there exists experience in enhancing saccharification during enzymatic hydrolysis (EH) [19]. Due to the amphiphilic nature of PEG, this surfactant can be adsorbed onto the lignocellulosic surface, facilitating the immobilization of enzymes on the substrate and avoiding the inhibition effect caused by the adsorption of cellulase on lignin and cellulose [20]. Alternatively, regarding metal adsorption, PEG has been successfully applied to improve the adsorption performance of acid orange II dye by the reduction in the agglomeration of the sorbent material particles, therefore allowing for the availability of more active surface sites [21]. The suitability of PEG for promoting both enzymatic lignocellulose digestion and metal binding prompted us to investigate the use of PEG in combining the capability of improving the saccharification process and the solid residue for adsorption purposes.

In fact, to the best of our knowledge, PEG has not been tested for improving the sugar conversion of BR, nor for producing sorbent materials with enhanced adsorbent properties from BR waste material. Here, the example of $Nd^{3+}$ is a representative of the rare earth metal family. In this context, the current research focuses on evaluating the residue from enzymatic hydrolysis with PEG as an adsorbent material of $Nd^{3+}$ ions from the aqueous phase. The evaluations include the parameters that affect the adsorption performance, such as the pH, particle size, initial Nd concentration and reaction time. We also tested to what extent the material can be reused. The present research contributes to the green chemistry concept through the incorporation of a renewable raw material based on a residue from a residue, seeking the avoidance of wastes and the use of degradable chemical products. This approach tends to improve the value chain and circular economy of the banana market, which, in some developing countries, is just limited to the exportation of banana fruit.

## 2. Materials and Methods

### 2.1. Chemicals

Neodymium (III) nitrate hexahydrate ($Nd(NO_3)_3 \cdot 6H_2O$, 99.9%), sodium hydroxide (NaOH, 98%) and nitric acid ($HNO_3$, 70%) were obtained from Alfa Aesar Inc. (Kandel, Germany). Polyethylene glycol 4000 (PEG), cellulase (enzyme blend of cellulases, ß-glucosidases and hemicellulose, Lot #SLBW7460) and citric acid (99.5%) were the products of Sigma Aldrich Inc. (St. Louis, MO, USA). Deionized water was used to prepare the solutions.

### 2.2. Adsorbent Material Preparation

The solid residue from the EH of BR treated with PEG was labeled as BR–PEG. This BR–PEG material was used as the adsorbent material. Prior to obtaining BR–PEG, the BR was collected from the Musa Cavendish plantation located in the south of Ecuador (3°15′ S, 80°51′ W). The neat BR was washed, chopped in pieces of around 2–5 cm, dried at 75 °C, milled and sieved. These particles were dried afterwards at 75 °C for 48 h.

The batch EH experiments were performed in a 1 L reactor (Tecnal, Piracicaba, Brazil). The EH consisted of adding 50 g of BR in 1 L of distilled water and autoclaving at 121 °C for 15 min. The solid residue was transferred to a citrate buffer solution of pH 4.8. Then, 1 g of PEG (2% *w*/*w*) was added and kept in the blend for 1 h at room temperature (25 ± 2 °C). Finally, 3.25 mL of the cellulase blend was added to the reactor (30 FPU/g—enzymatic activity of 206.58 FPU/mL). The hydrolysis experiments were performed by keeping the temperature at 50 °C at 150 rpm for 96 h. The obtained solid residue (BR–PEG) was dried at 75 °C, milled and sieved to three particle size fractions: (i) 150–180 µm, (ii) 500–600 µm and (iii) 850–1000 µm. To analyze the data using Design-Expert 11 software, the averages of these particle size values were used, according to Table 1. The experiments were conducted in triplicate.

**Table 1.** Factors and levels used in CCD experiments.

| Factor | Units | Levels | | |
|---|---|---|---|---|
| | | −1 | 0 | +1 |
| A: pH | pH | 2 | 3.7 | 5.5 |
| B: particle size | µm | 165.0 | 550.0 | 925.0 |

### 2.3. Material Characterization

The composition in terms of the quantity of cellulose, hemicellulose and lignin was analyzed by the ASTM D-1103-60, ASTM D-1104-56 and TAPPI 222 om-02 standard methodologies, respectively [22].

The pH point of zero charge ($pH_{pzc}$) was determined by measuring the pH change in electrolyte solutions loaded with BR–PEG. A total of 25 mg of BR–PEG was added in 25 mL of the electrolyte solutions (adsorbent dosage: 1 g/L) of NaCl 0.01 M with an initial pH ($pH_o$) of 3.0, 5.0, 7.0 and 9.0. These analyses were executed in triplicate.

A scanning electron microscope (SEM) coupled with an energy dispersive X-ray probe (EDX) was used to observe the morphology of the BR–PEG (Phenom XL, Eindhoven, The Netherlands). Prior to the observations, the BR–PEG was sputtered with a carbon film.

Fourier transform infrared (FTIR-ATR) spectroscopy was used to determine the surface functional groups (Spectrum Two IR, Perkin Elmer, Waltham, MA, USA).

Thermogravimetric analysis (TGA) was carried out from 25 to 800 °C, applying a heating rate of 10 K/min under a nitrogen atmosphere at a flow rate of 30 mL/min (TGA/DSC 3+, Mettler Toledo, Greifensee, Switzerland).

### 2.4. Adsorption Experiments

#### 2.4.1. pH and Particle Size Influence

To establish the optimal pH of the aqueous phase and the optimal particle size of BR–PEG to recover $Nd^{3+}$ ions, a response surface methodology (RSM) based on central composite design (CCD) with two factors and face-centered was conducted. The experiments were run in duplicate, and the central point was replicated five times. Table 1 shows the factors and levels evaluated.

The adsorption capacity ($q_e$) in mg/g was used as the response. The $q_e$ was calculated by the following Equation (1).

$$q_e = \frac{(C_i - C_e)V}{m} \quad (1)$$

where $C_i$ and $C_e$ are the initial and the equilibrium concentrations, respectively, in mg/L, $V$ is the volume in L and m is the mass of the adsorbent material in grams.

A second-order model was generated according to Equation (2).

$$y = \beta_o + \sum_{i=1}^{k} \beta_i x_i + \sum_{i=1}^{k} \beta_{ii} x_i^2 + \sum\sum_{i<j} \beta_{ij} x_i x_j + \epsilon \quad (2)$$

The adsorption experiments consisted of adding 25 mg of the sorbent material in 25 mL of a solution containing 50 mg/L of $Nd^{3+}$ ions, kept at 120 rpm in an orbital shaker for 24 h prior to the Nd analysis in a microwave plasma–atomic emission spectrometer (MP-AES 4100, Agilent Technologies, Santa Clara, CA, USA). The pH of the solutions was adjusted using convenient solutions of $HNO_3$ or NaOH with a concentration of 0.1 M or 0.01 M.

2.4.2. Kinetic Study

The adsorption kinetics reaction was evaluated by measuring the remaining concentration of $Nd^{3+}$ ions during the time. The experiments consisted of adding 500 mg of BR into 500 mL of a solution which was composed of 100 mg/L of $Nd^{3+}$ and was previously adjusted to pH 4.8. Nonlinear mathematical models including pseudo first-order (PFORE), pseudo second-order (PSORE) and Elovich equations, according to Equations (3)–(5), respectively [23], were fit to the experimental data to evaluate the kinetics of the process. The experiments were carried out in triplicate.

Pseudo first-order rate equation (PFORE):

$$\frac{dq}{dt} = k_1(q_e - q) \quad (3)$$

Pseudo second-order rate equation (PSORE):

$$\frac{dq}{(q_e - q)^2} = k_2 dt \quad (4)$$

Elovich equation:

$$q = \frac{1}{\beta} ln(1 + \propto \beta t) \quad (5)$$

where $q_e$ and $q$ are the equilibrium sorption capacity and the adsorption capacity (mg/g) at any time, respectively, $t$ is the time (min), $k_1$ and $k_2$ are the PFORE rate constant (1/min) and the PSORE rate constant (g/mg·min), respectively, $\alpha$ is the initial adsorption rate (mg/g·min) and $\beta$ is a desorption constant related to the extent of surface coverage and activation energy.

2.4.3. Equilibrium Study

The equilibrium isotherms were obtained by varying the initial $Nd^{3+}$ concentration ($C_i$) from 10 to 200 mg/L at an initial pH of 4.8. The equilibrium isotherms were modeled by applying the Langmuir, Freundlich and Dubinin–Radushkevich (R-D) nonlinear models, according to Equations (6)–(8), respectively. The experiments were carried out in triplicate.

Langmuir non-linear equation:

$$q_e = \frac{q_{max} b C_e}{1 + b C_e} \quad (6)$$

Freundlich non-linear equation:

$$q_e = K_F C_e^{1/n} \quad (7)$$

SIPS non-linear equation:

$$q_e = \frac{q_{max} K_s C_e^{1/ms}}{1 + K_s C_e^{1/ms}} \quad (8)$$

where $q_e$ is the adsorption capacity (mg/g), $C_e$ is the equilibrium concentration (mg/L), $q_{max}$ is the Langmuir and SIPS adsorption maximum capacity expressed in mg/g, $b$ is the Langmuir constant (L/mg), $K_F$ is the Freundlich constant ((mg/g)·(L/mg)$^{1/n}$), $n$ is the sorption intensity (dimensionless), $K_s$ is the SIPS equilibrium constant (L/mg) and $ms$ is the SIPS model exponent.

#### 2.4.4. Desorption Evaluation

Several adsorption–desorption cycles were carried out using $HNO_3$ 0.1 M as the eluent. The recovery of $Nd^{3+}$ was calculated by using Equation (9).

$$Recovery\ \% = \frac{C_D * V_D}{(C_i - C_e) * V_A} * 100 \quad (9)$$

where $C_D$ is the concentration (mg/L) of the desorbed $Nd^{3+}$ ions in the desorption cycle, $C_i$ and $C_e$ are the initial and equilibrium concentrations of $Nd^{3+}$ ions (mg/L) in the adsorption cycle, respectively, and $V_D$ and $V_A$ are the volumes used in the adsorption and elution cycles, respectively. The experiments were carried out in triplicate.

## 3. Results and Discussion

### *3.1. Material Characterization*

#### 3.1.1. SEM Observations

The surface morphology observations of BR before and after the sugar production process (BR–PEG) were carried out by SEM (Figure 1). The surface of neat BR presented heterogenous textural properties, combining flower-like rugosity and a smooth surface (Figure 1a). The smooth surface could be attributed to the cementing non-cellulosic layers of the cell wall of BR, including lignin, hemicelluloses and pectin compounds [24]. After the enzymatic hydrolysis with PEG (Figure 1b), the rugosity of the BR–PEG surface partially disappears, and consequently, fibers and smoothness are more appreciable. The fiber-like texture is presented in micro-sized bundles in some parts of the BR–PEG surface. The change in the morphology is caused by the enzymatic hydrolysis previously applied. In addition, the presence of PEG in the BR–PEG aids the dispersity of the fibers. PEG has demonstrated effective effects in minimizing the fiber agglomeration [25], which is also beneficial for the adsorption process because it further increases the available contact area of an adsorbent material.

**Figure 1.** Microscopical observations of materials (magnification: X500). (**a**) neat BR, (**b**) BR–PEG.

### 3.1.2. Compositional Analysis

Neat BR is mainly composed of cellulose (30.66%), followed by hemicelluloses (14.47%) and lignin (12.4%), as depicted in Table 2. As expected, after the enzymatic hydrolysis aided by PEG, some changes in the composition were determined—particularly, the lignin content increased to 62.96%, and the cellulose and hemicellulose concentrations decreased to 17.26% and 10.27%, respectively. The changes in the composition are related to the enzymatic hydrolysis, which consumes the cellulose and hemicelluloses, decreasing their contents but consequently causing an increase in the lignin content. Thus, the material is mainly composed of lignin.

**Table 2.** Structural composition of the evaluated materials.

| Composition | Neat BR | BR–PEG |
| --- | --- | --- |
| Hemicelluloses (%) | 14.47 | 10.27 |
| Cellulose (%) | 30.66 | 17.26 |
| Lignin (%) | 12.40 | 62.96 |

### 3.1.3. Thermogravimetric Analysis

The thermal degradation characteristics provide key information about the stability and composition of a sorbent material, because the thermal behavior depends on the inherent characteristic and molecular interactions of a solid material [24]. The thermogravimetric (TGA) and differential thermogravimetric (DTG) profiles of BR and BR–PEG (Figure 2) showed an initial small weight loss from 35 °C to 120 °C due to the evaporation of the absorbed water of the materials. BR showed an initial degradation from 170 °C to 310 °C, corresponding to the water evaporation and decomposition of extractives, hemicelluloses and lignin [24], and a final weight loss in the range between 300 °C to 380 °C, accounted for by the cellulose decomposition. Two stages of weight loss were observed in the BR–PEG sample: i) an initial and fast decomposition at 217 °C corresponding to the decomposition of PEG and the remaining hemicelluloses and ii) a second significant degradation of ~60% of mass weight from 220 °C to 360 °C due to the thermal degradation of the lignin, cellulose and polyether chain of PEG [26,27].

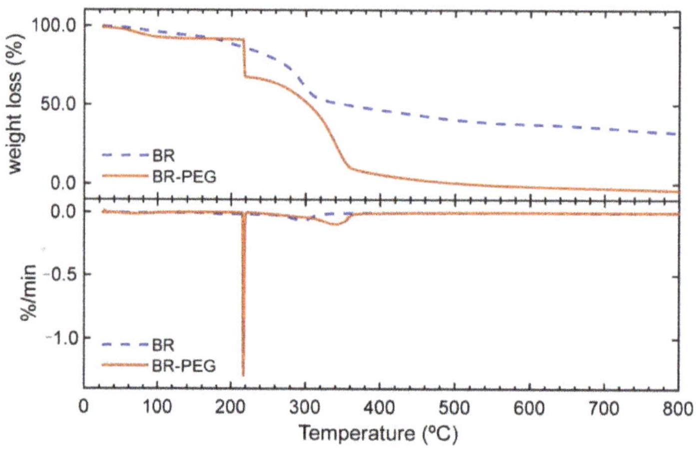


**Figure 2.** Thermogravimetric profile and DTG curves of neat BR and BR–PEG.

### 3.1.4. Infrared Analysis

FTIR analysis based on the determination of surface functional groups provided information about the changes in composition of BR, BR–PEG and BR–PEG after contact with $Nd^{3+}$ ions (BR–PEG–Nd) (Figure 3). A broad stretching vibration around 3300 cm$^{-1}$ belonging to the -OH groups is attributed to the lignocellulosic compounds such as hemicelluloses, pectins and cellulose present in the neat BR, BR–PEG and BR–PEG–Nd materials. A change in the stretching vibration from 2900 cm$^{-1}$ in neat BR to 2888 cm$^{-1}$ in BR–PEG is attributed to the hydrogen bonding caused by the incorporation of PEG in the material [28]. Various new peaks in BR–PEG in the fingerprint region between 1700 cm$^{-1}$ and 700 cm$^{-1}$ can be attributed to the lignin compounds [17]. The stretching vibration at 1032 cm$^{-1}$ related to C-O and O-C-O and the aromatic C-H groups related to the hemicelluloses, cellulose and lignin [12,29] remain in BR–PEG as the original neat BR, indicative of the presence of these compounds in both materials. The FTIR analysis indicates that PEG appears incorporated in BR–PEG; both materials maintain the hemicelluloses, cellulose and lignin. These compositional changes revealed by FTIR analysis agree with the chemical and TGA analysis discussed in Sections 3.1.2 and 3.1.3.

The BR–PEG material presented four shifts in the stretching vibrations after the contact with $Nd^{3+}$ ions during the adsorption process. A bathochromic shift from 1633 cm$^{-1}$ to 1655 cm$^{-1}$ was related to the participation in the adsorption of C=C bands from aromatic rings [17]. The stretching vibration at 1591 cm$^{-1}$, which was attributed to the carboxylic acids of BR [12], changed to 1607 cm$^{-1}$, which is indicative of the interaction of this negative $COO^-$ group with $Nd^{3+}$ ions. The peak at 1409 cm$^{-1}$, which was attributed to the methylene vibration of glucopyranosyl [30], experienced a change to 1422 cm$^{-1}$. The aromatic ring vibration, plus the C=O stretching vibration from the lignin, experienced a red change from 1386 cm$^{-1}$ to 1370 cm$^{-1}$. The shifts in these specific stretching vibrations of the BR–PEG surface are related to functional groups originating from the lignin, cellulose and hemicelluloses participating in the adsorption of $Nd^{3+}$ ions.

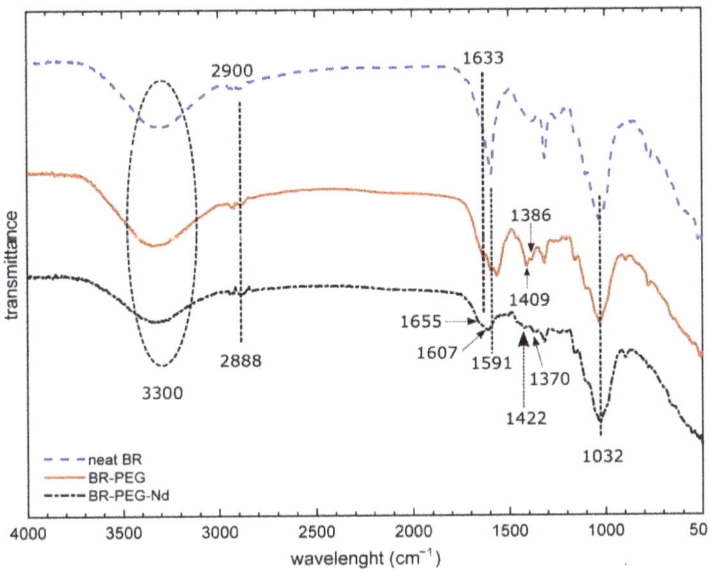


**Figure 3.** Infrared analysis of neat BR and BR–PEG.

### 3.2. Adsorption Experiments

#### 3.2.1. pH and Particle Size Factorial Experiments

The pH of the aqueous phase and the particle size of the adsorbent material had been considered as factors influencing the adsorption process [31,32]. This section presents the simultaneous evaluation of two factors such as the pH of the solution and the particle size of the BR–PEG applied to the $Nd^{3+}$ ion recovery. The optimal levels of these two factors were determined to set the conditions prior to performing the equilibrium isotherms, kinetics and desorption evaluation.

The results obtained after CCD experiments were adjusted among four mathematical models including linear, two-factor interaction (2FI), quadratic and cubic models (Table 3). The quadratic model was found to be the best model because the lack-of-fit *p*-value was statistically significant ($p < 0.05$) and the adjusted and predicted $R^2$ values showed acceptable values ($R^2$ closer to 1), except for the cubic model, which is discarded because the model was aliased. Although the linear and 2FI show statistical significance ($p < 0.05$), the $R^2$ (adjusted and predicted) is relatively farther than that shown by the quadratic model. Thus, as expected, the quadratic model was selected for further data analysis.

**Table 3.** Fit summary adjusting to different models.

| Model | Lack-of-Fit *p*-Value | Adjusted $R^2$ | Predicted $R^2$ | |
|---|---|---|---|---|
| Linear | <0.0001 | 0.7617 | 0.7097 | |
| 2FI | <0.0001 | 0.7459 | 0.6836 | |
| Quadratic | 0.0025 | 0.9829 | 0.9743 | Suggested |
| Cubic | 0.0132 | 0.9902 | 0.9840 | Aliased |

ANOVA Table (Table 4) shows the main effects and statistically significant interactions of the quadratic model. High F-values together with *p*-values less than 0.05 (95% of the confidence level) define which factor is statistically significant with respect to the model. The quadratic model result is statistically significant based on the high F-value of 207.67

and $p < 0.0001$, defining the accuracy of the model. In addition, the variables *A* (pH) and $A^2$ were significant based on high F-values and $p < 0.05$, while $B^2$ (particle-size) showed a lower effect in the model. On the other hand, the interaction of the variables *B* and *AB* did not show a significant effect on the adsorption phenomena (low F-values and $p > 0.05$).

**Table 4.** ANOVA Table for the quadratic model.

| Source | Sum of Squares | Degree of Freedom | Mean Square | F-Value | *p*-Value | |
|---|---|---|---|---|---|---|
| Model | 418.68 | 5 | 83.74 | 207.67 | <0.0001 | significant |
| *A*-pH | 334.00 | 1 | 334.00 | 828.35 | <0.0001 | |
| *B*-particle size | 0.0795 | 1 | 0.0795 | 0.1972 | 0.6643 | |
| *AB* | 0.0353 | 1 | 0.0353 | 0.0876 | 0.7720 | |
| $A^2$ | 82.79 | 1 | 82.79 | 205.33 | <0.0001 | |
| $B^2$ | 4.71 | 1 | 4.71 | 11.69 | 0.0046 | |
| Residual | 5.24 | 13 | 0.4032 | | | |

The model result from the quadratic model fitting is represented in Equation (10). Figure 4a shows the surface response.

$$q_e = -16.67 + 13.61 * A - 0.0084 * B + 0.0001 * A * B - 1.4 * A^2 + 7.18 * B^2 \tag{10}$$

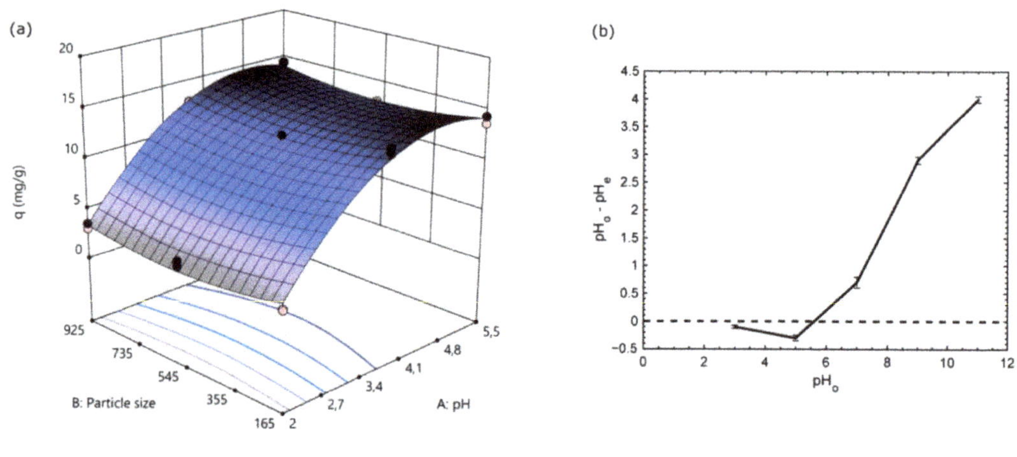


**Figure 4.** Effect of pH and particle size on $Nd^{3+}$ adsorption capacity. (**a**) surface response and (**b**) $pH_{pzc}$ profile of BR–PEG. (experimental conditions for a): temperature: 25 °C; sorbent dosage, SD: 1 g/L; agitation speed: 180 rpm).

The main effect influencing $q_e$ was the pH, followed by the pH squared, while the particle size and the interaction *AB* did not significantly affect the response. The optimal levels calculated by the Design Expert 11 software were: pH: 4.5 and particle size: 209.19 μm.

The squared behavior described by pH can be observed in the curvature presented in the pH axis of the modeled surface response depicted in Figure 4a. The lower pH tested (pH 2) presented a lower $q_e$ of around 3 mg/g, while pHs around 4.5 presented around the maximum $q_e$ of 14 mg/g, which is a ca. fourfold increase. The changes in $q_e$ along the pH evaluated can be explained by the variability of the electrostatic forces imposed by the functional groups available in BR–PEG. According to the $pH_{pzc}$ profile (Figure 4b), BR–PEG material is negatively charged between pH 3 and 5.6, and the maximum negativity matches with the optimal pH of 4.8, at which point the maximum $Nd^{3+}$ uptake was achieved. The adsorption is mainly caused by the electrostatic forces imposed by the surface functional groups of the material. In accordance with the FTIR analysis, the surface functional groups are related with carboxylic and carbonyl groups and lignin compounds [12,33]. The carboxylic group becomes anionic ($-COO^-$) at pHs < $pK_a$ of 2.0 to 5.0. Thus, negatively charged carboxylic groups are the active functional groups responsible for the REE adsorption and corroborate the expected high affinity of rare earths to the oxygenated surface functionalities [34] present in BR–PEG material.

### 3.2.2. Kinetics

The kinetic evaluation provides the information regarding the reaction time for capturing an adsorbate in the solid adsorbent material, the reaction rate at which the adsorption occurs and the probable adsorption mechanisms governing the phenomena [23]. The behavior of the $Nd^{3+}$ ion concentration over the time during the adsorption is shown in Figure 5a, and the mathematical modeling of the kinetic process using three non-linear kinetics model fittings of PFOR, PSOR and Elovich is presented in Figure 5b.

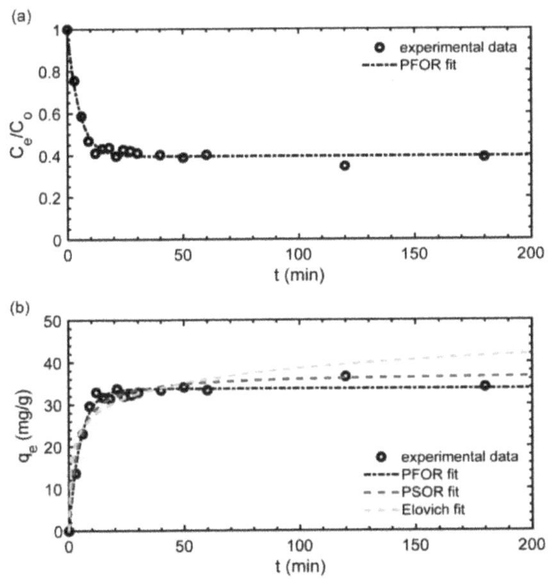


**Figure 5.** Kinetics modeling of $Nd^{3+}$ ion recovery. (**a**) reaction time vs. concentration, (**b**) reaction time vs. $q_e$.. (experimental conditions: temperature: 25 °C; sorbent dosage, SD: 1 g/L; particle size: 150–180 μm; agitation speed: 180 rpm; $pH_o$: 4.5; $C_o$: 50 mg/L).

The typical kinetics steps regarding the adsorption stages can be identified in Figure 5a. During the first stage (film diffusion), from 0 to 10 min, the adsorbate is transported speedily from the liquid phase to the primary external surface of the BR–PEG, achieving around 30 mg/g of adsorption capacity. Step 2 (pore or intraparticle diffusion) takes approximately from 10 to 20 min, in which the $Nd^{3+}$ ions are transported from the external BR–PEG surface into the material pores, where experimental adsorption capacities between 31 mg/g and 36 mg/g were reached. Finally, Step 3 (surface reaction) occurs after 20 min until reaching the material saturation. The total reaction time can be 30 min (Figure 5a), which can be considered competitive compared with that of the raw BR material [12] or can be considered fast compared with the natural material chitosan–iron's reaction time of 5 h [8]. The rapid adsorption is indicative of strong affinities of the material for the adsorbate [35], which corroborates the favorable adsorption determined by the $R_L$ factor in Section 3.2.3.

Moreover, short reaction times imply a reduced volume reactor during the process scaling and consequently play a key role in reducing cost investments.

Three mathematical models—PFOR, PSOR and Elovich—were fit to the experimental kinetic data (Figure 5b and Table 5). The PFOR model and PSOR presented close values of regression coefficients ($R^2$) of 0.98 and 0.96 for PFOR and PSOR, respectively, followed by the Elovich model ($R^2$: 0.89). Considering the closeness of the $R^2$ values, a statistic error function RMSE (root mean sum-of-squares error) is used to determine which is the best fitting model for describing the kinetic experimental data. The RMSE error function provides a quantification of the error between the model parameter and experimental values and can be used to discriminate when the $R^2$ values are too close [23]. Thus, PFOR presents a lower RMSE than the PSOR model (1.38 for PFOR and 1.91 for the PSOR model) and can be considered a model suitable for describing the experimental kinetic data. The PFOR model suggests that the rate of adsorbate occupation is of the first order and is related to the surface sites available, and the adsorption rate is controlled by the diffusion of the adsorbate, while the PSOR and Elovich models are representative of chemisorption [23,36]. Thus, physical sorption is assumed to be the dominant adsorption mechanism to the $Nd^{3+}$ ions on BR–PEG. The PFOR model has been used to describe the adsorption kinetics of a banana peel adsorbing Cd(II) [37] and Cr(VI) [38].

**Table 5.** Kinetic parameters of BR–PEG.

| PFOR | | | | PSOR | | | | Elovich | | | |
|---|---|---|---|---|---|---|---|---|---|---|---|
| $k_1$ (1/min) | $q_e$ (mg/g) | $R^2$ | *RMSE* | $k_2$ (g/mg ∗ min) | $q_e$ (mg/g) | $R^2$ | *RMSE* | α | β | $R^2$ | *RMSE* |
| 0.19 | 33.75 | 0.98 | 1.38 | 0.0079 | 37.26 | 0.96 | 1.91 | 108.5 | 0.19 | 0.89 | 3.11 |

3.2.3. Equilibrium Isotherms

Equilibrium isotherms represent the capacity of sorbent materials to capture the target element or molecules at different initial concentrations. Figure 6 shows the equilibrium isotherm representation of Nd adsorption on BR–PEG material at different Nd concentrations. The data were fit to three mathematical models: Langmuir, Freundlich and SIPS.

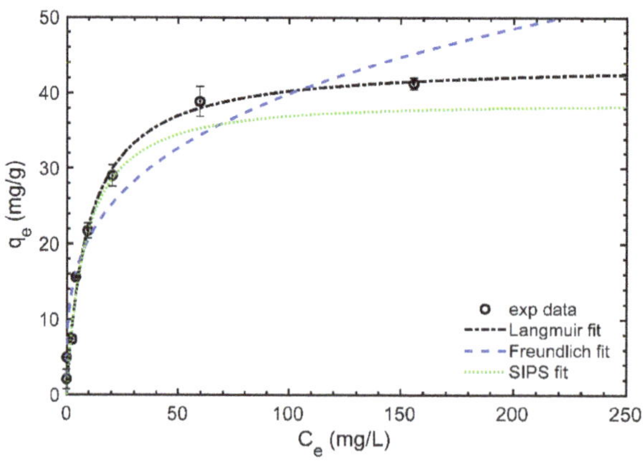


**Figure 6.** Equilibrium isotherm of $Nd^{3+}$ ion recovery. (Experimental conditions: temperature: 25 °C; sorbent dosage, SD: 1 g/L; particle size: 150–180 μm; agitation speed: 180 rpm; $pH_0$: 4.5.)

The BR–PEG equilibrium isotherm presents a type I shape (convex upward) [39], showing a deep slope at the beginning until a $q_e$ of 21 mg/g, followed by a horizontal plateau reaching the highest $q_e$ values at around 40 mg/g. Table 6 shows the parameters obtained after fitting the experimental data on the mathematical models evaluated using the Langmuir, Freundlich and SIPS models.

**Table 6.** Equilibrium isotherm modeling parametrization.

| REE | Langmuir | | | | Freundlich | | | SIPS | | | |
|---|---|---|---|---|---|---|---|---|---|---|---|
| | $q_{max}$ | $b$ | $R^2$ | $R_L$ at Ci: 100 mg/L | $K_F$ | $n$ | $R^2$ | $q_{ms}$ | $K_s$ | $ms$ | $R^2$ |
| | mg/g | L/mg | | | $(mg^{1-1/n}/g\ L^{1/n})$ | | | mg/g | L/g | | |
| Nd | 44.11 | 0.10 | 0.97 | 0.0022 | 10.78 | 3.52 | 0.92 | 44.2 | 0.10 | 1.01 | 0.96 |

The Langmuir equation better describes the experimental data ($R^2$ = 0.97) among the three non-linear mathematical models applied. Langmuir theory assumes that adsorption occurs in definite localized sites covering a homogeneous adsorbent surface [40]. Therefore, once $Nd^{3+}$ ions occupy a particular surface functional group, those sites cannot be held again. In addition, the $R_L$ dimensionless factor indicates a favorable adsorption ($0 < R_L < 1$) [39]. The Langmuir model has been probed as an adequate model for describing equilibrium isotherms where lignocellulosic residues were tested in metal binding, including various parts of banana wastes such as banana rachis for adsorbing critical rare earths [12], banana stalk for removing Pb(II) [41] or banana peel activated carbon for adsorbing Cu(II), Ni(II) and Pb(II) [42]. In addition, BR–PEG showed a better adsorption capacity than other biopolymers such as chitosan, which needs modifications to perform improved adsorption capacities. For instance, chitosan was modified with $Fe(OH)_3$, improving its adsorption capacity from 3.77 mg/g to 11.51 mg/g for neat chitosan and chitosan–$Fe(OH)_3$ [8].

The $q_{max}$ calculated by the Langmuir model resulted in 44.1 mg/g, which is compared with similar materials applied in $Nd^{3+}$ ions recovery (Table 7). The BR–PEG results are moderate compared with those of similar materials derived from banana waste including peel (47.0 mg/g) and pseudo stem (66.4 mg/g) and are less than the $q_{max}$ showed by its neat BR material banana rachis (104.0 mg/g) [12]. In addition, BR–PEG presented a better adsorption capacity compared with other non-modified or natural materials such as bone powder (10.9 mg/g) [43], clinoptilolite (1.8 mg/g) [44], *C. marxianus yeast* (12 mg/g) [45] or chitosan–iron modified material (13.8 mg/g) [8].

Neat BR showed a better adsorption capacity than its residual material BR–PEG. However, the application of BR–PEG represents a route of valorization involving a previous process for producing fermentable sugars and, parallelly, the adsorbent BR–PEG material. This approach provides extra economic profit, because the fermentable sugars obtained during the enzymatic hydrolysis with PEG can be transformed into the most valuable products (i.e., ethanol, ethylene, etc.), representing an advantageous alternative instead of using neat BR directly. In addition, BR in its raw form is able to be used once [12] due to the disintegration of the material, attributed to the solubilization of water-soluble compounds. The reusability of BR–PEG and its advantages compared to neat BR are presented in Section 3.2.4.

**Table 7.** Nd (III) recovery with different materials.

| Sorbent | pH | $q_{max}$ (mg/g) | Authors |
|---|---|---|---|
| Bone powder | - | 10.9 | [43] |
| Kluyveromyces marxianus yeast | 1.5 | 12 | [45] |
| Chitosan/iron (III) hydroxide | 6 | 13.8 | [8] |
| Banana peel waste | 4.5 | 47.03 | [12] |
| Banana pseudo steam waste | 4.5 | 66.46 | [12] |
| Banana rachis waste | 4.5 | 104 | [12] |
| BR–PEG | 4.5 | 44.11 | Present work |

In addition, experiments focused on the evaluation of selectivity in solutions containing a mixture of Nd/Fe ions were performed (Figure S1). The initial concentrations simulate a leachate from an Nd/Fe/B magnet, according to the leaching experiments performed by Riaño et al. [46]. The initial concentration of Nd and Fe followed the ratio Fe/Nd: ~6, resulting in the initial concentrations of Fe: 6.69 mmol/L and Nd: 1.10 mmol/L, and prepared in acetate buffer to avoid iron precipitation. The chemical distribution diagrams (Figure S2) show the soluble Fe and Nd ions present in the working solution. The results, in terms of adsorption capacity, indicate no selectivity for Nd ions over Fe ions. Fe ions are preferred 32% more than Nd ions. However, according to the results presented in terms of the yield of extraction (%), the Nd ions were extracted 27% more than the Fe ions, which is because the initial concentration of Fe is higher than that of Nd. BR–PEG material can be used during a concentration stage where the Nd ions, together with some number of Fe ions, can be separated by the adsorbent material, extracting most of the rare earth from the starting solution. That approach is under evaluation, focusing on the selectivity towards rare earths and proposing process strategies towards rare earth extraction and purification.

### 3.2.4. Desorption Evaluation

BR–PEG was reused for five adsorption–desorption cycles to assess its performance of adsorption and desorption using $HNO_3$ 0.1 M as the eluent solution (Figure 7). The performance of the adsorption of BR–PEG material remains relatively constant, with over 73% of recovery efficiency throughout the five reusages. The recovery during the desorption showed recoveries >60% during the first two cycles and reached the best recovery in the third cycle, with up to 76%.

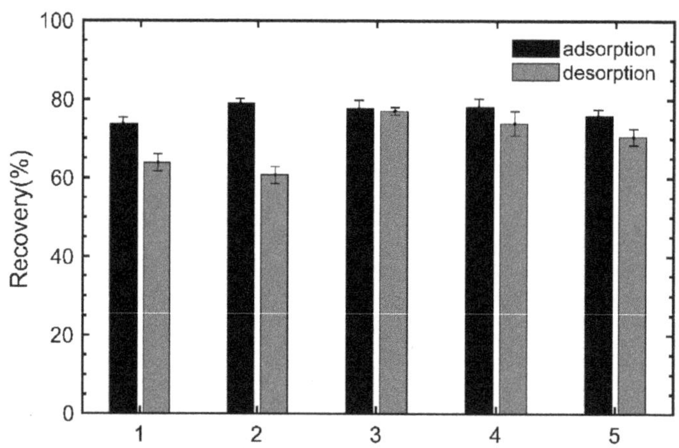


**Figure 7.** Desorption performance of BR–PEG after several reusages. (Experimental conditions: temperature: 25 °C; sorbent dosage, SD: 1 g/L; particle size: 150–180 μm; agitation speed: 180 rpm; $pH_o$: 4.5.)

BR–PEG presents stability after the adsorption–desorption of $Nd^{3+}$ ions. FTIR spectra (Figure S3) comparing BR–PEG, BR–PEG after $Nd^{3+}$ ions adsorption (BR–PG–Nd) and BR–PEG after the desorption of $Nd^{3+}$ ions (BR–PEG desorbed) were performed to identify changes in the composition after being desorbed with $HNO_3$ 0.1 M. It is observed that the main functionalities remain similar after one adsorption–desorption cycle. Stretching vibrations located at 3300 $cm^{-1}$, 1633 $cm^{-1}$, 1409 $cm^{-1}$, 1386 $cm^{-1}$ and 1032 $cm^{-1}$ and related to the OH, C=C, C=O and C-O from lignin, cellulose and hemicellulose did not show changes in their location compared with the starting BR–PEG material, demonstrating the stability in relation to the usage.

BR–PEG presents stability after the adsorption of $Nd^{3+}$ ions. By comparing the FTIR spectra (Figure 3) between BR–PEG and BR–PG–Nd, it is observed that the main functionalities remain similar after the adsorption (peaks at 3300 $cm^{-1}$, 1316 $cm^{-1}$, 1031 $cm^{-1}$, 1010 $cm^{-1}$ and 1151 $cm^{-1}$) related to the cellulose, hemicellulose and lignin composition of BR–PEG, demonstrating the stability in relation to the usage.

The capacity of an adsorbent material to be reused increases its competitive value against adsorbent materials [47]. BR–PEG material is demonstrated to be a stable material after five reusages. This is to say that, during the whole usage, BR–PEG can adsorb around 200 mg/g (~40 mg/g multiplied by five times) compared with ~100 mg/g for neat BR. The possibility of using BR–PEG several times is a remarkable advantage in relation to neat BR and contributes to the total cost of producing BR–PEG and the circular economy of banana agribusiness. The stability of BR–PEG material is attributed to the presence of PEG in its composition.

## 4. Conclusions

BR–PEG residual material from a saccharification process was probed as an adsorbent material for $Nd^{3+}$ ions recovery from aqueous solutions. BR–PEG showed competitive adsorption capacities and kinetics compared with similar natural-sourced materials. The main mechanisms suggest that adsorption is carried out by electrostatic attraction induced by surface functional groups. The mathematical modeling of equilibrium isotherms suggests that the adsorption was produced in a monolayer. BR–PEG can be considered a sustainable alternative for recovering $Nd^{3+}$ ions due to the fact that this material is valorized from a waste, contributing to the circular economy of the banana market.

**Supplementary Materials:** The following supporting information can be downloaded at: https://www.mdpi.com/article/10.3390/polym15071666/s1, Figure S1: Recovery in a mixture Nd/Fe performed by BR-PEG; Figure S2: Chemical distribution diagram to the Nd/Fe mixture used in the selectivity experiments. (a) Nd ions, (b) Fe ions; Figure S3: FTIR analysis after one cycle of adsorption-desorption.

**Author Contributions:** Conceptualization, A.M.S., M.B. and B.L.; methodology, S.P., H.D. and M.M.; formal analysis, B.L.; resources, A.M.S. and M.B.; writing—original draft preparation, B.L.; writing—review and editing, A.M.S., M.B., S.P., M.M. and H.D.; funding acquisition, A.M.S. All authors have read and agreed to the published version of the manuscript.

**Funding:** This research was funded by the Spanish Ministry of Economy and Competitiveness, MINECO (Project CTM2017-83581-R and Project PID2021-127028OB-100).

**Institutional Review Board Statement:** Not applicable.

**Data Availability Statement:** Not applicable.

**Conflicts of Interest:** The authors declare no conflict of interest.

## References

1. Binnemans, K.; Jones, P.T.; Blanpain, B.; Van Gerven, T.; Yang, Y.; Walton, A.; Buchert, M. Recycling of rare earths: A critical review. *J. Clean. Prod.* **2013**, *51*, 1–22. [CrossRef]
2. de Vargas Brião, G.; da Silva, M.G.C.; Vieira, M.G.A. Reusable and efficient clay material for the fixed-bed neodymium recovery. *Sustain. Chem. Pharm.* **2022**, *25*, 100623. [CrossRef]
3. Junne, T.; Wulff, N.; Breyer, C.; Naegler, T. Critical materials in global low-carbon energy scenarios: The case for neodymium, dysprosium, lithium, and cobalt. *Energy* **2020**, *211*, 118532. [CrossRef]
4. Liu, Q.; Sun, K.; Ouyang, X.; Sen, B.; Liu, L.; Dai, T.; Liu, G. Tracking Three Decades of Global Neodymium Stocks and Flows with a Trade-Linked Multiregional Material Flow Analysis. *Environ. Sci. Technol.* **2022**, *56*, 11807–11817. [CrossRef] [PubMed]
5. U.S. Department of Energy. *Critical Materials Strategy*; Secretary of Energy: Washington, DC, USA, 2011.
6. Rim, K.T.; Koo, K.H.; Park, J.S. Toxicological evaluations of rare earths and their health impacts to workers: A literature review. *Saf. Health Work* **2013**, *4*, 12–26. [CrossRef] [PubMed]
7. Pavon, S.; Kutucu, M.; Coll, M.T.; Fortuny, A.; Sastre, A.M. Comparison of Cyanex 272 and Cyanex 572 for the separation of Neodymium from a Nd/Tb/Dy mixture by pertraction. *J. Chem. Technol. Biotechnol.* **2018**, *93*, 2152–2159. [CrossRef]

8. Demey, H.; Lapo, B.; Ruiz, M.; Fortuny, A.; Marchand, M.; Sastre, A.M. Neodymium recovery by chitosan/iron(III) hydroxide [ChiFer(III)] sorbent material: Batch and column systems. *Polymers* **2018**, *10*, 10020204. [CrossRef]
9. Anastopoulos, I.; Bhatnagar, A.; Lima, E. Adsorption of rare earth metals: A review of recent literature. *J. Mol. Liq.* **2016**, *221*, 954–962. [CrossRef]
10. De Quadros Melo, D.; De Oliveira Sousa Neto, V.; De Freitas Barros, F.C.; Raulino, G.S.C.; Vidal, C.B.; Do Nascimento, R.F. Chemical modifications of lignocellulosic materials and their application for removal of cations and anions from aqueous solutions. *J. Appl. Polym. Sci.* **2016**, *133*, 1–22. [CrossRef]
11. Crini, G.; Lichtfouse, E. *Green Adsorbents for Pollutant Removal*; Springer Nature: Cham, Switzerland, 2018; ISBN 9783540228608.
12. Lapo, B.; Bou, J.; Hoyo, J.; Carrillo, M.; Peña, K.; Tzanov, T.; Sastre, A.M. A potential lignocellulosic biomass based on banana waste for critical rare earths recovery from aqueous solutions. *Environ. Pollut.* **2020**, *264*, 114409. [CrossRef]
13. Kema, G.; Drenth, A.; Dita, M.; Jansen, K.; Vellema, S.; Stoorvogel, J. Fusarium wilt of banana, a recurring threat to global banana production. *Front. Plant Sci.* **2021**, *11*, 628888. [CrossRef] [PubMed]
14. Sawarkar, A.N.; Kirti, N.; Tagade, A.; Tekade, S.P. Bioethanol from various types of banana waste: A review. *Bioresour. Technol. Rep.* **2022**, *18*, 101092. [CrossRef]
15. Acevedo, S.; Díaz, A.; Flórez-Lopez, E.; Grande-tovar, C. Recovery of banana waste-loss from production and processing: A contribution to a circular economy. *Molecules* **2021**, *26*, 5282. [CrossRef] [PubMed]
16. Guerrero, A.B.; Ballesteros, I.; Ballesteros, M. The potential of agricultural banana waste for bioethanol production. *Fuel* **2018**, *213*, 176–185. [CrossRef]
17. Costa, S.; Rugiero, I.; Uria, C.L.; Pedrini, P.; Tamburini, E. Lignin degradation efficiency of chemical pre-treatments on banana rachis destined to bioethanol production. *Biomolecules* **2018**, *8*, 141. [CrossRef]
18. Ayala, H.; Kaiser, D.; Pavón, S.; Molina, E.; Siguenza, J.; Bertau, M.; Lapo, B. Valorization of cocoa's mucilage waste to ethanol and subsequent direct catalytic conversion into ethylene. *J. Chem. Technol. Biotechnol.* **2022**, *97*, 2171–2178. [CrossRef]
19. Capanema, N.S.V.; Mansur, A.A.P.; de Jesus, A.C.; Carvalho, S.M.; de Oliveira, L.C.; Mansur, H.S. Superabsorbent crosslinked carboxymethyl cellulose-PEG hydrogels for potential wound dressing applications. *Int. J. Biol. Macromol.* **2018**, *106*, 1218–1234. [CrossRef]
20. Li, J.; Li, S.; Fan, C.; Yan, Z. The mechanism of poly(ethylene glycol) 4000 effect on enzymatic hydrolysis of lignocellulose. *Colloids Surf. B Biointerfaces* **2012**, *89*, 203–210. [CrossRef]
21. Mandal, S.; Kalaivanan, S.; Mandal, A.B. Polyethylene glycol-modified layered double hydroxides: Synthesis, characterization, and study on adsorption characteristics for removal of acid orange II from aqueous solution. *ACS Omega* **2019**, *4*, 3745–3754. [CrossRef]
22. TAPPI. Acid-Insuble Lignin in Wood and Pulp. 2006. Available online: https://www.tappi.org/content/sarg/t222.pdf (accessed on 22 March 2023).
23. Tan, K.L.; Hameed, B.H. Insight into the adsorption kinetics models for the removal of contaminants from aqueous solutions. *J. Taiwan Inst. Chem. Eng.* **2017**, *74*, 25–48. [CrossRef]
24. Deepa, B.; Abraham, E.; Cordeiro, N.; Mozetic, M.; Mathew, A.P.; Oksman, K.; Faria, M.; Thomas, S.; Pothan, L.A. Utilization of various lignocellulosic biomass for the production of nanocellulose: A comparative study. *Cellulose* **2015**, *22*, 1075–1090. [CrossRef]
25. De Lima, G.F.; De Souza, A.G.; Rosa, D.S. Effect of adsorption of polyethylene glycol (PEG), in aqueous media, to improve cellulose nanostructures stability. *J. Mol. Liq.* **2018**, *268*, 415–424. [CrossRef]
26. Shi, Y.; Wang, G. Influence of molecular weight of PEG on thermal and fire protection properties of PEPA-containing polyether flame retardants with high water solubility. *Prog. Org. Coat.* **2016**, *90*, 390–398. [CrossRef]
27. Loof, D.; Hiller, M.; Oschkinat, H.; Koschek, K. Quantitative and qualitative analysis of surface modified cellulose utilizing TGA-MS. *Materials* **2016**, *9*, 415. [CrossRef] [PubMed]
28. Tehrani, Z.; Nordli, H.R.; Pukstad, B.; Gethin, D.T.; Chinga-Carrasco, G. Translucent and ductile nanocellulose-PEG bionanocomposites—A novel substrate with potential to be functionalized by printing for wound dressing applications. *Ind. Crops Prod.* **2016**, *93*, 193–202. [CrossRef]
29. Florian, T.D.M.; Villani, N.; Aguedo, M.; Jacquet, N.; Thomas, H.G.; Gerin, P.; Magali, D.; Richel, A. Chemical composition analysis and structural features of banana rachis lignin extracted by two organosolv methods. *Ind. Crops Prod.* **2019**, *132*, 269–274. [CrossRef]
30. Xu, F.; Yu, J.; Tesso, T.; Dowell, F.; Wang, D. Qualitative and quantitative analysis of lignocellulosic biomass using infrared techniques: A mini-review. *Appl. Energy* **2013**, *104*, 801–809. [CrossRef]
31. Neris, J.B.; Luzardo, F.H.M.; da Silva, E.G.P.; Velasco, F.G. Evaluation of adsorption processes of metal ions in multi-element aqueous systems by lignocellulosic adsorbents applying different isotherms: A critical review. *Chem. Eng. J.* **2019**, *357*, 404–420. [CrossRef]
32. Rápo, E.; Tonk, S. Factors affecting synthetic dye adsorption; desorption Studies: A review of results from the last five years. *Molecules* **2021**, *26*, 5419. [CrossRef]
33. Liu, H.; Sun, J.; Leu, S.-Y.; Chen, S. Toward a fundamental understanding of cellulase-lignin interactions in the whole slurry enzymatic saccharification process. *Biofuels Bioprod. Biorefining* **2016**, *10*, 648–663. [CrossRef]
34. Das, N.; Das, D. Recovery of rare earth metals through biosorption: An overview. *J. Rare Earths* **2013**, *31*, 933–943. [CrossRef]

35. Tran, H.N.; You, S.J.; Nguyen, T.V.; Chao, H.P. Insight into the adsorption mechanism of cationic dye onto biosorbents derived from agricultural wastes. *Chem. Eng. Commun.* **2017**, *204*, 1020–1036. [CrossRef]
36. Li, Y.; Liu, J.; Yuan, Q.; Tang, H.; Yu, F.; Lv, X. A green adsorbent derived from banana peel for highly effective removal of heavy metal ions from water. *RSC Adv.* **2016**, *6*, 45041–45048. [CrossRef]
37. Anwar, J.; Shafique, U.; Salman, M.; Dar, A.; Anwar, S. Bioresource technology removal of Pb(II) and Cd(II) from water by adsorption on peels of banana. *Bioresour. Technol.* **2010**, *101*, 1752–1755. [CrossRef] [PubMed]
38. De Langhe, E.; Vrydaghs, L.; Maret, P.; Perrier, X.; Denham, T. Why bananas matter: An introduction to the history of banana domestication. *Ethnobot. Res. Appl.* **2009**, *7*, 165–177. [CrossRef]
39. Al-Ghouti, M.A.; Da'ana, D.A. Guidelines for the use and interpretation of adsorption isotherm models: A review. *J. Hazard. Mater.* **2020**, *393*, 122383. [CrossRef]
40. Allen, S.J.; Mckay, G.; Porter, J.F. Adsorption isotherm models for basic dye adsorption by peat in single and binary component systems. *J. Colloid Interface Sci.* **2004**, *280*, 322–333. [CrossRef]
41. Ogunleye, O.O.; Ajala, M.A.; Agarry, S.E. Evaluation of Biosorptive Capacity of Banana (Musa paradisiaca) Stalk for Lead(II) Removal from Aqueous Solution. *J. Environ. Prot.* **2014**, *05*, 1451–1465. [CrossRef]
42. Van Thuan, T.; Quynh, B.T.P.; Nguyen, T.D.; Ho, V.T.T.; Bach, L.G. Response surface methodology approach for optimization of $Cu^{2+}$, $Ni^{2+}$ and $Pb^{2+}$ adsorption using KOH-activated carbon from banana peel. *Surf. Interfaces* **2017**, *6*, 209–217. [CrossRef]
43. Butnariu, M.; Negrea, P.; Lupa, L.; Ciopec, M.; Negrea, A.; Pentea, M.; Sarac, I.; Samfira, I. Remediation of rare earth element pollutants by sorption process using organic natural sorbents. *Int. J. Environ. Res. Public Health* **2015**, *12*, 11278–11287. [CrossRef]
44. Vasylechko, V.; Stechynska, E.; Stashkiv, O.; Gryshchouk, G.; Patsay, I. Sorption of neodymium and gadolinium on transcarpathian clinoptilolite. *Acta Phys. Pol. A* **2018**, *133*, 794–797. [CrossRef]
45. Vlachou, A.; Symeopoulos, B.D.; Koutinas, A. A comparative study of neodymium sorption by yeast cells. *Radiochim. Acta* **2009**, 437–441. [CrossRef]
46. Riaño, S.; Petranikova, M.; Onghena, B.; Vander Hoogerstraete, T.; Banerjee, D.; Foreman, M.R.S.; Ekberg, C.; Binnemans, K. Separation of rare earths and other valuable metals from deep-eutectic solvents: A new alternative for the recycling of used NdFeB magnets. *RSC Adv.* **2017**, *7*, 32100–32113. [CrossRef]
47. Attar, K.; Demey, H.; Bouazza, D.; Sastre, A.M. Sorption and desorption studies of Pb(II) and Ni(II) from aqueous solutions by a new composite based on alginate and magadiite materials. *Polymers* **2019**, *11*, 340. [CrossRef] [PubMed]

polymers

MDPI

*Article*

# Ecofriendly Elimination of Ni (II) Using Fabricated Nanocomposite Based on Chitosan/Silver Nanoparticles/Carbon Nanotubes

**Eid M. S. Azzam [1,2,*], Walaa I. Elsofany [1,3], Fahad Abdulaziz [1], Hind A. AlGhamdi [4] and Abdullah Y. AL alhareth [1]**

1 Department of Chemistry, College of Science, University of Ha'il, Ha'il 81451, Saudi Arabia; wa.ibrahim@uoh.edu.sa (W.I.E.); fah.alanazi@uoh.edu.sa (F.A.); abdulla_medshil@hotmail.com (A.Y.A.a.)
2 Egyptian Petroleum Research Institute, Cairo 11727, Egypt
3 Photochemistry Department, Chemical Industries Research Division, National Research Centre, 33 EL Buhouth St., Dokki, Giza 12622, Egypt
4 Chemistry Department, College of Science, Imam Abdulrahman Bin Faisal University, Dammam 31441, Saudi Arabia; hasalghamdi@iau.edu.sa
* Correspondence: ei.ibrahim@uoh.edu.sa or eazzamep@yahoo.com

**Abstract:** Nickel ions are hazardous heavy metals that are non-biodegradable and can lead to allergic sensitivity and dermatitis. Nanomaterials are chosen for their effective elimination of impurities from water structures based entirely on the variety of therapy and degree of purification. The target of this work was the combination of the properties of biopolymers such as chitosan, silver nanoparticles (SNPs), and carbon nanotubes (CNTs) in one ecofriendly compound for Ni (II) uptake from the aqueous solution. To attain this target, the endeavor was made by creating a nanocomposite based on chitosan/SNPs/CNTs. The characterization of the structure of the fabricated nanocomposite (Chit-SNPs-CNTs) was carried out using different techniques. The removal of Ni (II) was examined by studying the adsorption of Ni (II) ions onto the fabricated nanocomposite by batch adsorption using UV, XRD, XPS, and ICP techniques. Moreover, we investigated the effect of the contact time, pH of the solution, and mass of the adsorbent on the efficiency of the adsorption of Ni (II). The results show that the adsorption capacity of Ni (II) increased by increasing the contact time with a neutral pH. The maximum removal of Ni (II) ions (99.70%) was found using 0.3 g of the (Chit-SNPs-CNTs) nanocomposite. In addition, the results indicate that the fabricated nanocomposite has a high adsorption effectivity, which is associated to the function of the chitosan, SNPs, and CNTs in upgrading the adsorption efficiency. Finally, the results in the existing work indicate that the ecofriendly nanocomposite organized here gave excessive effectivity closer to the elimination of Ni (II).

**Keywords:** biopolymer; silver nanoparticles; carbon nanotubes; ecofriendly nanocomposites; wastewater; Ni (II)

## 1. Introduction

Water is an integral but restricted useful resource that all dwelling matters want to access. The viability of clean water is the most necessary factor for life on the earth. Because of the intense pressure introduced by the increasing human population, industrialization, civilization, environmental changes, and agricultural activities, water shortage is a world problem that impacts even water-rich regions. The difficulty in obtaining the right of entry to smooth water is unavoidable, and will take widespread learning about how to increase new, much-less-high-priced strategies of purification while considering electricity utilization and environmental effects [1]. Water pollution consists of all kinds of liquid pollution, such as river and ocean contamination. Liquid pollution, as the phrase implies, occurs in quick liquid-containing regions, such as bays, streams, rivers, lakes, and subterranean water. It entails the discharge of toxic materials, pathogenic microorganisms, substances that want a large amount of oxygen to break down, and substances that are without problems regarding solubility, radiation, etc., that are deposited on the back side and whose accumulations will influence the fitness of aquatic ecosystems. Heavy metals such as lead (Pb), arsenic (As), mercury (Hg), chromium (Cr), especially hexavalent chromium, nickel (Ni), barium (Ba), cadmium (Cd), cobalt (Co), selenium (Se), and vanadium (V), as well as oils and grease, pesticides, etc., are some of the pollutions that are extraordinarily harmful, toxic, and toxic even in the ppb (parts per billion) range [2]. Nickel ions are hazardous heavy metals that are non biodegradable and can lead to allergic sensitivity and dermatitis. The highest approved amount of Ni in US effluents from electroplating procedure wastewater is 4.1 mg/L, whereas that in ingesting water needs to be much less than 0.1 mg/L, in accordance with World Health Organization rules. Industrial tactics along with electroplating, battery production, mining, steel polishing, and forging are the major producers of nickel air pollution in water [3]. The desalination of sea and brackish water; extensive reuse of wastewater; disinfection and decontamination of water, i.e., biosorption and nanoadsorption for contaminant removal; nanophoto catalysis for the chemical degradation of contaminants; nanosensors for contaminant detection; and unique membrane-utilized sciences along with reverse osmosis, nanofiltration, ultrafiltration, and electrodialysis are all examples of nanotechnology applications in water treatment that are being developed and used [1]. Based on their physical and surface characteristics, nanomaterials are chiefly divided into countless classes. Carbon nanotubes (CNTs), metallic nano-adsorbents ($Al_2O_3$, ZNO, $TiO_2$, and $CeO_2$ nanoparticles), metal nanoparticles (Au and Ag), combined oxide nanoparticles (Fe-Ti nanoparticles), polymer nano-adsorbents, nanofibers, and nanoclays are examples of nanomaterials. In addition, they make use of nanoscale holes determined in zeolite filtration membranes and nanocatalysts [4].

Three-dimensional (3D) carbon nanomaterial assemblies are of great interest in emerging applications [5]. Carbon nanotubes (CNTs), a novel adsorbent that was first developed in 1991 by using Iijima, have been tested to currently be of greater quality than activated carbon. In the literature, there are, however, distinctly few investigations on the absorption of heavy metals with CNTs [6]. The extremely good adsorption conducts carbon nanotubes (CNTs) toward several hazardous contaminants. These adsorbents are very fine in casting off unique pollution due to their excessive adsorption effectivity and high adsorption rate. These traits make CNTs the present day substance for the purification of water contaminated with heavy steel ions and dyes [4]. The most time-honored biopolymer after cellulose is chitosan. It has been utilized in water purification systems for over thirty years. In recent years, a new type of unique polymeric material with intrinsic microporosity (PIM) has emerged as a practical material for a range of technologies, such as membrane separation, catalysis, energy storage, sensors, and so on [7]. Chitosan, a polysaccharide that takes place naturally and is located in crabs, shrimps, lobsters, coral jellyfish, butterflies, ladybugs, and one-of-a-kind creatures, is the deacetylated structure of chitin. Because it possesses free amino groups, chitosan is a higher first-rate chelating agent than chitin [8]. Since chitosan has free amino groups, it can chelate substances more effectively than chitin. Biocompatibility, biodegradability, and an improved medication absorption are among chitosan's unique properties. Chitosan should therefore be viewed as a young polymer that is useful for analytical chemistry and water treatment [8]. For wastewater treatment, chitosan (clay-containing silver nanoparticles)-based nanocomposites were utilized [8]. However, to enhance the absorption potential and separation rate, the sketch and synthesis of novel adsorbents relying on chitosan nanocomposites are still needed [8]. In the present work, we fabricated an ecofriendly and simple nanocomposite based on chitosan, CNTs, and silver nanoparticles (SNPs) for the uptake of Ni (II) ions from wastewater. We investigated the sorption of Ni (II) ions using different techniques, such as FTIR, UV-Vis, EDX, XRD, XPS, SEM, TEM, ICPE, and atomic adsorption. The sorption of Ni (II) ions was studied in this work by a variation in different parameters, such as the contact time, pH of the solution, and mass of the fabricated nanocomposite (Chit-CNTs-SNPs) as the absorbent.

## 2. Materials

All of the chemicals used in this project came from the German Sigma-Aldrich company (St. Louis, MO, USA).

### *2.1. Synthesis of Chitosan from Chitin*

The preparation of chitosan in this work was achieved by deacetylation of chitin, which was supplied from the shrimp peel as shown in Scheme 1. The chitin reacted with 12.5 M NaOH at a solid/liquid ratio of 1:15 (g/mL) and was stirred for 24 h. The mixture was cooled and frozen for 24 h. After that, the mixture's temperature was raised to 115 °C, and the reaction continued while being stirred for 6 h. The crude product of chitosan was filtrated followed by washing with distilled water until neutral pH. Finally, the produced chitosan was dried at 70 °C [9].

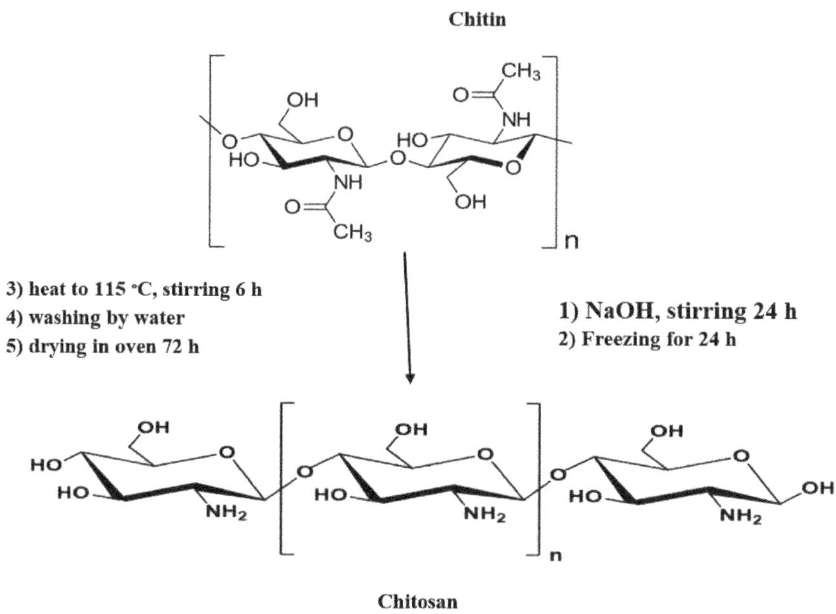


**Scheme 1.** Preparation of chitosan from chitin.

### *2.2. Synthesis of Silver Nanoparticles (SNPs)*

Using trisodium citrate to reduce $AgNO_3$, an SNP colloidal solution was created. The process was carried out as follows: the solution was stirred for 30 min while adding 5 mL of 1% trisodium citrate dropwise after heating 50 mL of $1 \times 10^{-3}$ M $AgNO_3$ until boiling. This produced a pale-yellow color of SNPs [8].

### *2.3. Nano Formation of Chitosan with SNPs*

The nano formation of the chitosan with SNPs (Chit-SNPs) was produced as follows: acetic acid solution of chitosan (5 mL) was added to 20 mL of SNPs aqueous solution. The mixed solution of chitosan and SNPs was stirred until the color changed to colorless [10].

### *2.4. Fabrication of the Nanocomposite (Chit-CNTs-SNPs)*

The nanocomposite (Chit-CNTs-SNPs) was prepared using CNTs and Chit-SNPs as follows: the ethanolic solution of CNTs (1 g/100 mL) was poured into 50 mL solution of Chit-SNPs prepared in step (2.3) and was stirred for 24 h. The output material was filtered, and water-cleaned. The pure nanocomposite (Chit-CNTs-SNPs) was created after overnight drying under vacuum [11].

### *2.5. Removal of Ni (II) Ions Using the Nanocomposite (Chit-CNTs-SNPs)*

The removal of Ni (II) ions using the nanocomposite (Chit-CNTs-SNPs) was carried out by the following steps.

#### 2.5.1. Preparation of Ni (II) Solutions

The nickel (II) sulfate hexahydrate (100 mg) was dissolved in 1 L of distilled water to prepare Ni (II) solution (100 mg/L) [3].

#### 2.5.2. Effect of Time on Removal of Ni Ions

In this work, Ni (II) solution (100 mg/L) and 0.1 g of adsorbent-prepared nanocomposite (Chit-CNTs-SNPs) at various periods (20, 40, 60, 80, 100, 120, 140, and 160 min) with pH 7 at ambient temperature were used to investigate the influence of time on the removal of Ni ions. To bring the adsorbent nanocomposite and the Ni ions into equilibrium, the solution was agitated for 90 min. A UV-VIS double-beam PC scanning spectrophotometer (LABOMED, INC., UV-2950, Los Angeles, CA, USA) was used to evaluate the solid portion of the filtered Ni (II) solutions as well as the leftover filtrates [3].

#### 2.5.3. Effect of pH on Removal of Ni Ions

A total of 0.1 g of the adsorbent (Chit-CNTs-SNPs) was combined with a series of Ni (II) solutions (100 mg/L). Using a standard acid solution of 0.01 M $H_2SO_4$ and an alkaline solution of 0.125 M NaOH, the pH of the solution was altered to be at pH levels of 1, 2, 10, and 11. Before taking any measurements, the pH meter was calibrated using buffers with pH values of 4.0, 7.0, and 10. The pH meter (Mettler-Toledo AG 8603 made by Mettler-Toledo Group, Schwerzenbach, Switzerland) was used to measure all pH measurements. To remove the solid component, the solutions were filtered. Then, using a UV-VIS double-beam PC scanning spectrophotometer (LABOMED, INC., UV-2950, Los Angeles, CA, USA), the residual Ni (II) concentrations were found [3].

#### 2.5.4. Nanocomposite Mass Effect on Sorption of Ni Ions

The mass effect of the nanocomposite as adsorbent for Ni ions was investigated by mixing of different weights (0.1, 0.2, 0.3, and 0.4 g) of the nanocomposite (Chit-CNTs-SNPs) separately to a series of 100 mL of Ni solutions (100 mg/L) at pH 7.0 while stirring for 90 min to reach the equilibrium. The solutions were filtered to separate the solid part of the nanocomposite. The remaining Ni (II) concentrations were detected using UV–VIS double-beam PC scanning spectrophotometer (LABOMED, INC., UV-2950, USA). We used Equations (1,2) to calculate the amount of adsorption qe (mg/g) and the percentage of removal (% removal) [3].

$$q_e = (C_0 - C_e) \times V/m \quad (1)$$

$$\% \text{ removal} = (C_0 - C_e)/C_0 \times 100\% \quad (2)$$

where $C_0$ and $C_e$ are the initial Ni (II) concentration and the concentration at equilibrium in mg/L, m is the mass of the adsorbent, and V is the volume of solution.

## 3. Experimental Analyses

### *3.1. Infrared Spectrometer (FTIR) and Ultraviolet Absorption Measurements (UV)*

The FTIR measurements for the samples used in this study were carried using Thermos Nicolet 6700 FT-IR optical spectrometer (Mundelein, IL 60060 USA). The sample (2 mg) was grinded with 100 mg of KBr with grinding. The sample was converted to pellets by pressing into pills with a compressor. The wavenumber range of the spectra used was 4000–400 $cm^{-1}$, with resolution of 4.0 $cm^{-1}$ at an angle of incidence 80° relative to the surface normal. In this work, we used a UV–VIS double-beam PC scanning spectrophotometer (Labomed, Inc., UV-2950, Los Angeles, CA, USA). The distilled water was used as solvent for all measurements.

### 3.2. Scanning Electron Microscope, Energy-Dispersive X-Ray Spectroscopy, and Transmission Electron Microscope (TEM)

A VEGA 3 TESCAN scanning electron microscope (Tescan, Brno, Czech Republic) with a secondary electron (SE) detector and an energy-dispersive spectrometer (EDS) detector, Element Silicon Drift Detector AMETEK MATERIALS ANALYSIS DIVISION, USA, alongside an electron backscatter diffraction (EBSD) system, was employed for "in situ" chemical composition analysis, morphology observation, and structure determination of the prepared nanocomposite (Chit-CNTs-SNPs), where the measurement was carried out at a selected voltage of 20 KV with a working distance of 8–10 mm between the specimen and the detector without any polishing or use of conductive coating. The compositional characterization diffraction mapping of the nanocomposite (Chit-CNTs-SNPs) was carried out using EDX analysis. An investigation of the chitosan-SNPs nanostructure was conducted using TEM. The TEM images were measured by TEM model (Joel JeM–2100, Tokyo, Japan).

### 3.3. X-Ray Diffraction (XRD) and X-Ray Photoelectronic Spectroscopy (XPS)

The morphology composition of the fabricated nanocomposite (Chit-CNTs-SNPs) was investigated using X-ray diffraction (XRD-7000S, Shimadzu, Japan) operating with a Cu anticathode ($\lambda$ = 1.5406 Å) at 40 kV and 30 mA, using a continuous scan mode of $2^\circ$/min with $2(\theta)$ range from $20^\circ$ to $100^\circ$. The XPS measurements for the nanocomposite (Chit-CNTs-SNPs) were performed using X-ray photoelectron spectroscopy, K-Alpha Plus, Thermo Fisher Scientific, Waltham, MA, USA, equipped with Al K$\alpha$ multi-focused monochromatic.

### 3.4. Inductively Coupled Plasma Emission (ICPE) and Atomic Adsorption

The residual amount of Ni ions in nickel sulfate hexahydrate solution (100 mg/L) after the removal of Ni (II) from this solution using the nanocomposite (Chit-CNTs-SNPs) was determined using inductively coupled plasma emission spectroscopy (ICPE-9800, Shimadzu, Japan). In this work, the Perkin Elmer Analyst AA400 model flame atomic adsorption spectrometer (FAAS) (PerkinElmer Co., Norwalk, CT, USA) was used to determine the uptake of the Ni (II) ions absorbed by the nanocomposite (Chit-CNTs-SNPs).

## 4. Results and Discussion

### 4.1. Chemical Structure of the Synthesized Chitosan

The chemical structure of the synthesized chitosan in Scheme 1 was established using FTIR.

We can observe the FTIR of the synthesized chitosan as shown in Figure 1. The bands in the region 3289–3362 cm$^{-1}$ are related to the N–H and O–H stretching, as well as the intramolecular hydrogen bonds. These bands appear at 3291–3361 cm$^{-1}$ as mentioned in a previous publication [12]. The bands that appear at around 2931 and 2876 cm$^{-1}$ are the bands of C–H symmetric and asymmetric stretching, respectively, which are compatible with the position of these bands at 2921 and 2877 cm$^{-1}$ mentioned in a previous work [12]. It is known that these bands are a typical characteristic of polysaccharides [12]. In addition, the band at 1575 cm$^{-1}$ corresponds to the N–H bending of the primary amine [12]. The observed band at 1163 cm$^{-1}$ can be considered as the asymmetric stretching of the C–O–C bridge. Moreover, the bands at 1071 and 1016 cm$^{-1}$ correspond to C–O stretching [12]. The bands observed at 1648.79 and 1374.36 cm$^{-1}$ show the presence of C=O stretching and C-H deformation, respectively [13]. All above bands are related to the spectra of chitosan as shown in other publications [12,13].

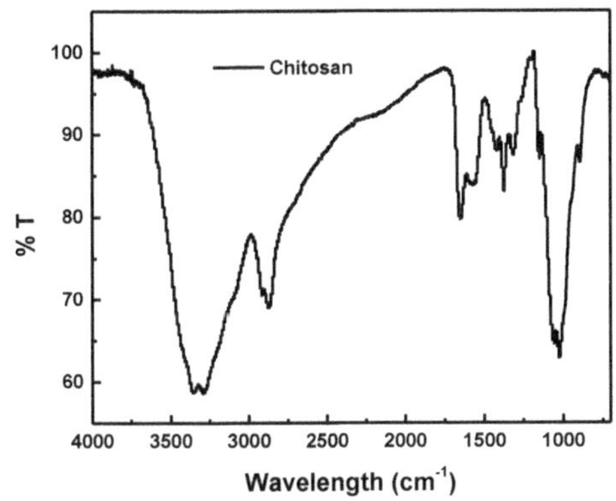


**Figure 1.** FTIR of the prepared chitosan.

*4.2. Chitosan Assembling on Silver Nanoparticles (SNPs)*

The assembling of chitosan molecules on SNPs was confirmed using UV absorbance spectrum and TEM as shown in Figures 2 and 3. Figure 2 shows the absorbance spectra of SNPs individually and after the assembling of chitosan molecules. The absorbance peak for SNPs appears at 422 nm because of the surface plasmon absorption of SNPs. This peak disappears after the assembling of chitosan molecules on SNPs as shown in Figure 3. The reason for this phenomenon is related to the neutralization of the charges surrounding the SNPs and reduction in the plasmon charge on it [14]. Further confirmation of the assembling of chitosan molecules on SNPs was achieved using TEM images as shown in Figure 3. TEM micrographs of the individual SNPs solution and the SNPs in the presence of chitosan molecules are represented in Figure 3. The nanostructure of SNPs in Figure 3 can be seen as taking a spherical shape and polycrystalline structure, with an average size range from 19.1 to 21.44 nm. After assembling chitosan molecules on the SNPs, the individual nanoparticles converted to nanoshells. This behavior leads to restricting the aggregation of SNPs due to the formation of nanoshells after the assembling of chitosan molecules into SNPs (see Figure 3). Moreover, the assembling of chitosan molecules increased the stability of the SNPs and reduced their aggregation. Therefore, the presence of $NH_2$ groups in chitosan enhances the assembling of the chitosan molecules on SNPs via chemosorption due to the presence of a lone pair of electrons on the $N_2$ atom in $NH_2$ groups. In addition, due to the effect of $NH_2$ groups in chitosan molecules on the charge of SNPs, the surface charge of the SNPs is reduced [15]. The above factors prevented the aggregation of SNPs after the formation of nanoshells with chitosan molecules as shown in Figure 3.

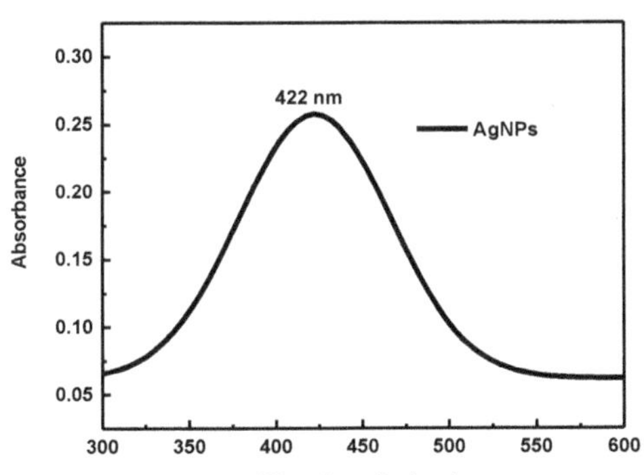


(a)

**Figure 2.** *Cont.*

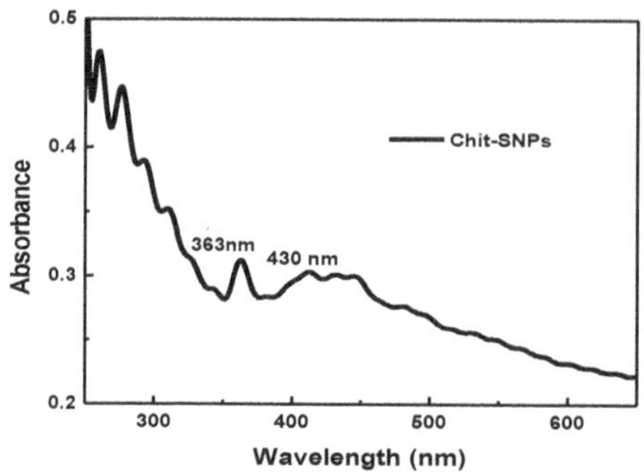


(b)

**Figure 2.** The UV-VIS spectra of SNPs (**a**) and Chit-SNPs (**b**).

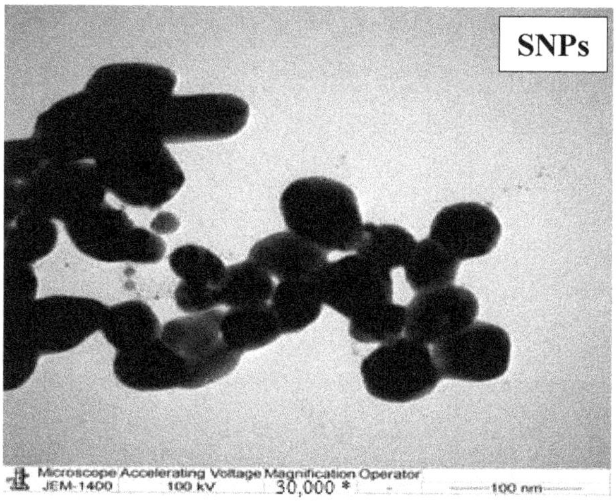


(**a**)

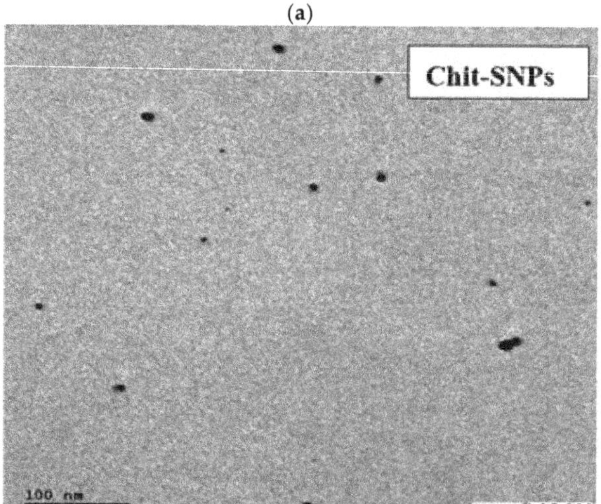


(**b**)

**Figure 3.** The TEM images of SNPs (**a**) and Chit-SNPs (**b**).

*4.3. Establishment of the Morphology and Composition of the Prepared Nanocomposite (Chit-SNPs-CNTs)*

The morphology and composition of the prepared nanocomposite (Chit-SNPs-CNTs) was studied using some techniques as follows.

4.3.1. FTIR Spectroscopy of the (Chit-SNPs-CNTs)

Figure 4 represents the FTIR spectra of the CNTs and Chit-CNTs-SNPs. We used the FTIR to confirm the formation of the Chit-CNTs-SNPs nanocomposite by testing the vibration frequency changes in their functional groups within the range of 4000–400 $cm^{-1}$. The FTIR of the individual CNTs showed absorption bands at 2860 and 2917 $cm^{-1}$ for symmetric/asymmetric $CH_2$ groups stretching, respectively. These groups may be located at the defect sites on the sidewall surface [16]. After the formation of the nanocomposite (Chit-CNTs-SNPs), the FTIR of CNTs in Figure 4 is completely changed and new bands appear at 2162, 2015, and 1979 $cm^{-1}$, which are related to the stretching vibration of SNPs and the CN group of chitosan, respectively [17].

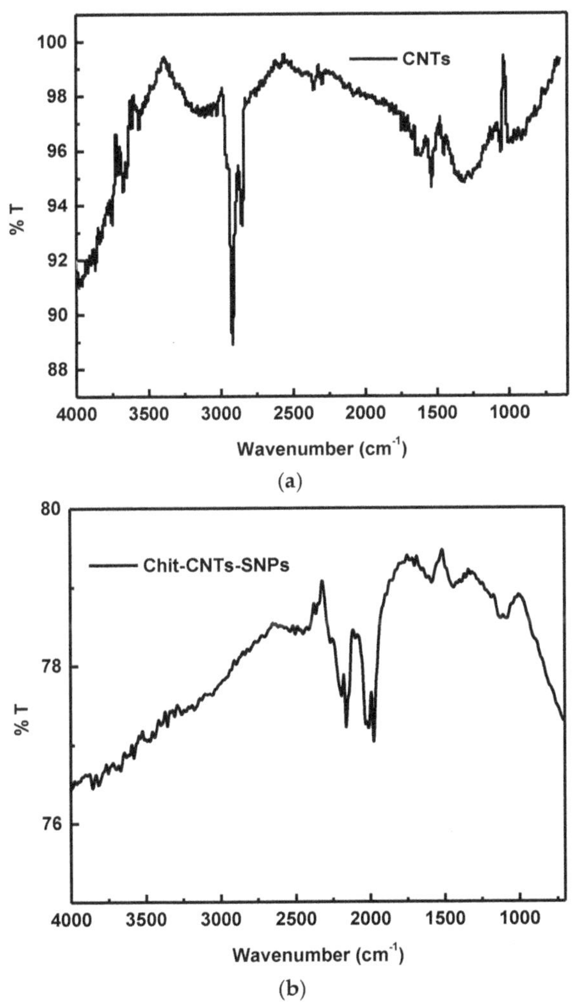


**Figure 4.** The FTIR spectra of the CNTs and Chit-CNTs-SNPs nanocomposite.

4.3.2. Energy-Dispersive X-ray (EDX)

The EDX chart in Figure 5 for CNTs represents the C (K) peak for the CNTs. After the addition of SNPs coated with chitosan, new peaks for Ag (M) and Ag (L) appeared with the C (K) peak of CNTs as shown in Figure 5, which gives an indication of the formation of the (Chit-CNTs-SNPs) nanocomposite.

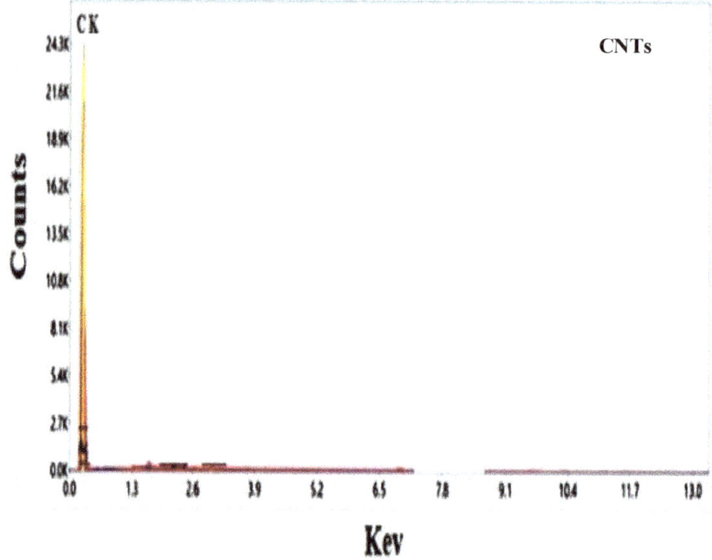


(a)

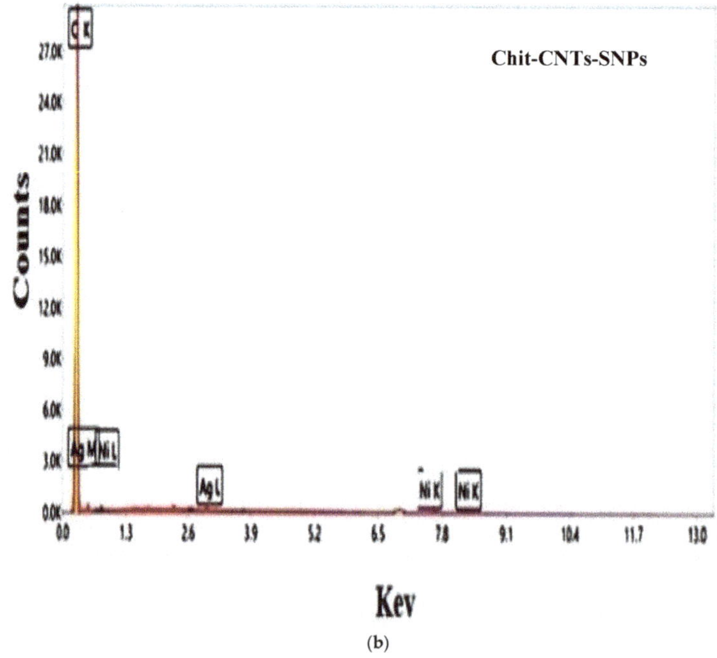


(b)

**Figure 5.** *Cont.*

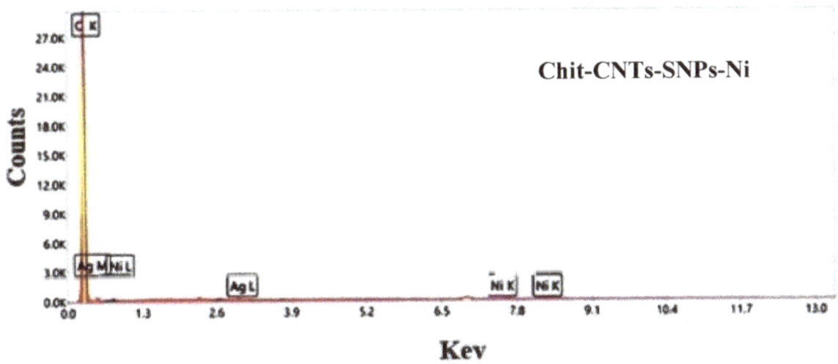


(c)

**Figure 5.** THE EDX of CNTs (**a**), Chit-CNTs-SNPs (**b**), and Chit-CNTs-SNPs-Ni (**c**).

4.3.3. Scanning Electron Microscope (SEM)

In this study, we used an SEM to study the surface morphology of the CNTs and the prepared nanocomposite (Chit-CNTs-SNPs) as shown in Figure 6. It can be noticed from the SEM images of the CNTs in Figure 6 that the surface of the CNTs contains smooth and wide blanks between the layers. It is clear from Figure 6 that the blanks between the CNTs layers decreased after the addition of the SNPs coated with chitosan. This may be due to the filling of these blanks by the SNPs coated by the chitosan molecules in the prepared nanocomposite (Chit-CNTs-SNPs) [6].

(**a**)

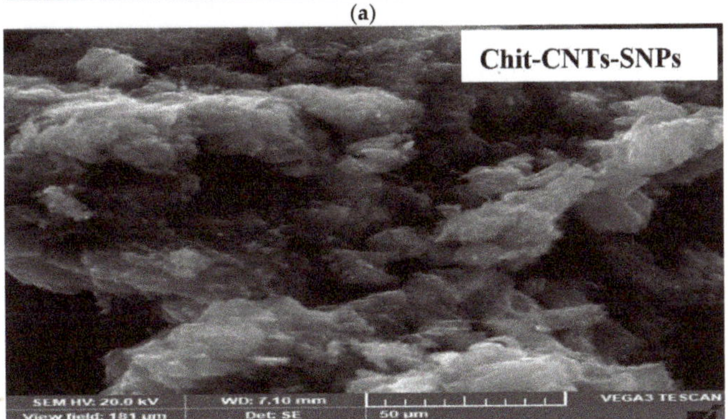


(**b**)

**Figure 6.** *Cont.*

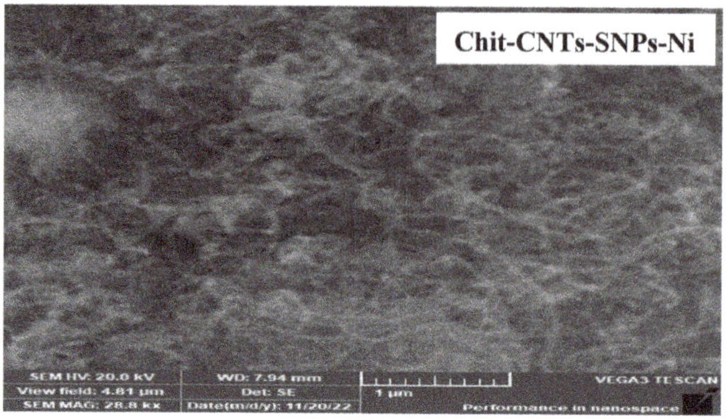


(c)

**Figure 6.** SEM images of CNTs (**a**), Chit-CNTs-SNPs (**b**), and Chit-CNTs-SNPs-Ni (**c**).

4.3.4. X-ray Diffraction (XRD)

In addition to the SEM and EDX, we used XRD for a further study of the morphology of the fabricated nanocomposite (Chit-CNTs-SNPs) as represented in Figure 7. The XRD pattern CNTs show a (002) peak at ~26.3° and (101) peak at 43.7°. These peaks arise due to the tubular structure of the carbon atoms [18]. The intensity of this peak decreased after the addition of SNPs coated by chitosan molecules to form the nanocomposite (Chit-CNTs-SNPs) as shown in Figure 7. This may be due to the interpenetration of the SNPs coated by chitosan between the CNTs layers, which confirms the formation of the fabricated nanocomposite (Chit-CNTs-SNPs).

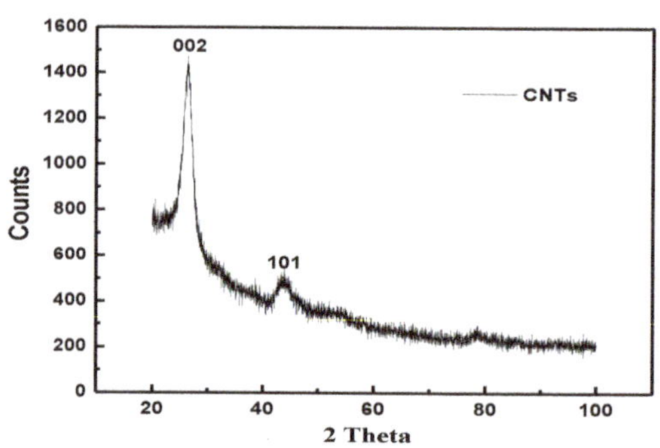


(a)

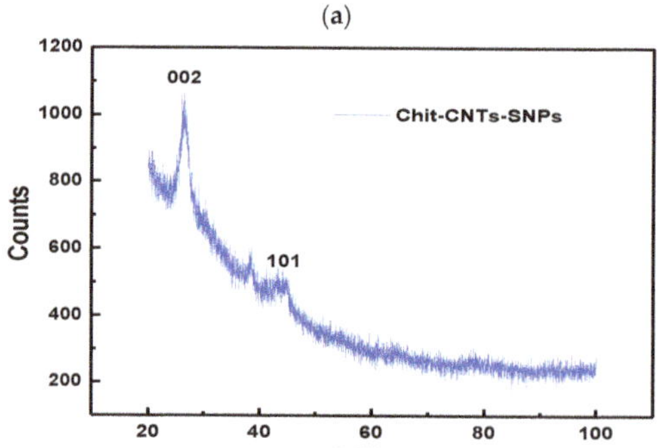


(b)

**Figure 7.** *Cont.*

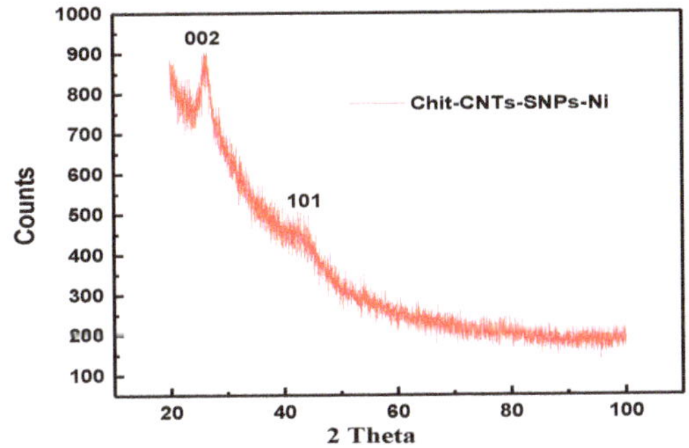


(c)

**Figure 7.** XRD of CNTs (**a**), Chit-CNTs-SNPs (**b**), and Chit-CNTs-SNPs-Ni (**c**).

4.3.5. X-ray Photoelectronic Spectroscopy (XPS)

The composition of the fabricated nanocomposite (Chit-CNTs-SNPs) was also confirmed using the XPS technique. Figure 8 represents XPS spectra of the nanocomposite (Chit-CNTs-SNPs). When focused into the region characteristic for the carbon, the C1s peak of the CNTs was resolved at 284.5 eV and the O1s peak of $O_2$ was detected at 532.2 eV as shown in Figure 8 [19]. After the addition of SNPs coated with chitosan molecules to the CNTs for the formation of the nanocomposite (Chit-CNTs-SNPs), new peaks appeared as represented in Figure 8. One peak appeared at 400.54 eV (N 1s) for $N_2$ atoms, and two peaks appeared at 368.02 (Ag3d 5/2) and 373.23 eV (Ag3d 3/2) with respect to the silver nanoparticles (SNPs).

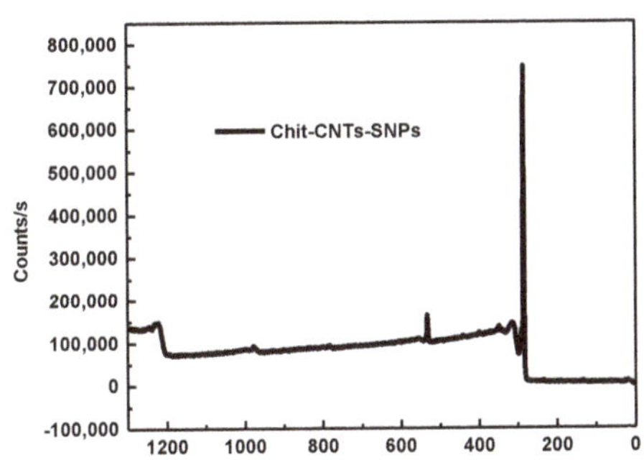


(a)

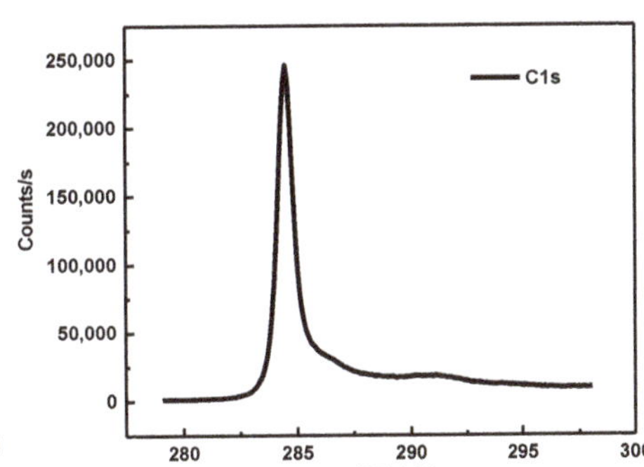


(b)

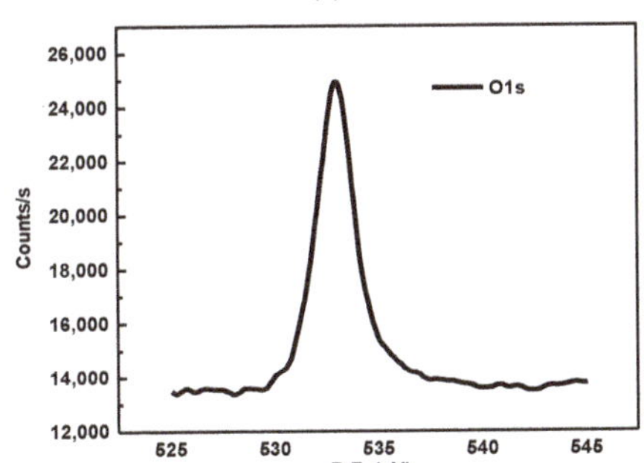


(c)

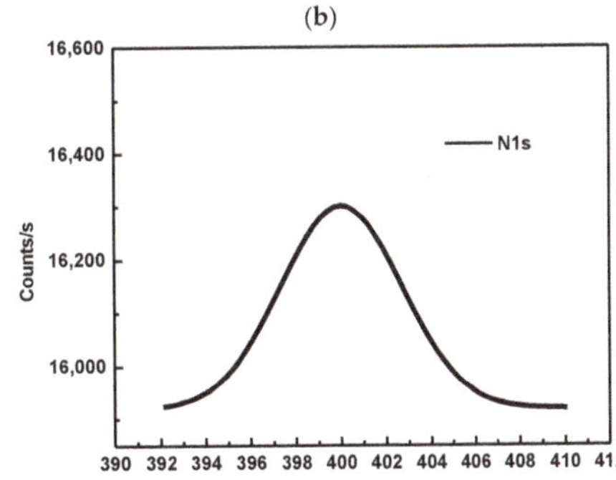


(d)

**Figure 8.** *Cont.*

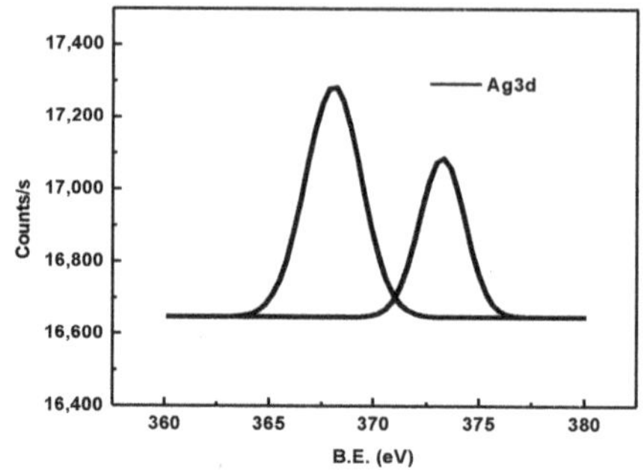


(e)

**Figure 8.** XPS spectra of Chit-CNTs-SNPs nanocomposite. (**a**) Chit-CNTs-SNPs; (**b**) C1s; (**c**) O1s; (**d**) N1s; (**e**) Ag3d.

### *4.4. The Uptake of Ni (II) Ions from Nickel Sulphate Solution*

The following techniques were used in this work to establish the uptake of Ni (II) ions from the nickel sulfate hexahydrate ($NiSO_4{\cdot}6H_2O$) solution onto the fabricated nanocomposite (Chit-CNTs-SNPs): the EDX results in Figure 5 were used to confirm the removal of Ni (II) ions using the fabricated nanocomposite (Chit-CNTs-SNPs). It is clear in Figure 5 that some peaks related to Ni (L) and Ni (K), in addition to the other peaks for Ag (M) and Ag (L), appeared with the C (K). The presence of the Ni peaks in Figure 5 establishes the sorption of Ni ions onto the surface of the nanocomposite (Chit-CNTs-SNPs). The SEM image in Figure 6 shows the surface morphology of the nanocomposite after the sorption of Ni ions (Chit-CNTs-SNPs-Ni). It is clear from this image that the surface morphology is completely changed after the uptake of Ni (II) onto the surface of the Chit-CNTs-SNPs nanocomposite. The surface contains bumps, small dark gray spots, and tight thickness due to the uptake of Ni (II) ions. Herein, we used XRD in this work for the investigation of the sorption of Ni (II) onto the fabricated nanocomposite (Chit-CNTs-SNPs) as shown in Figure 7. As mentioned in Figure 7, the XRD pattern CNTs show a (002) peak at $\sim$26.3° and (101) peak at 43.7°. The intensity of these peaks of CNTs became weaker, as shown in Figure 7, which may be due to the sorption of Ni ions onto the surface of the Chit-CNTs-SNPs nanocomposite. XPS analysis was used here to gain further investigation into the sorption of Ni (II) ions as shown in Figure 8. Comparing between the XPS of Chit-CNTs-SNPs (see Figure 8) and the XPS after sorption of Ni in Figure 9, it is noticed that new peaks appear at binding energy (BE) values of 856.6 and 870.2 eV related to Ni2p, which consists of two spin-orbit doublets, Ni2$p_{3/2}$ and Ni2$p_{1/2}$, respectively [20]. These results confirm the uptake of the Ni ions on the surface of the Chit-CNTs-SNPs nanocomposite. The uptake of Ni (II) using 0.1 g of the prepared nanocomposite (Chit-CNTs-SNPs) was investigated in this work using ICPE and atomic adsorption as represented in Table 1. The result of atomic adsorption shows that 72.70 ppm of Ni (II) ions were uptaken from a 100 ppm $NiSO_4.6H_2O$ solution. In addition, the ICPE data in Table 1 confirm that 27.30 ppm of Ni (II) ions remain in the $NiSO_4.6H_2O$ solution. These results in Table 1 confirm the uptake of Ni (II) ions using the Chit-CNTs-SNPs nanocomposite.

**Table 1.** Removal % of Ni II ions adsorption capacity ($q_e$) using UV absorbance at different times using the fabricated nanocomposite (Chit-SNPs-CNTs) at pH = 7 and ambient temperature.

| **Time (min)** | **20 min** | **40 min** | **60 min** | **80 min** | **100 min** | **120 min** | **140 min** | **160 min** |
|---|---|---|---|---|---|---|---|---|
| Removal % | 50% | 55% | 60% | 65% | 75% | 82% | 82% | 82% |
| $C_e$ (g/l) | 0.050 | 0.045 | 0.040 | 0.035 | 0.025 | 0.018 | 0.018 | 0.018 |
| $q_e$ (mg/g) | 250 | 275 | 300 | 325 | 375 | 410 | 410 | 410 |

Moreover, the uptake of Ni (II) ions using the fabricated nanocomposite (Chit-CNTs-SNPs) was investigated experimentally using the UV-VIS technique at different contact times, pHs of the Ni (II) solution, and masses of the adsorbent (Chit-CNTs-SNPs) nanocomposite as follows.

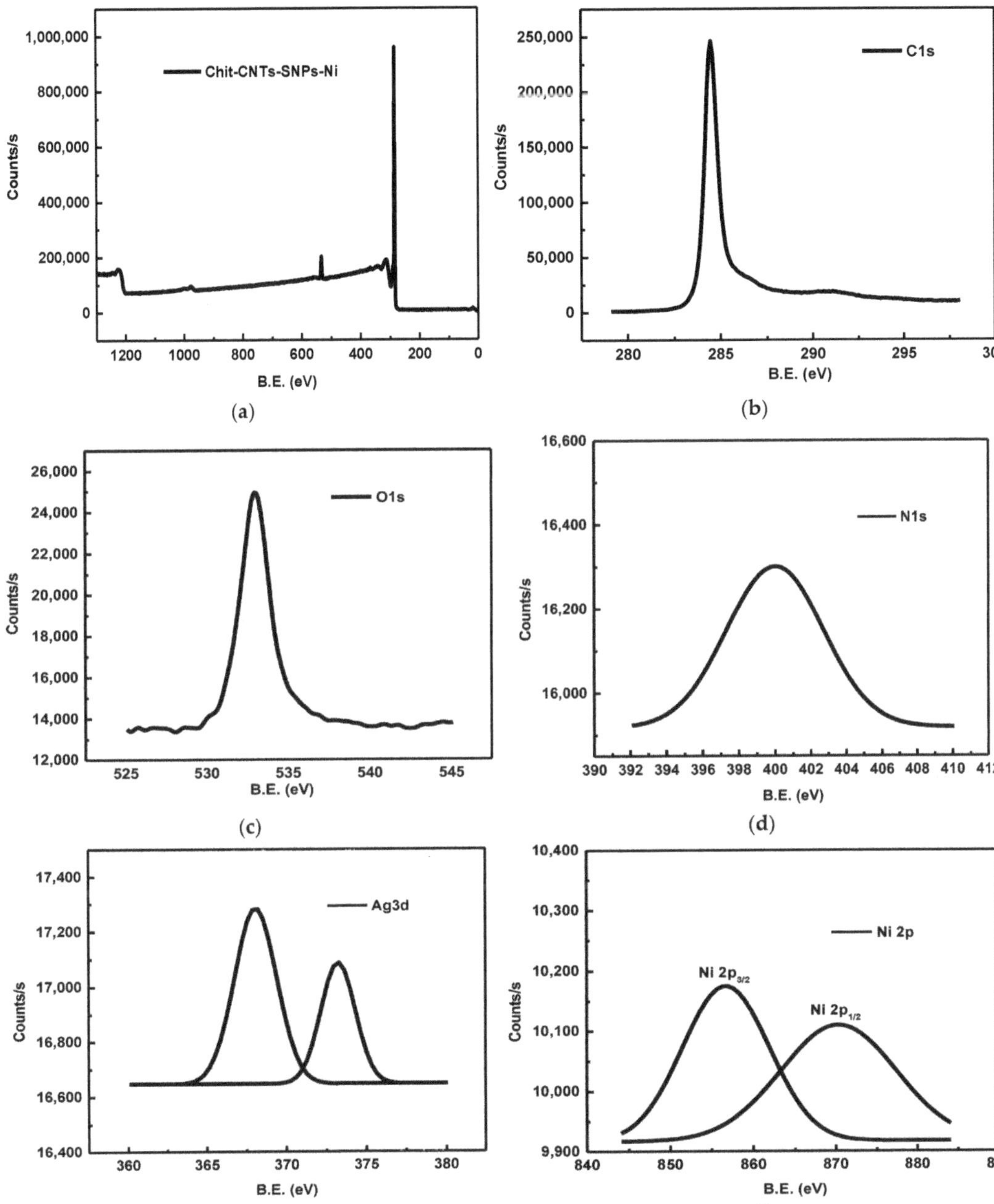


**Figure 9.** XPS spectra of Chit-CNTs-SNPs-Ni. (**a**) Chit-CNTs-SNPs-Ni; (**b**) C1s; (**c**) O1s; (**d**) N1s; (**e**) Ag3d; (**f**) Ni2p.

### 4.5. Variation in Contact Time

The data in Figure 10 and Table 1 represent the relationship between the contact time and the absorbent (Chit-CNTs-SNPs) nanocomposite's adsorption capacity (qe). The contact time varied at 20, 40, 60, 80, 100, 120, 140, and 160 min, with an initial metal concentration of nickel (II) sulfate hexahydrate of 100 mg/L. The analysis of the data in Figure 10 and Table 1 shows that the adsorption rate of Ni (II) ions using the nanocomposite (Chit-CNTs-SNPs) increases from a 50% removal of Ni (II) and $q_e$ = 250 mg of Ni (II) per gram of (Chit-CNTs-SNPs) at 20 min until 82% and $q_e$ = 410 mg of Ni (II) per gram of (Chit-CNTs-SNPs) at 120 min. These results in Figure 10 and Table 1 make it abundantly evident that the chelation between the Ni ions and the $NH_2$ group in chitosan may play a role in the adsorption of Ni (II) using the (Chit-CNTs-SNPs) nanocomposite. The rapid diffusion of Ni (II) ions from the solution to the outer surface (Chit-CNTs-SNPs) as a result of the interaction between CNTs and SNPs in the nanocomposite (Chit-CNTs-SNPs) may also be connected to this [3].

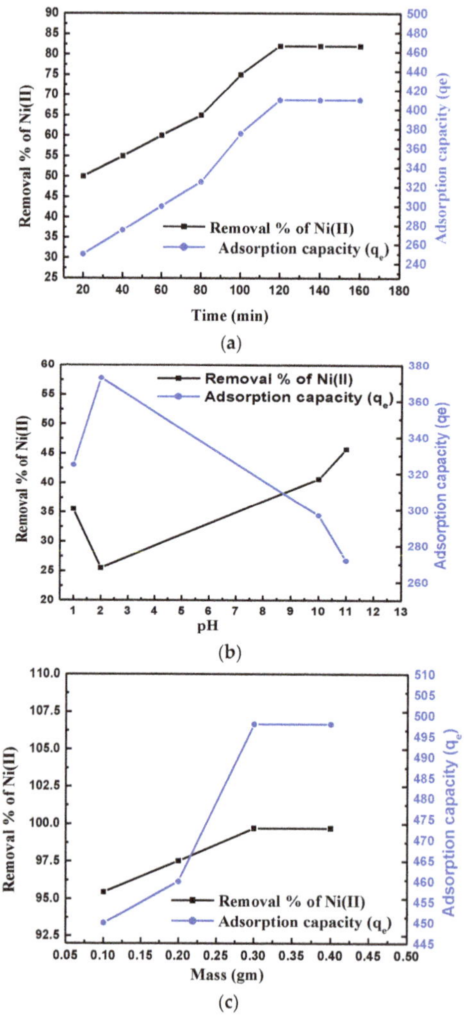


**Figure 10.** Effect of contact time (**a**), pH (**b**), and the mass of nanocomposite (Chit-CNTs-SNPs) (**c**) on the sorption Ni ions.

### 4.6. Variation in pH

It is known from the previous publications that the variation in pH of the metal ion solution has an effect on the adsorption capacity ($q_e$) of an absorbent material such as the (Chit-CNTs-SNPs) nanocomposite [3]. Figure 10 and Table 2 show the results of the investigation into how the pH of the starting solution affects the removal percentage of Ni (II) and adsorption capacity of the synthesized nanocomposite (Chit-CNTs-SNPs). The high percentage (45.70%) of the Ni (II) ions adsorbed by the (Chit-CNTs-SNPs) nanocomposite was found at pH = 11 as represented in Figure 10 and Table 2. The $NH_2$ groups of chitosan in the nanocomposite (Chit-CNTs-SNPs) convert to $-NH^{3+}$ cation at a lower pH (1 and 2), which deactivates the chelation between the $NH_2$ group and Ni (II) ions. Moreover, the conversion of the $NH_2$ groups to the $-NH^{3+}$ cation causes electrostatic repulsion between Ni (II) ions and $-NH^{3+}$ cation, which restricts the sorption of Ni (II) ions using the nanocomposite (Chit-CNTs-SNPs) [3]. At a high pH (10 and 11), the Ni (II) starts to deposit as $Ni(OH)_2$, which causes a decrease in the adsorption capacity ($q_e$ = 297 at pH = 10 and 272 at pH = 11) as shown in Figure 10 and Table 2 [3].

**Table 2.** Removal % of Ni II ions adsorption capacity ($q_e$) using UV absorbance at different pH using the fabricated nanocomposite (Chit -CNTs-SNPs) after 1 h at ambient temperature.

| pH of Ni (II) Solution | pH = 1 | pH = 2 | pH = 10 | pH = 11 |
|---|---|---|---|---|
| Removal % | 35.50% | 25.50% | 40.60% | 45.70% |
| Ce (g/l) | 0.035 | 0.0255 | 0.0406 | 0.0457 |
| qe (mg/g) | 325 | 373 | 297 | 272 |

### 4.7. Variation in the Nanocomposite Mass

The removal % of Ni (II) and the adsorption capacity were considered in this work by a mass variation in the fabricated nanocomposite (Chit-CNTs-SNPs) from 0.1 to 0.4 g at a constant pH and contact time as shown in Table 3 and Figure 10. It is clear from the data in Table 3 and Figure 10 that the maximum removal of Ni (II) ions reaches (99.70%) and the adsorption capacity (qe) equals (498 mg/g) using 0.3 g of the (Chit-CNTs-SNPs) nanocomposite. Comparing these data in Table 4 and Figure 10 with the results given in the previous publications [3] gives an indication that the sorption of Ni (II) ions using the fabricated nanocomposite (Chit-CNTs-SNPs) in this work has a greater efficiency for the removal of heavy metal ions than the other composites in the previous publications.

**Table 3.** Removal % of Ni (II) ions and adsorption capacity ($q_e$) using UV absorbance at different weights using the fabricated nanocomposite (Chit-CNTs-SNPs) after 1 h at pH = 7 and ambient temperature.

| Mass of (Chit-CNTs-SNPs) | 0.1 gm | 0.2 gm | 0.3 gm | 0.4 gm |
|---|---|---|---|---|
| Removal % | 95.44% | 97.53% | 99.70% | 99.70% |
| $C_e$ (g/l) | 0.0955 | 0.0080 | 0.0003 | 0.0003 |
| $q_e$ (mg/g) | 450 | 460 | 498 | 498 |

**Table 4.** Comparison between the maximum adsorption capacity (mg/g) of the fabricated nanocomposite (Chit-CNTs-SNPs) and some absorbents for heavy metals removal.

| Absorbent | Maximum Adsorption Capacity (mg/g) | Reference |
|---|---|---|
| Chitosan–zeolite composites | 14.75–51.32 | 3 |
| Semi-IPN hydrogel based on chitosan and gelatin | 153.9 | 3 |
| Xanthate-modified magnetic chitosan | 34.5 | 3 |

**Table 4.** *Cont.*

| Absorbent | Maximum Adsorption Capacity (mg/g) | Reference |
|---|---|---|
| Chitosan nanofibrils | 186.6 | 3 |
| Chitosan (Ch)/clay, chitosan (Ch)/AgNPs/clay, and chitosan (Ch)/AuNPs/clay | 181.5 | 3 |
| Chit-CNts-SNPs | 498 | Present work |

### *4.8. The Mechanism for the Uptake Process of Ni (II) Ions*

The fabrication of the nanocomposite (Chit-CNTs-SNPs), which is used to adsorb Ni (II) ions, depends on a number of variables. One of these elements is the coordination between the metal cation and amino group in chitosan, which causes chelation [6,21]. The amino groups of the chitosan molecule, which function as a Lewis base, can donate electron pairs to the metal cation, such as Ni (II), which acts as a Lewis acid [3]. As the nanocomposite (Chit-CNTs-SNPs) contains ions exchangeable with silver nanoparticles (SNPs) in addition to the metal chelation with the $NH_2$ group of chitosan, the removal of the Ni (II) ions in this work depends on both the ion-exchange mechanism and the chelation [3,21]. The information in Table 4 demonstrates that the synthesized nanocomposite (Chit-CNTs-SNPs), which was used as an absorbent in this study, was more effective in absorbing heavy metal ions than other absorbents in earlier studies [3].

## 5. Conclusions

Recently, greater focus has been placed on creating environmentally acceptable nanocomposites that function as adsorbents, particularly those that rely on biopolymers, CNTs, and nanoparticles. Here, we created an environmentally friendly nanocomposite (Chit-CNTs-SNPs) for absorbing Ni (II). We looked into how the mass of the absorbent, pH of the solution, and contact time affected the removal of Ni (II) ions. According to the results of this work, 0.3 g of the Chit-CNTs-SNPs nanocomposite had the highest removal rate (99.70%) of Ni (II). The results also show that qe = 498 mg/g was the critical value for this work's maximal Ni (II) ion uptake. The Chit-CNTs-SNPs nanocomposite is also very active in capturing Ni (II) ions from wastewater treatment.

**Author Contributions:** Conceptualization, A.Y.A.a., E.M.S.A. and E.M.S.A.; methodology, E.M.S.A. and A.Y.A.a.; software, E.M.S.A.; validation, E.M.S.A. and A.Y.A.a.; formal analysis, E.M.S.A., A.Y.A.a. and H.A.A.; investigation, E.M.S.A.; resources, E.M.S.A., A.Y.A.a. and F.A.; data curation, W.I.E. and H.A.A.; writing—original draft preparation, E.M.S.A., A.Y.A.a. and W.I.E.; writing—review and editing, E.M.S.A., F.A. and W.I.E.; visualization, E.M.S.A. and H.A.A.; supervision, E.M.S.A.; project administration, E.M.S.A., F.A. and W.I.E.; funding acquisition, E.M.S.A., W.I.E. and F.A. All authors have read and agreed to the published version of the manuscript.

**Funding:** This research was funded by Deputy for Research & Innovation, Ministry of Education through Initiative of Institutional Funding at University of Ha'il—Saudi Arabia grant number IFP-22 089.

**Institutional Review Board Statement:** Not applicable.

**Data Availability Statement:** The data presented in this study are available on request from the corresponding author. The data are not publicly available due to privacy.

**Acknowledgments:** This research has been funded by Deputy for Research & Innovation, Ministry of Education through Initiative of Institutional Funding at University of Ha'il–Saudi Arabia through project number IFP-22 089.

**Conflicts of Interest:** The authors declare no conflict of interest.

## References


1. Kumar, S.; Ahlawat, W.; Bhanjana, G.; Heydarifard, S.; Nazhad, M.M.; Dilbaghi, N. Nanotechnology-based water treatment strategies. *J. Nanosci. Nanotechnol.* **2014**, *14*, 1838–1858. [CrossRef] [PubMed]
2. Verma, R.; Dwivedi, P. Heavy metal water pollution—A case study. *Recent Res. Sci. Technol.* **2013**, *5*, 98–99.
3. Azzam, E.M.; Elsofany, W.I.; Alrashdi, G.K.; Alenezi, K.M.; Alsukaibi, A.K.; Latif, S.; Abdulaziz, F.; Atta, A.M. New route for removal of Cu (II) using fabricated nanocomposite based on cationic surfactant/Ag-nanoparticles/silica gel. *Arab. J. Chem.* **2022**, *15*, 103897. [CrossRef]
4. Jain, N.; Kanu, N.J. The potential application of carbon nanotubes in water treatment: A state-of-the-art-review. *Mater. Today Proc.* **2021**, *43*, 2998–3005. [CrossRef]
5. Wang, X.; Yin, H.; Sheng, G.; Wang, W.; Zhang, X.; Lai, Z. Fabrication of self-entangled 3D carbon nanotube networks from metal–organic frameworks for Li-ion batteries. *ACS Appl. Nano Mater.* **2018**, *7*, 7075–7082. [CrossRef]
6. Kandah, M.I.; Meunier, J.L. Removal of nickel ions from water by multi-walled carbon nanotubes. *J. Hazard. Mater.* **2007**, *146*, 283–288. [CrossRef] [PubMed]
7. Guo, H.; Li, H.; Jing, C.; Wang, X. Soluble polymers with intrinsic porosity for efficient removal of phenolic compounds from water. *Micropor. Mesopor. Mater.* **2021**, *1*, 111068. [CrossRef]
8. Azzam, E.M.; Eshaq, G.H.; Rabie, A.M.; Bakr, A.A.; Abd-Elaal, A.A.; El Metwally, A.E.; Tawfik, S.M. Preparation and characterization of chitosan-clay nanocomposites for the removal of Cu (II) from aqueous solution. *Int. J. Biol. Macromol.* **2016**, *89*, 507–517. [CrossRef] [PubMed]
9. De Queiroz Antonino, R.S.; Lia Fook, B.R.; de Oliveira Lima, V.A.; de Farias Rached, R.Í.; Lima, E.P.; da Silva Lima, R.J.; Peniche Covas, C.A.; Lia Fook, M.V. Preparation and characterization of chitosan obtained from shells of shrimp (*Litopenaeus vannamei* Boone). *Mar. Drugs* **2017**, *15*, 141. [CrossRef] [PubMed]
10. Azzam, E.M.; Sami, R.M.; Alenezi, K.M.; Haque, A.; El Moll, H.; Soury, R.A.; Ismail, A.R. Inhibition of Sulfate-Reducing Bacteria by Para-amino-N-((1-Alkylpyridin-1-Ium Bromide)-4-Yl) Benzamide Surfactants and Surfactant-Coated Silver Nanoparticles. *J. Surfactants Deterg.* **2022**, *25*, 125–131. [CrossRef]
11. Azzam, E.M.; Fathy, N.A.; El-Khouly, S.M.; Sami, R.M. Enhancement the photocatalytic degradation of methylene blue dye using fabricated CNTs/TiO2/AgNPs/Surfactant nanocomposites. *J. Water Process Eng.* **2019**, *28*, 311–321. [CrossRef]
12. Queiroz, M.F.; Teodosio Melo, K.R.; Sabry, D.A.; Sassaki, G.L.; Rocha, H.A.O. Does the use of chitosan contribute to oxalate kidney stone formation? *Mar. Drugs* **2014**, *13*, 141–158. [CrossRef] [PubMed]
13. Vijayalakshmi, K.; Devi, B.M.; Sudha, P.N.; Venkatesan, J.; Anil, S. Synthesis, characterization and applications of nanochitosan/sodium alginate/microcrystalline cellulose film. *J. Nanomed. Nanotechnol.* **2016**, *7*, 419.
14. Azzam, E.M.S.; El-Farargy, A.F.M.; Abd-Elaal, A.A. Enhancement the detection of $Ni^{2+}$ and $Zn^{2+}$ ions using nanostructure of synthesized dithiol surfactants with gold nanoparticles. *J. Ind. Eng. Chem.* **2014**, *20*, 3905–3912. [CrossRef]
15. Dąbrowski, A.; Hubicki, Z.; Podkościelny, P.; Robens, E. Selective removal of the heavy metal ions from waters and industrial wastewaters by ion-exchange method. *Chemosphere* **2004**, *56*, 91–106. [CrossRef] [PubMed]
16. Do Amaral Montanheiro, T.L.; Cristóvan, F.H.; Machado, J.P.B.; Tada, D.B.; Durán, N.; Lemes, A.P. Effect of MWCNT functionalization on thermal and electrical properties of PHBV/MWCNT nanocomposites. *J. Mater. Res.* **2015**, *30*, 55–65. [CrossRef]
17. Velmurugan, N.; Kumar, G.G.; Han, S.S.; Nahm, K.S.; Lee, Y.S. Synthesis and characterization of potential fungicidal silver nano-sized particles and chitosan membrane containing silver particles. *Iran. Polym. J.* **2009**, *18*, 383–392.
18. Abdulrazzak, F.H.; Alkiam, A.F.; Hussein, F.H. Behavior of X-Ray Analysis of Carbon Nanotubes. In *Perspective of Carbon Nanotubes*; IntechOpen: London, UK, 2019.
19. Azzam, E.M.S.; Abd El-aal, A.A.; Shekhah, O.; Arslan, H.; Wöll, C. Fabrication of spr nanosensor using gold nanoparticles and self-assembled monolayer technique for detection of $Cu^{2+}$ in an aqueous solution. *J. Dispers. Sci. Technol.* **2014**, *35*, 717–724. [CrossRef]
20. Al-Kuhaili, M.F.; Ahmad, S.H.A.; Durrani, S.M.A.; Faiz, M.M.; Ul-Hamid, A. Application of nickel oxide thin films in NiO/Ag multilayer energy-efficient coatings. *Mater. Sci. Semiconduct. Process.* **2015**, *39*, 84–89. [CrossRef]
21. Gandhi, M.R.; Meenakshi, S. Preparation and characterization of silica gel/chitosan composite for the removal of Cu (II) and Pb (II). *Int. J. Biol. Macromol.* **2012**, *50*, 650–657. [CrossRef] [PubMed]

polymers

MDPI

*Article*

# Designing Sustainable Polymer Blends: Tailoring Mechanical Properties and Degradation Behaviour in PHB/PLA/PCL Blends in a Seawater Environment

**Leonardo G. Engler [1,2], Naiara C. Farias [1], Janaina S. Crespo [1,2], Noel M. Gately [3], Ian Major [1], Romina Pezzoli [3] and Declan M. Devine [1,*]**

1 PRISM Research Institute, Technological University of the Shannon: Midlands Midwest, Athlone Campus, University Road, N37 HD68 Athlone, Ireland; a00278634@student.tus.ie (L.G.E.)
2 Postgraduate Program in Materials Science and Engineering, University of Caxias do Sul, Francisco Getúlio Vargas Street, 1130, Caxias do Sul 95070-560, Brazil
3 Applied Polymer Technologies Gateway, Technological University of the Shannon: Midlands Midwest, Athlone Campus, University Road, N37 HD68 Athlone, Ireland
* Correspondence: ddevine@ait.ie

**Abstract:** Biodegradable polyesters are a popular choice for both packaging and medical device manufacture owing to their ability to break down into harmless components once they have completed their function. However, commonly used polyesters such as poly(hydroxybutyrate) (PHB), poly(lactic acid) (PLA), and polycaprolactone (PCL), while readily available and have a relatively low price compared to other biodegradable polyesters, do not meet the degradation profiles required for many applications. As such, this study aimed to determine if the mechanical and degradation properties of biodegradable polymers could be tailored by blending different polymers. The seawater degradation mechanisms were evaluated, revealing surface erosion and bulk degradation in the blends. The extent of degradation was found to be dependent on the specific chemical composition of the polymer and the blend ratio, with degradation occurring via hydrolytic, enzymatic, oxidative, or physical pathways. PLA presents the highest tensile strength (67 MPa); the addition of PHB and PCL increased the flexibility of the samples; however, the tensile strength reduced to 25.5 and 18 MPa for the blends 30/50/20 and 50/25/25, respectively. Additionally, PCL presented weight loss of up to 10 wt.% and PHB of up to 6 wt.%; the seawater degradation in the blends occurs by bulk and surface erosion. The blending process facilitated the flexibility of the blends, enabling their use in diverse industrial applications such as medical devices and packaging. The proposed methodology produced biodegradable blends with tailored properties within a seawater environment. Additionally, further tests that fully track the biodegradation process should be put in place; incorporating compatibilizers might promote the miscibility of different polymers, improving their mechanical properties and biodegradability.

**Keywords:** poly(lactic acid) (PLA); poly(hydroxybutyrate) (PHB); polycaprolactone (PCL); thermal degradation; seawater degradation; biodegradable blends

## 1. Introduction

As our dependence on plastics continues to grow globally, the amount of plastic waste that finds its way into our oceans, rivers, and landfills is largely a result of human behaviour. This trend has catastrophic consequences, as plastic particles are now being incorporated into the food chain of animals, including humans, with unpredictable consequences [1,2]. Given that oceans account for approximately 97% of the world's water supply and cover nearly 70% of the Earth's surface, it is likely that a significant portion of this plastic waste ultimately ends up in our oceans, particularly when not properly disposed of [3].

Biodegradable polymers are a viable approach to addressing the issue of plastic pollution in marine ecosystems. In contrast to conventional plastics, these polymers are capable of rapidly decomposing over a brief span of a few months through natural mechanisms, such as enzymatic digestion, hydrolysis, oxidation, or mechanical degradation, and are hence environmentally friendly. Nevertheless, not all biodegradable polymers degrade in seawater; for instance, PHB and PCL are known to biodegrade in seawater [4,5], whereas PLA is not [6,7]. Therefore, it is crucial to study the blending process and characterisation of these polymers to enable the comprehensive seawater degradation of these materials.

Seawater solutions contain a higher content of salts than freshwater, with chloride, sodium, sulphate, magnesium, calcium, and potassium making up 99% of the ionic salts present [8]. The temperature of seawater varies between 30 °C on the surface in summer and −1 °C on the seabed during winter, depending on the season and geographical region, while the pH can range from 7.5 to 8.4 [1,9]. Additionally, previous studies have shown that the natural dynamic movement of waves in oceans can promote degradation by mechanical fragmentation of the biodegradable polymers studied, such as PLA, PCL, and PHAs [6,10].

Reproducible investigations of seawater degradation in the natural environment are challenging owing to the varying climatic conditions in different regions. However, laboratory experiments can standardise the experimental parameters, increasing the likelihood of reproducible results [1]. Similarly, when blending PLA with more flexible materials, such as PCL and PHB, the goal is to enhance flexibility while preserving the mechanical properties of the blends. Although this approach shows promise, it is essential to overcome the limitations of multi-material mixtures using techniques such as plasticisation, copolymerisation, and blending [11,12]. The use of different polymers to produce ternary blends is an interesting alternative that can be used to tailor the desired properties of materials. In the context of biomedical polymeric devices, understanding their degradation process is crucial for ensuring their stability and functionality. By examining the degradation mechanisms and behaviour of these ternary blends, researchers can gain valuable insights into how the materials perform under real-world conditions. This knowledge enables the development of more robust and reliable biomedical devices that can withstand the challenges of the human body and contribute to improved patient outcomes.

The degradation process of biomedical polymeric devices is a multifaceted phenomenon that is subject to variations in the environmental conditions and kinetics, which depend on the processing settings, temperature, sample geometry, and packing properties; e.g., moulded parts are more densely packed than extruded components, which are susceptible to die swelling (the expansion due to polymer chain realignment) once they exit the extruder [13–15], therefore presenting a more complex degradation process than in seawater. Similarly, the degradation of such medical devices can also occur via four distinct mechanisms, namely hydrolytic, oxidative, enzymatic, and physical degradation [1]. Hydrolytic degradation is triggered by the interaction between the polymeric chains and the water present in the surrounding tissues. On the other hand, oxidative degradation results from the release of oxidants by tissues, which acts as a biological defence mechanism against the implant. Enzymatic degradation also involves a biological response, but it varies from patient to patient and from tissue to tissue over time. Finally, physical degradation occurs as a result of water swelling and mechanical tension, leading to changes in the glass transition temperature and geometry of the swollen polymers. This, in turn, affects the mechanical properties of the device and could potentially lead to fractures, particularly in situations where friction is involved in the motion under pressure [1,16].

Narancic et al. (2018) [15] previously examined the marine biodegradability of 25 mg extruded samples of different biodegradable polymers and found that only PHB and thermoplastic starch (TPS) were marine degradable. To overcome this challenge, the novelty of this work is to determine whether, through the production of ternary blends containing common biodegradable polymers, PLA, and PCL, a commercially relevant marine biodegradable polymer blend could be produced. Furthermore, most medical devices and packaging types are produced by injection moulding which manufactures parts that have tightly packed polymer chains compared to extruded parts and are more challenging from a biodegradation point of view.

The selection of polymer concentrations in ternary blends of PHB, PLA, and PCL, is also challenging and can be justified based on various factors. Firstly, each polymer in the blend contributes distinct properties, with PHB offering biodegradability and excellent mechanical properties, PLA providing stiffness and strength, and PCL imparting flexibility and toughness to the blends [13,14]. Accordingly, this study proposes that the low flexibility of PLA can be mitigated by blending it with more pliable materials such as PCL and PHB through a blending process, while simultaneously tailoring its degradation properties. The selected blends exhibit enhanced flexibility and can be applied in various fields, including packaging and medical devices, particularly in the production of biodegradable straws as an alternative to single-use plastics. Additionally, optimal blends will serve in future studies as feedstock material for the development of flexible biodegradable ureteral stents.

## 2. Materials and Methods

### 2.1. Materials

Poly-(lactic acid) (PLA 4043D) was purchased from NatureWorks (Minnetonka, MN, USA), it has a melt flow index (MFI) of 4.9 g·10 $min^{-1}$, and a density of 1.24 g·$cm^{-3}$. Poly-(3-hydroxybutyrate) (PHB P226) was purchased from Biomer (Krailling, Germany), it is highly crystalline with an MFI of 16.3 g·10 $min^{-1}$ and a density of 1.25 g·$cm^{-3}$. Polycaprolactone (PCL CAPA 6500) was purchased from Ravago Chemicals (Barnsley, UK), it has a high molecular weight ($M_w$ 50,000 g·$mol^{-1}$), and MFI of 7.0 g·10 $min^{-1}$, all MFI tests were performed at a temperature of 175 °C and a load of 2.16 kg, following standard methods. All polymers were used as received and they were dried with forced hot air circulation at 45 °C for 24 h before processing, to avoid any moisture residue that could promote the hydrolysis degradation of the polyesters.

### 2.2. Material Extrusion

The compounding of materials was carried out on a Leistritz Micro 27 twin-screw co-rotating extruder (Leistritz Group, Nuremberg, Germany) with a 27 mm screw diameter and a 38:1 length-to-diameter ratio, with a constant screw speed of 150 rpm and temperature profile from feeder to die of 110/120/130/140/150/160/170/180 °C, and two additional heating zones at the head of the extruder, a flange and a strand die, at 190 and 200 °C, respectively. The extruder was fed at 4 kg·$h^{-1}$, and the compounded samples were named as PHB, PLA, PCL, 50/50/0, 50/0/50, 50/25/25 and 30/50/20 (wt.% PHB/PLA/PCL). The filament produced was pelletised to 3 mm granulates after cooling down in a conveyor belt.

### 2.3. Material Moulding

Tensile specimens with dimensions (type V): length 63.50 mm, width 9.50 mm, gauge length 9.50 mm, gauge width 3.18 mm, and thickness 3.20 mm, were moulded according to ASTM D638 on a BabyPlast® 6/12 (Rambaldi, Italy) injection moulding process, equipped with a 14 mm diameter piston. The machine possesses three temperature-controlled zones: plasticising zone, chamber and nozzle represented as zone 1, 2 and 3, respectively. The injection moulding parameters were optimised for each sample and are described in Table 1. At least 5 specimens for each sample were produced.

### 2.4. Seawater Degradation

For the seawater degradation study, a marine broth solution based on ASTM D6691 was prepared and placed into a 22 L unstirred thermostatic bath Clifton™ (Oldmixon Cres, Weston, UK), allowing the temperature constant to maintain at 30 °C for up to 8 weeks of the experimental testing. The marine broth was prepared by weighing all the individual components of the solution in a high precision analytical scale (±0.1 mg), before dissolving them in distilled water under constant stirring for 3 h to promote solution homogeneity. A commercial fish tank pump was introduced to the thermostatic bath to promote the aeration of the solution and prevent components of the solution from depositing at the bottom of the bath.

**Table 1.** Injection moulding process parameters of PHB, PLA, and PCL and their respective blends with different compositions (wt.% PHB/PLA/PCL).

| Sample | Temperature (°C) | | | Shot Size (mm) | Cooling | | Pressure (Bar) | |
|---|---|---|---|---|---|---|---|---|
| | Zone 1 | Zone 2 | Zone 3 | | T (°C) | t (s) | 1st | 2nd |
| PHB | 185 | 175 | 170 | 35 | 50 | 25 | 40 | 30 |
| PLA | 210 | 200 | 180 | 40 | 35 | 35 | 80 | 70 |
| PCL | 170 | 160 | 150 | 40 | 35 | 35 | 100 | 90 |
| 50/50/0 | 160 | 150 | 140 | 37 | 35 | 35 | 90 | 85 |
| 50/0/50 | 150 | 140 | 130 | 36 | 35 | 35 | 70 | 65 |
| 50/25/25 | 150 | 140 | 130 | 36 | 35 | 35 | 70 | 65 |
| 30/50/20 | 160 | 150 | 140 | 36 | 35 | 35 | 90 | 85 |

Moulded tensile samples were weighed and placed into the seawater solution distanced at a minimum of 2 cm from each other, to assess degradation during 0, 14, 28, and 56 days with 3 minimum replicates for each sample. Each collected sample was washed and dried before testing.

### 2.5. Weight Variance

The weight variance of samples was measured from the beginning of the degradation tests and after 14, 28, and 56 days of seawater degradation. After each collection date, the samples were removed from the seawater environment, washed with current water, and dried for 24 h in an oven at 40 °C. Their final weight was measured and recorded using a Sartorius Entris analytical balance (Sartorius AG, Göttingen, Germany), prior to any subsequent testing.

### 2.6. Mechanical Properties

Tensile testing was conducted in accordance with ASTM D3039 standard procedure using a ZwickRoell Z010 universal testing machine (ZwickRoell Ltd., Ulm, Germany) equipped with a 5 kN load cell. The tests were carried out at a strain rate of 5 mm·min$^{-1}$, at room temperature and the tensile properties were obtained directly from the equipment analysis software. A minimum of 3 replicates were evaluated per group, and the dimensions of each sample were recorded prior to testing. The stress–strain curves were obtained from the software TestXpert II (ZwickRoell, Ulm, Germany), and utilised for the calculation of the Young Modulus.

### 2.7. Infrared Analysis

The infrared analysis was performed using a Spectrum One (Perkin Elmer, Waltham, MA, USA) with a universal attenuated total reflectance (ATR) sample adapter. All data were obtained at room temperature (20 °C) in the spectral range between 4000–650 cm$^{-1}$ against a background of air, using 10 scans per sample cycle and a resolution of 4 cm$^{-1}$. The compression force was kept constant at 80 N and the data were collected from the spectrum software used to perform the analysis. The chemical groups were evaluated in regions by performing the ratio area of the groups related to the main chain (–CH3 and –CH2) and the oxygenated groups (–C=O, –C–O). The areas of –CH3, –CH2 stretches; $A_1$ (3050–2800 cm$^{-1}$); –C=O: $A_2$ (1840–1600 cm$^{-1}$); and –C–O: $A_3$ (1250–1000 cm$^{-1}$) were determined using the software Origin (OriginLab©). Results were expressed as the mean and standard deviation.

### 2.8. Thermal Analysis

By using a thermal analyser (TGA-50, Shimadzu Co., Kyoto, Japan) with a dynamic nitrogen atmosphere of 50 mL·min$^{-1}$, the blends were characterised by thermogravimetric analysis to evaluate the thermal stability of the blends produced. Samples weighing between 8–12 mg were heated from room temperature (20 °C) to 700 °C at a rate of 10 °C·min$^{-1}$, providing information about the weight loss plotted against the temperature for each sample. Additionally, a derivative curve (DTG) was produced to indicate the temperatures at which maximum rates of weight loss occurred.

The thermal properties of the samples were investigated in a differential scanning calorimeter DSC Pyris 4000 (Perkin Elmer, Waltham, MA, USA). Samples weighing between 4–6 mg were crimped in non-perforated aluminium pans with reference to an empty crimped pan. The analysis was carried out with a nitrogen flow rate of 30 mL·min$^{-1}$ to prevent the sample degradation. The samples were heated from −60 to 190 °C at a rate of 10 °C·min$^{-1}$ holding the temperature for 2 min before cooling down to −60 °C and holding the temperature for 2 min again, this process is repeated twice for each sample. The thermal properties of the samples were analysed with the equipment software Pyris Series DSC 4000 (Perkin Elmer, Waltham, MA, USA) and the data analysed based on the second heating cycle of the test. The crystallinity of the samples was calculated with the values in Table 4 by applying the crystallinity index equation,

$$C_I = \frac{\Delta Hm}{\Delta Hm_0 \times f} \times 100\% \quad (1)$$

where $C_I$ is the crystallinity index, ΔHm is the experimental melting enthalpy, f is the polymer weight fraction and $\Delta Hm_0$ is the theoretical melting enthalpy of the pure material (considering it as 100% crystalline). The theoretical melting enthalpy for PHB, PLA and PCL are 146 J·g$^{-1}$, 93.7 J·g$^{-1}$, and 135 J·g$^{-1}$, respectively [17–19].

Dynamic mechanical thermal analysis (DMTA) was run to provide information about thermal transitions ($T_g$ values) that were not possible to identify in the previous thermal analysis, due to the overlapping of these properties in the blends of this study. DMTA was carried out in single-cantilever mode using a Perkin Elmer (Waltham, MA, USA) DMA 8000. Rectangular samples of size 20 × 10 × 3 mm$^3$ were subjected to a temperature ramp from −80 up to 100 °C at a constant heating rate of 3 °C·min$^{-1}$. A frequency of 1 Hz was used for all tests and the maximum flexural deformation (γ) was set at 10 μm.

### 2.9. Morphology Characterisation

Tensile samples were subjected to a cryofracture process using liquid nitrogen to observe the internal morphology of the samples without any deformation in the inner layers of the sample. A field emission scanning electron microscope (Tescan Mira, Oxford Instruments, Cambridge, UK) (5 kV) was used to assess the degradation areas of samples after 56 days in the seawater environment.

## 3. Results and Discussion

### 3.1. Seawater Degradation

As reported in previous studies [20], the degradation of a biomaterial may occur through four different mechanisms: hydrolytic degradation (scission of polymeric chains by hydrolytic activity), enzymatic degradation (scission of polymeric chains by enzymatic activity), oxidative degradation (radical attack supported by peroxide-producing inflammatory reactions, when in vivo), and physical degradation (depending on the physical activity of the environment, swelling-deswelling, etc.) [21].

Surface erosion and bulk degradation represent two distinct mechanisms of hydrolytic degradation, although their occurrence may not always be mutually exclusive [22]. Bulk degradation is the predominant mechanism of degradation in most polyester-based polymers, and when the rate of water diffusion is greater than the rate of hydrolysis (Dt >> Ht), sample saturation by the degradation media and a non-linear mass loss over time occur [20]. Additionally, as a result of the accumulation of degradation products, it is common to have autocatalytic reactions in situ, which leads to areas of accelerated degradation that might compromise or cause the structure to collapse [14]. Surface erosion, on the other hand, commonly presents a linear loss of mass over time and is characterised by a faster rate of hydrolysis than water diffusion (Dt ≤ Ht). This effect reduces the accumulation of degradation products since they are free to diffuse away from the polymeric matrix; therefore, no autocatalytic reactions occur [20].

The phenomenon of bulk degradation in poly(α-hydroxy-esters) samples with a thickness (Lc) up to 7.4 cm was previously reported in studies with temperatures ranging between 15 and 37 °C, which reported an increased rate of PHA and PLA degradation in alkali environments. Additionally, the onset of percent crystallinity change during hydrolysis was observed at approximately 37 °C, corresponding to the temperature at which the hydrolysis of PLA chains starts to occur noticeably [22–25]. This has been demonstrated through experimental results as well, as depicted in Figure 1. Over a period of 56 days, an observable increase in the weight of some samples is noted, attributable to the formation of fissures and cracks, which permit the infiltration of aqueous solutions into the PLA-based samples, leading to a transient rise in their mass, as observed in samples of 100 wt.% of PLA, 50/50/0, and 30/50/20 (wt.% PHB/PLA/PCL). The same investigations also demonstrated that microbial activity exerts a marginal to negligible impact on the hydrolysis phase of degradation, which involves the cleavage of ester linkages in PLA samples [24]. PHB and PCL samples exhibited notable mass variations throughout the 56 days of seawater degradation, whereby PCL demonstrated a weight reduction of up to 10%. Furthermore, the specimens containing higher amounts of these two polymers exhibited comparable behaviours, as observed for the samples 50/0/50 and 50/25/25 (wt.% PHB, PLA, and PCL) in Figure 1. It is assumed that surface erosion serves as the main degradation mechanism for these samples, as evidenced by the constant decrease in weight over time. Bagheri et al. (2017) [7] investigated the biodegradability of PHB, PLA, PCL, and poly(lactic-co-glycolic acid) (PLGA) in a seawater environment at a constant temperature of 25 °C for over a year; only the latter underwent complete degradation within this timeframe, while the other polymers exhibited minimal degradation. Specifically, PHB demonstrated only 6% weight loss, whereas PLA and PCL displayed less than 1% weight loss. These findings align with the supposition posited by previous researchers [1] that temperature significantly affects the degradation process. In addition, Volova et al. (2010) [4] reported that PHB and PHBV films experienced 46% mass loss after being submerged in seawater at 29 °C for 160 days, with complete sample degradation occurring after 350 days, as reported in a similar study [1].

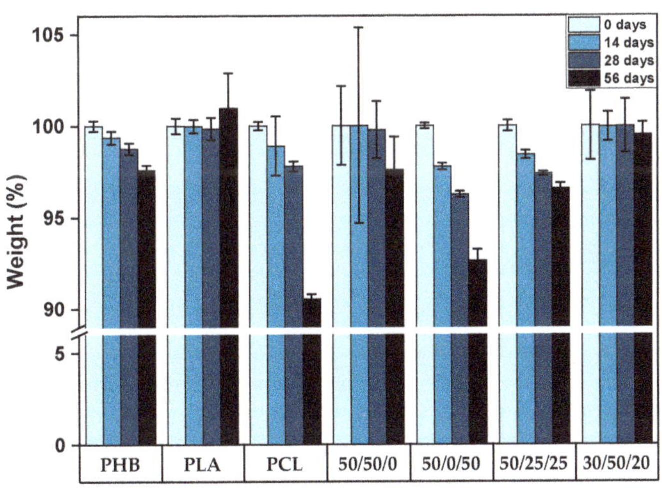


**Figure 1.** Weight variance (%) of PHB/PLA/PCL and their respective blends with different compositions (wt.% PHB/PLA/PCL) after 0, 14, 28 and 56 days of seawater degradation.

PLA degradation is a complex phenomenon involving several steps. Initially, its polymer chains undergo hydrolysis, which breaks down the bonds between the individual monomers that constitute the polymer. Water and other solutions permeate the PLA matrix and initiate the hydrolysis reaction, causing the disintegration of the polymer chains. Once the polymer chains are broken down into smaller fragments, the resulting lactic acid oligomers are metabolised by microorganisms present in the environment. This process involves the enzymatic degradation of lactic acid oligomers by microorganisms, which consume the fragments and use them as a source of carbon and energy [26,27].

In an oxygen-rich environment, the biodegradable polymers PLA and PCL undergo degradation primarily into $CO_2$, CO, water and short-chain acids. However, blending these polymers with PHB can have a significant impact on the degradation process. The addition of PHB promotes the initial degradation of crotonic acid (2-butenoic acid), followed by the degradation of the aforementioned products [16]. This occurs due to the presence of microbial communities that preferentially consume the crotonic acid in the blended polymer mixture. As the crotonic acid is consumed, the degradation of the other components is accelerated, leading to the ultimate breakdown of the polymer blend into its constituent components [28].

As the average temperature and microorganism density in seawater are relatively low, the biodegradation of PLA in this environment occurs slowly. In a previous degradation study [6] using thin films of PLA (0.05 mm of thickness), after 10 weeks, samples in a real seawater solution at 25 °C showed an insignificant change in the average molar mass ($M_w$). Additionally, PLA films of about 0.32 mm thickness presented no mass loss in a seawater solution, even after a year of exposure. Nevertheless, studies comparing the degradation of PCL and PHB films (thickness of 0.1 mm) at 25 °C [29] in seawater reported weight loss of 100 and 41%, respectively, after 4 weeks. The same study conducted with freshwater presented weight loss of only 25 and 9% for PCL and PHB, respectively, after 10 weeks. confirming the significance of the biopolymer type, geometry of the samples, and also the conditions of the environment, such as temperature and type and concentrations of marine microbes in the seawater [1,7,29].

Similarly, research carried out in a marine pelagic environment using 25 mg extruded films of blends produced with PBS, PHB, PLA and PCL reported promising relative biodegradation (wt.%) over a period of 56 days [15]. However, when comparing the relative biodegradation of PHB/PCL (60/40) samples reported in the previous study, which demonstrated approximately 15 wt.% weight loss in 25 mg films, with the blend 50/0/50 (PHB/PLA/PCL) used in our current research involving injection-moulded parts, we observed a higher loss of weight. These findings suggest that our approach, which utilises the industrial production method of injection moulding to create more densely packed components, yields more rapid biodegradation in the produced blends compared to extruded components.

Some studies reported delayed disintegration of the blends due to the selectivity of degradation of specific polymeric chains. The significant effect of hydrolysis on these polymeric chains was reported as being responsible for changes in the mechanical properties, resulting in softening of the blends, including PLA samples [7].

Aliphatic polyesters are known to biodegrade via the hydrolytic breakdown of their ester bonds [30]. The by-products of hydrolytic biodegradation lead to a more acidic environment, which raises a concern for implantable devices, as a common response would be local tissue inflammation. To lessen the acidity, researchers reported adding basic salts in the media or within the polyester samples [31,32].

The chemical hydrolysis rate increases in amorphous areas around the polymers due to the facilitated diffusion of water into the polymeric matrix compared to semi- and crystalline polymers, which present more organised structures causing diffusion limitation, even at temperatures higher than the Tg. Moreover, the diffusion of polymers with Tg values lower or similar to the temperature used in the degradation studies is primarily governed by the amorphous area of the polymeric matrix, where chemical hydrolysis is dominant [10,31].

According to previous research, the amount of PCL added to a polymeric matrix can be altered to control the rate at which the copolymers degrade [33]. These studies explored the addition and crosslinking of different PCL contents to fine-tune the degradation behaviour of copolymers, providing a greater degree of control over the release of therapeutic agents or other functional components. This has significant implications for the development of biomaterials that require precise degradation profiles as well as for the engineering of implantable medical devices and drug delivery systems [31,34].

*3.2. Mechanical Properties*

The PHB/PLA/PCL-based specimens were tested at a low strain rate of 5 mm·min$^{-1}$, and the results are shown in Figure 2 after 0 and 56 days of seawater degradation. The tensile properties were acquired using equipment software after each test. However, to confirm the accuracy of this information, sample data were used to calculate the mechanical properties of the polymer blends.

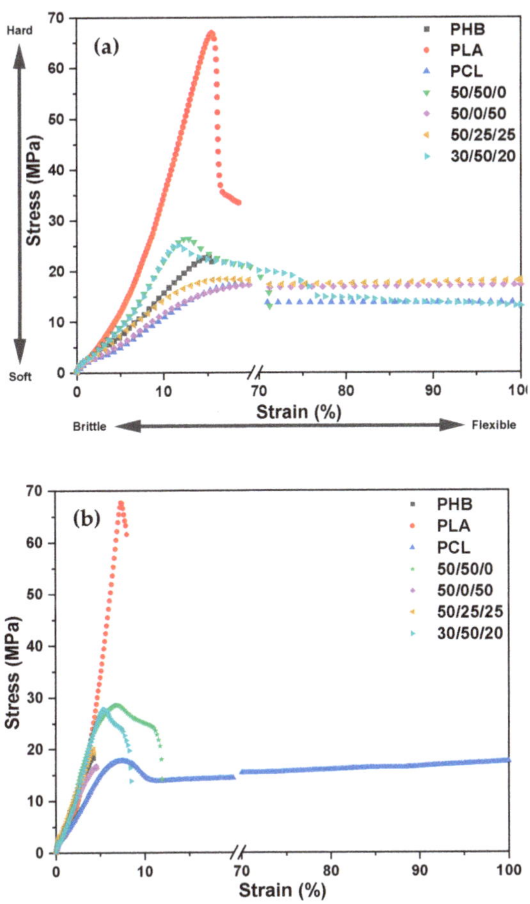


**Figure 2.** Stress-Strain curves of PHB/PLA/PCL and their respective blends with different compositions (wt.% PHB/PLA/PCL) after 0 days in (**a**), and after 56 days of seawater degradation, in (**b**). Individual curves of Stress vs. Strain after 0 and 56 days for each material are plotted in Figure S1.

The tensile results of neat PLA presented a higher yield stress resistance due to its higher intrinsic brittleness, while the remaining samples, including PHB, PCL, and blends of PHB, PLA, and PCL, presented a notable increase in the ultimate strain, evidenced by the higher flexibility of these materials. Similar results were reported in previous studies using PLA and PHB blends [35]. Notwithstanding, previous studies on the toughness of PLA [36] reported that the tensile strength and elastic modulus of ternary blends may differ slightly compared with neat PLA, implying that these mechanical properties are mainly influenced by the content of the PLA phase in the blends. And because PLA is more susceptible to thermal and hydrolytic degradation, a reduction in molecular weight due to the degradation process might lead to lower mechanical properties in the final material [37,38]. This is evidenced in Figure 2b, where all samples became more brittle after 56 days of seawater degradation.

Neat PLA is quite brittle when compared to PHB and PCL, as observed from the high stress and low strain relationship seen in Figure 2; it has been reported to have a low yield elongation of about 7% [39], and a modulus of about 3.4 GPa [11], whereas its tensile strength was found to be approximately 70 MPa, as seen in Figure 3 and supported by previous investigations [12]. These properties may limit its application in certain fields that require materials with higher ductility and toughness, such as flexible medical devices, or some packaging [40]. Previous investigations focusing on PLA/PCL blends have documented a reduction in sample stiffness with increasing PCL content, while conversely observing an increase in the elongation at break within the same samples. Furthermore, these studies reported a notable enhancement in impact strength with higher PCL concentrations. Specifically, the impact strength exhibited an improvement of approximately 200% with the incorporation of 30 wt.% PCL and a significant increase of approximately 350% when 40 wt.% PCL was added [41].

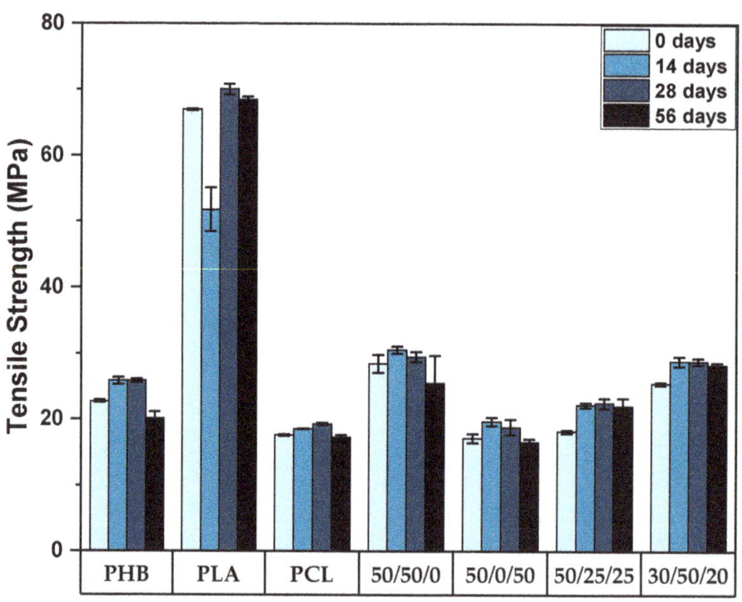


**Figure 3.** Tensile strength of PHB/PLA/PCL and their respective blends with different compositions (wt.% PHB/PLA/PCL) after 0, 14, 28 and 56 days of seawater degradation.

Tensile strength is an important mechanical property of polymers because it reflects the ability of any material to withstand pulling or stretching forces, which is commonly assessed for applications such as medical devices and packaging. The results provided in Figure 3 corroborate that PLA stands as the polymer with the highest tensile strength resistance in this study, and blends containing PLA also presented better performance in the tensile tests, as expected. As one of the drawbacks related to PLA is the low intrinsic elongation at break, the addition of PHB and PCL to tune the overall flexibility of blends is a promising solution for brittle samples, such as PLA. Previous investigations added different amounts of PCL to improve the ductile properties of PLA, which works similarly as plasticizers; however, the usual outcome is lower tensile strength [42,43]. As aforementioned, unblended PLA samples presented tensile strength results of 67 MPa, while with the addition of PHB and PCL, values of about 18 MPa (a decrease of 73.1%) and 25.5 MPa (a decrease of 61.9%) were obtained for the samples 50/25/25 and 30/50/20 (wt.% PHB/PLA/PCL), respectively, for samples not exposed to the seawater degradation environment. Furthermore, seawater exposure can lead to the absorption of the seawater solution by the polymeric blends, which may vary according to the hydrophilic nature of each individual polymer. This can cause swelling, softening, and plasticization of the blends, potentially leading to a decrease in mechanical properties such as tensile strength and stiffness, as reported in previous investigations [44]. Additionally, exposure to high temperatures, UV radiation, and mechanical stress can accelerate the degradation process.

### 3.3. Infrared Analysis

To better understand the main differences in the chemical structure and type of bonding, all samples were analysed at different times of seawater degradation after 0, 14, 28, and 56 days. A previous study suggested that PLA degradation is primarily driven by the generation of acetaldehyde and lactide monomers, although small amounts of carbon dioxide and carbon monoxide have also been detected [45]. The main degradation product of PHB has been identified as crotonic acid (2-butenoic acid) by numerous researchers. This assertion is supported by the characteristic peaks of the compound at 3056, 1722, and 1152 $cm^{-1}$, which are attributed to C–H, C=O, and C–O stretching vibrations, respectively [46]. In addition, other studies have indicated the release of ester groups during PHB degradation [47]. PLA hydrolysis degradation and formation of oligomeric lactic acids (OLAs) were supported by the development of a double peak related to $CH_3$ stretching at 1454 $cm^{-1}$ and another double peak at 1743 $cm^{-1}$ related to the C=O stretching vibrations [23,44]. The degradation of PCL is characterised by the evolution of ε-caprolactone (approximately 1736 $cm^{-1}$) and 5-hexenoic acid and its di- and trimers (approximately 3475 $cm^{-1}$); however, carbon dioxide and some traces of carbon monoxide have also been previously identified [45]. All FTIR peaks are shown in Figure S2.

The results presented in Table 2 show the ratio of the peak areas of the neat polymers PHB, PLA, and PCL, as well as their respective blends, after 0, 14, 28, and 56 days of seawater degradation. The ratios were determined based on the axial deformation of the bands $A_1$ = 3050–2800 $cm^{-1}$ (–$CH_3$, –$CH_2$), $A_2$ = 1840–1600 $cm^{-1}$ (–C=O), and $A_3$ = 1250–1000 $cm^{-1}$ (–C–O), which correspond to the specific chemical groups in the polymers of this study. Additionally, these results revealed that at the longest seawater degradation time (56 days), the neat PLA and PHB polymers exhibited a reduction in the peak ratio of areas $A_2/A_1$ and $A_3/A_1$ related to the C=O and C–O stretching absorption bands in relation to the nonpolar peaks (–$CH_3$, –$CH_2$), suggesting the evolution of the degradation of these polymeric chains, as confirmed by previous investigations [23,44]. The ratio of peak areas of PCL infrared samples remained slightly the same, suggesting that seawater degradation did not affect the chemical structure of the polymer over the time of these investigations, and thus it also affected blends containing PCL.

**Table 2.** Ratio of peak areas of FTIR spectra of neat polymers and their respective blends (wt.% PHB/PLA/PCL) at 0, 14, 28, and 56 days.

| | Ratio | Days | | | |
|---|---|---|---|---|---|
| | | 0 | 14 | 28 | 56 |
| PHB | $(A_2/A_1)$ | 5.17 | 3.91 | 14.13 | 3.78 |
| | $(A_3/A_1)$ | 5.28 | 3.49 | 15.27 | 3.66 |
| PLA | $(A_2/A_1)$ | 11.68 | 8.37 | 10.52 | 9.09 |
| | $(A_3/A_1)$ | 31.00 | 20.48 | 27.24 | 23.19 |
| PCL | $(A_2/A_1)$ | 1.46 | 1.41 | 1.44 | 1.33 |
| | $(A_3/A_1)$ | 1.65 | 2.24 | 1.94 | 2.02 |
| 50/50/0 | $(A_2/A_1)$ | 5.35 | 5.19 | 3.18 | 5.63 |
| | $(A_3/A_1)$ | 10.05 | 8.56 | 4.76 | 9.35 |
| 50/0/50 | $(A_2/A_1)$ | 2.35 | 2.08 | 2.22 | 4.05 |
| | $(A_3/A_1)$ | 2.43 | 2.15 | 1.88 | 4.72 |
| 50/25/25 | $(A_2/A_1)$ | 3.58 | 3.03 | 3.37 | 3.23 |
| | $(A_3/A_1)$ | 4.44 | 3.96 | 3.96 | 3.96 |
| 30/50/25 | $(A_2/A_1)$ | 4.03 | 4.46 | 4.13 | 4.49 |
| | $(A_3/A_1)$ | 5.58 | 6.48 | 5.84 | 7.10 |

Axial deformation of bands: $A_1$ = 3050–2800 cm$^{-1}$ (–$CH_3$, –$CH_2$); $A_2$ = 1840–1600 cm$^{-1}$ (–C=O); $A_3$ = 1250–1000 cm$^{-1}$ (–C–O).

For the polymer blends, the ratios of the peak areas varied depending on the blend ratio and degradation time. For the 50/50/0 blend, the ratio of the peak areas for A2/A1 decreased after 28 days but increased again at 56 days, indicating a complex degradation pattern. The same behaviour was observed for the peak ratios of A3/A1. For the 50/0/50 blend, the ratio of the peak areas for A2/A1 and A3/A1 increased over time, indicating that the degradation of both PHB and PCL occurred preferentially in the non-polar structures. For 50/25/25 and 30/50/25, the ratios of the peak areas remained relatively constant over the 56-day period, suggesting that the degradation of these blends was not significantly affected by seawater.

### *3.4. Thermal Transitions and Degradation of PHB/PLA/PCL Blends*

All neat polymers investigated (PHB, PLA, PCL) degrade in a one-step stage with typical representative temperatures $T_5$, $T_{10}$, $T_{max}$ and the residual weight summarised in Table 3, for both neat polymers and blends. Blends containing either two or three polymers exhibited more than a one-step degradation process, and it is worth noting that the degradation of PLA is slightly delayed within the blends compared to neat PLA, as can be seen by how the $T_5$ and $T_{10}$ change in different blends [11,45]. Emphasising the blend with the lowest amount of wt.% PHB (30/50/20), presenting the thermal degradation of 312.3 °C ($T_5$).

Previous studies found that the presence of PCL in similar blends delayed the degradation of PLA, meaning that PCL promotes the thermal stability of the blends. Studies using blends of 50/50 and 70/30 (wt.% PCL/PLA) compositions reported an increase in the characteristic decomposition peak temperature of PLA to 325 °C for 50/50 blends and 334 °C for 70/30 blends [11,46]. Similar results were found using blends of PLA and PCL with only 30 wt.% PCL. The latter suggested that the immiscibility between PLA and PCL is evidenced by the presence of a two-step degradation process and that the amount of weight loss is proportional to the polymer content in each blend during degradation [48–50]. According to Arrieta et al., ternary blends such as PHB/PLA/PCL tend to degrade in a three-step process, where each process corresponds to the individual polymer and the mass loss for each step is proportional to the polymeric content in the blend; the same behaviour is observed in the blends in this study, as reported in the blends 50/25/25 and 30/50/20 (wt.% PHB/PLA/PCL) in Figure 4 [51].

**Table 3.** Thermal degradation properties of PHB/PLA/PCL blends obtained by TG/DTG characterisation. Curves of TG/DTG and the respective deconvolution of each thermal degradation are represented in Figure S3.

| Material | $T_5$ (°C) | $T_{10}$ (°C) | $T_{max}$ (°C) | | | Residual Weight (%) |
|---|---|---|---|---|---|---|
| | | | PLA | PHB | PCL | |
| PHB | 389.9 | 403.7 | - | 435.2 ± 0.1 | - | 0.26 ± 0.1 |
| PLA | 247.1 | 292.2 | 339.2 ± 0.0 | - | - | 1.82 ± 0.1 |
| PCL | 403.7 | 420.2 | - | - | 472.0 ± 0.1 | 4.26 ± 0.2 |
| 50/50/0 | 290.6 | 332.4 | 348.0 ± 0.1 | 429.3 ± 0.1 | - | 1.86 ± 0.2 |
| 50/0/50 | 265.3 | 326.5 | - | 344.2 ± 0.1 | 430.4 ± 0.1 | 1.20 ± 0.1 |
| 50/25/25 | 252.2 | 313.3 | 343.6 ± 0.1 | 421.5 ± 0.5 | 464.5 ± 0.5 | 4.21 ± 0.2 |
| 30/50/20 | 312.3 | 334.4 | 343.5 ± 0.1 | 380.6 ± 0.4 | 426.1 ± 0.1 | 3.27 ± 0.3 |

$T_5$, $T_{10}$ and $T_{max}$ being the temperatures at 5%, 10% and maximum weight loss rate.

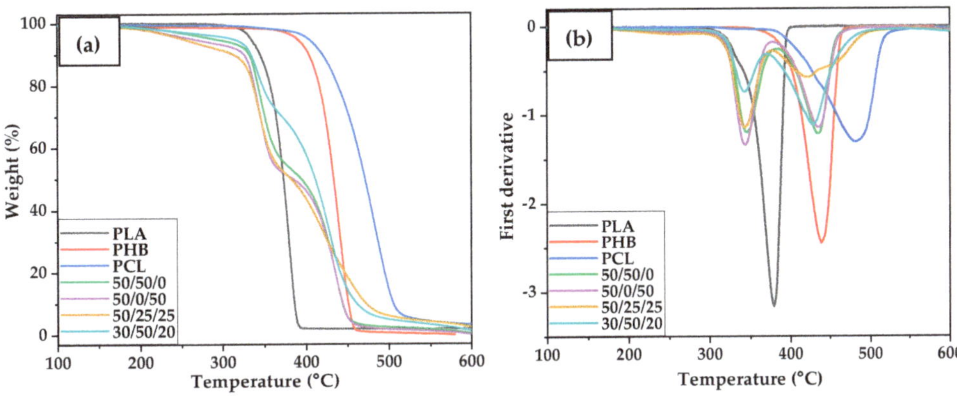


**Figure 4.** Comparative plot of the thermal degradation of PHB, PLA, and PCL samples and their respective ternary blends with different compositions (wt.% PHB/PLA/PCL), (**a**) TGA (thermograms), and (**b**) DTG (first derivative curves).

The DSC thermograms for blends containing both PLA and PCL, such as 50/25/25 and 30/50/20 (wt.% PHB/PLA/PCL) revealed that the melting peak of PCL overlaps with the $T_g$ of PLA at approximately 60 °C. For this reason, conventional DSC is unable to detect fine variations in the $T_g$ of PLA in blends containing PCL. To mitigate this challenge, DMTA analysis of samples was used to detect the $T_g$ points of PHB, PLA, and PCL; the values found are presented in Table 4. Due to the diluting effect in the blend, the cold crystallisation process is visible as a slightly broad peak at the same location as in the neat PLA. This shows that PCL has little effect on the crystal structure of PLA, strongly supporting their poor miscibility [11]. The DSC thermograms of neat polymers and their respective blends are shown in Figure S4.

Table 4 shows that the glass transition temperature ($T_g$) of PLA is around 67.2 °C after 56 days of seawater degradation, which is slightly higher than its original $T_g$ of 66.7 °C before degradation, similar effect happened to the $T_g$ of PCL, whereas the $T_g$ of PHB increased from 55.3 °C to 62.9 °C after the same period. This suggests that seawater degradation did not have a significant effect on the $T_g$ of PLA and PCL, but caused the PHB chains to become more rigid, possibly due to the formation of cross-links between the chains [48,51].

**Table 4.** Differential scanning calorimetry (1st heating curves) for neat polymers and their respective blends (wt.% PHB/PLA/PCL) after 0 and 56 days of seawater degradation.

| Sample Name | d | $T_g$ (°C) | $T_m$ (°C) | $T_c$ (°C) | $\Delta H_c$ (J/g) | $\Delta H_m$ (J/g) | Crystallinity Index (%) |
|---|---|---|---|---|---|---|---|
| PHB | 0 | 55.3 | 167.0 | 110.6 | −76.8 | 67.2 | 46.0 |
| | 56 | 62.9 | 166.8 | 111.2 | −72.2 | 65.2 | 44.7 |
| PLA | 0 | 66.7 | 149.9 | 59.2 | −5.5 | 6.8 | 7.3 |
| | 56 | 67.2 | 150.1 | 59.4 | −4.9 | 7.0 | 7.5 |
| PCL | 0 | −43.1 | 61.3 | 28.1 | −65.3 | 61.3 | 45.4 |
| | 56 | −41.7 | 65.0 | 28.7 | −71.0 | 82.2 | 60.9 |
| 50/50/0 | 0 | 53.8 | PLA (141.8), PHB (169.8) | PLA (97.2), PHB (106.4) | PLA (−1.3), PHB (−32.2) | PLA (10.8), PHB (27.4) | PLA (11.5), PHB (18.8) |
| | 56 | 55.7 | PLA (145.2), PHB (169.8) | PLA (97.9), PHB (107.5) | PLA (−1.8), PHB (−20.0) | PLA (9.9), PHB (29.7) | PLA (10.6), PHB (20.3) |
| 50/0/50 | 0 | 29.4 | PCL (58.3), PHB (169.0) | PCL (29.4), PHB (105.6) | PCL (−35.6), PHB (−36.7) | PCL (34.4), PHB (34.9) | PCL (25.5), PHB (23.9) |
| | 56 | 29.8 | PCL (63.7), PHB (171.4) | PCL (29.8), PHB (107.1) | PCL (−34.2), PHB (−34.9) | PCL (39.7), PHB (34.3) | PCL (29.4), PHB (23.5) |
| 50/25/25 | 0 | 58.8 | PCL (58.8), PLA (143.5), PHB (168.8) | PCL (29.4), PHB (106.2) | PCL (−18.6), PHB (−29.0) | PCL (15.8), PLA (4.0), PHB (27.1) | PCL (11.7), PLA (4.3), PHB (18.6) |
| | 56 | 60.1 | PCL (60.1), PLA (143.8), PHB (168.7) | PCL (29.8), PHB (106.6) | PCL (−21.1), PHB (−31.7) | PCL (26.6), PLA (4.5), PHB (30.4) | PCL (19.7), PLA (4.8), PHB (20.8) |
| 30/50/20 | 0 | 59.8 | PCL (59.5), PLA (144.5), PHB (172.8) | PCL (30.2), PHB (105.0) | PCL (−11.4), PHB (−16.8) | PCL (9.3), PLA (9.2), PHB (13.7) | PCL (6.9), PLA (9.8), PHB (9.4) |
| | 56 | 61.4 | PCL (61.4), PLA (146.4), PHB (171.4) | PCL (30.8), PHB (105.6) | PCL (−15.1), PHB (−21.1) | PCL (21.7), PLA (11.8), PHB (17.8) | PCL (16.1), PLA (12.6), PHB (12.2) |

Furthermore, the $T_m$ values of both PLA and PHB remained relatively constant, indicating that seawater degradation did not significantly affect the crystallinity of these polymers. In contrast, the $T_m$ of PCL increased from 61.3 °C to 65.0 °C after degradation, and it also increased its crystallinity from 45.4% to 60.7% after 56 days of seawater degradation, due to the amorphous regions of the polymer being more susceptible to hydrolysis degradation.

The data also demonstrated that the thermal properties of the polymeric blends were similar to those of their respective neat polymers, varying according to the ratio of each polymer within the blend. For instance, the $T_g$ of the 50/50/0 blend of PLA and PHB was similar to that of neat PHB, but slightly lower than that of neat PLA. Seawater degradation also had a similar effect on the thermal properties of the blends, with the crystallinity indices of neat PLA and PHB polymers remaining slightly constant after degradation, whereas the crystallinity of PCL increased. A similar effect was observed for the respective blends, which varied according to their polymer ratios.

### 3.5. Morphology of the Samples

The examination of fractured specimens after tensile testing revealed the brittle fracture of samples of neat PLA and PHB at 0 and after 56 days of seawater degradation. Both polymers presented lower strain before failure in the stress–strain curves, as reported in Figure 2. A ductile fracture preceded by significant plastic deformation was observed, associated with extensive stress whitening that occurred throughout nearly the entire gauge zone of the specimen, which diminished in the same specimens after 56 d of seawater degradation. This was observed in the blends containing 50 wt.% of PLA, such as both

50/50/0 and 30/50/20 formulations, as shown in Figure 5. Furthermore, it was observed that blends with weight ratios of 50/0/50 and 50/25/25 (wt.% PHB/PLA/PCL) exhibit a greater occurrence of white spots, which are indicative of surface degradation accompanied by a loss of ductility and transition to brittle fracture behaviour [35], after 56 days of seawater degradation.

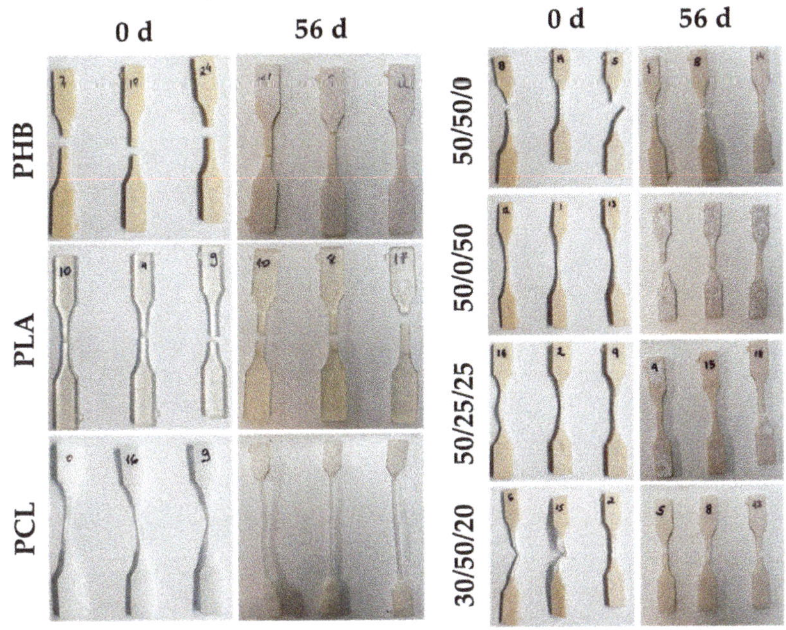


**Figure 5.** Comparative of the different morphologies of PHB, PLA, and PCL samples and their respective ternary blends with different compositions (wt.% PHB/PLA/PCL), after tensile testing at 0 and after 56 days of seawater degradation.

SEM photomicrographs revealed the formation of cracks and fissures on the surfaces of the PLA, which is also a typical indication of polymer degradation [52,53]. The formation of cracks on the surface of PLA samples is an important factor in promoting the diffusion of OLAs into the degradation environment [54]. These cracks increased the surface area of PLA, which was exposed to an environment that promoted degradation, thereby facilitating the diffusion of OLAs and other degradation products. The increased diffusion of OLAs from PLA also promotes local bacterial activity, as the microorganisms present in the environment can more readily access OLAs and use them as a source of carbon and energy. This localised bacterial activity can further accelerate the degradation of PLA, leading to a quicker breakdown of the polymer chains and the release of carbon dioxide and other by-products [24].

As depicted in Figure 6, the SEM analysis of the studied blends revealed uneven surface topography and disordered internal structures, which are indicative of phase separation within the blend. These observations are consistent with previous investigations and with the behaviour of immiscible polymer blends [55]. This is in contrast to the smooth and uniform fracture surfaces characteristic of neat PLA and PHB, which present brittle fracture behaviour and might be classified as rigid fillers in the blends, whereas the presence of PCL improves the overall ductility of the blends, as evidenced by SEM photomicrographs and mechanical properties from previous investigations [56–59].

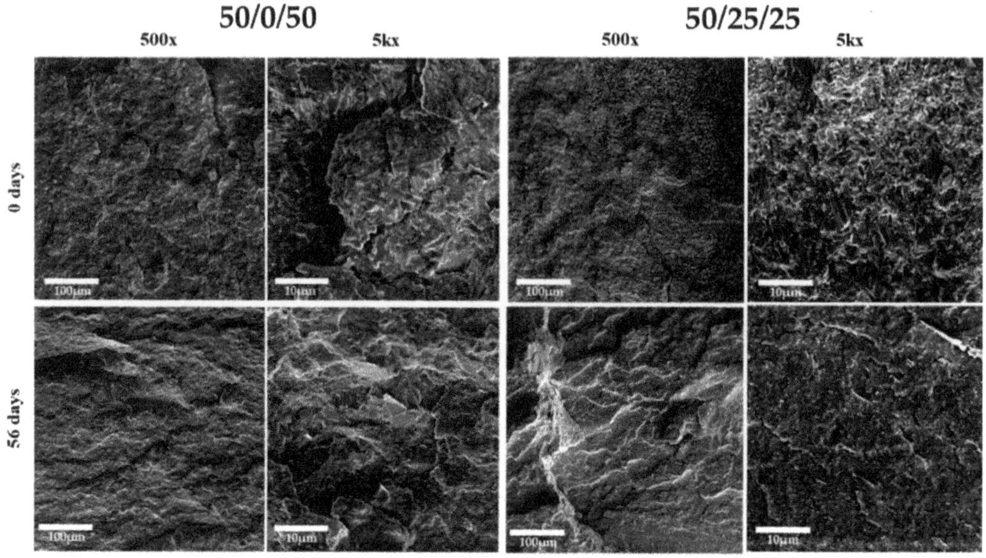


**Figure 6.** SEM photomicrographs of blends 50/0/50 (**left**) and 50/25/25 (**right**) (wt.% PHB/PLA/PCL) at 0 and after 56 days of seawater degradation.

In previous studies on ternary blends of PHB, PLA, and PCL, researchers reported that neat PLA has a brittle fracture with a typical smooth surface presenting several crack fronts [11]. Regarding PLA/PHB blends (60/40 wt.%), SEM photomicrographs also showed a morphology indicating phase separation between the polymers, and were confirmed by thermal analysis [35]. In other studies, the immiscibility of the blends was evidenced by the bead-like shape of the blends containing PCL, with the PCL dispersed phase containing spherical shapes ranging from 1 to 10 µm [39,60].

## 4. Conclusions

The present investigation utilised a methodology to fabricate and assess blends composed of biodegradable materials such as PHB, PLA, and PCL, resulting in the production of materials with customised mechanical properties. The evaluation of seawater degradation mechanisms revealed the occurrence of surface erosion and bulk degradation in the blends, where the weight variance was initially positive in certain blends, indicating the presence of bulk degradation with the formation of cracks and fractures that permitted the permeation of fluids and salts, subsequently resulting in an increase in weight. However, this effect tends to increase the degradation rate over time by augmenting the surface area of degradation and the local autocatalysis effect. These findings have significant implications for understanding the degradation behaviour of these polymers in a seawater environment.

In light of the findings of this study, the development of tailored mechanical properties for PHB/PLA/PCL blends is suggested by adjusting the content of each biopolyester. The tensile results indicated that all samples exhibited increased brittleness after 56 days of seawater degradation, with PLA being the most rigid polymer among the investigated materials. The blending process of the samples in this study with PCL promoted the flexibility of the blends, especially the ones with the highest content of PLA. As corroborated by previous studies, the addition of different amounts of PCL causes a decrease in the stiffness of PLA-based samples by improving their elongation at break and impact strength properties, which facilitates their use in packaging and medical device applications. Furthermore, neat polymers underwent degradation in a one-step stage, whereas blends containing two or three polymers underwent a multistep degradation process, indicating a degree of immiscibility between the polymers. Morphological analysis of the samples revealed brittle fracture in PLA and PHB and ductile fracture preceded by significant plastic deformation in blends containing 50 wt.% of PLA, such as 50/50/0 and 30/50/20 formulations, accompanied by extensive stress whitening across the gauge zone. In addition, blends with weight ratios of 50/0/50 and 50/25/25 (wt.% PHB/PLA/PCL) exhibited a higher frequency of white spots, indicating surface degradation, loss of ductility, and transition to brittle fracture behaviour. Based on the findings of this study, it is recommended to develop a biopolyester blend with a high PCL content to promote ductility, in combination with a suitable ratio of PHB and PLA to achieve the desired mechanical properties.

**Supplementary Materials:** The following supporting information can be downloaded at: https://www.mdpi.com/article/10.3390/polym15132874/s1, Figure S1: Stress vs. Strain of neat PHB/PLA/PCL and their respective blends at the initial state and after 56 days of seawater degradation: (**a**) PHB, (**b**) PLA, (**c**) PCL, (**d**) B2: 50/50/0, (**e**) B3: 50/0/50, (**f**) B4: 50/25/25, and (**g**) B5: 30/50/20 (wt.% PHB/PLA/PCL).; Figure S2: FTIR of neat polymers PHB/PLA/PCL and their respective blends: (**a**) PHB, (**b**) PLA, (**c**) PCL, (**d**) B2: 50/50/0, (**e**) B3: 50/0/50, (**f**) B4: 50/25/25, and (**g**) B5: 30/50/20 (wt.% PHB/PLA/PCL).; Figure S3: TGA and DTG of the neat polymers PHB/PLA/PCL and their respective blends (**left**: TG and DTG/**right**: deconvolution of the DTG peaks): (**a**) PHB, (**b**) PLA, (**c**) PCL, (**d**) B2: 50/50/0, (**e**) B3: 50/0/50, (**f**) B4: 50/25/25, and (**g**) B5: 30/50/20 (wt.% PHB/PLA/PCL).; Figure S4: DSC of neat polymers PHB/PLA/PCL and their respective blends: (**a**) PHB, (**b**) PLA, (**c**) PCL, (**d**) B2: 50/50/0, (**e**) B3: 50/0/50, (**f**) B4: 50/25/25, and (**g**) B5: 30/50/20 (wt.% PHB/PLA/PCL).

**Author Contributions:** Conceptualization, L.G.E., N.C.F., R.P. and D.M.D.; methodology, L.G.E., N.C.F., R.P. and D.M.D.; investigation, L.G.E. and N.C.F.; writing—original draft preparation, L.G.E.; writing—review and editing, N.C.F., J.S.C., N.M.G., R.P. and D.M.D.; supervision, J.S.C., I.M., N.M.G., R.P. and D.M.D.; project administration, D.M.D.; funding acquisition, I.M., N.M.G. and D.M.D. All authors have read and agreed to the published version of the manuscript.

**Funding:** This manuscript emanated from research funded by Enterprise Ireland under the capital call of 2019, grant number CE 20190068, the TTSI technology fund grant number SI20203032 and Technology Gateway programme TG-2017-1014, the AIT President's Seed Fund (PSF2020-DD), and the European Research Agency under the European Union, Horizon-CL6-Circbio-01-GA number 101059923.

**Institutional Review Board Statement:** Not applicable.

**Data Availability Statement:** The data that support the findings of this study are available from the corresponding author upon reasonable request.

**Conflicts of Interest:** The authors declare no conflict of interest.

## References

1. Kliem, S.; Kreutzbruck, M.; Bonten, C. Review on the biological degradation of polymers in various environments. *Materials* **2020**, *13*, 4586. [CrossRef] [PubMed]
2. Leslie, H.A.; van Velzen, M.J.M.; Brandsma, S.H.; Vethaak, A.D.; Garcia-Vallejo, J.J.; Lamoree, M.H. Discovery and quantification of plastic particle pollution in human blood. *Environ. Int.* **2022**, *163*, 107199. [CrossRef]
3. Chahine, M.T. The hydrological cycle and its influence on climate. *Nature* **1992**, *359*, 373–380. [CrossRef]
4. Volova, T.G.; Boyandin, A.N.; Vasiliev, A.D.; Karpov, V.A.; Prudnikova, S.V.; Mishukova, O.V.; Boyarskikh, U.A.; Filipenko, M.L.; Rudnev, V.P.; Bá Xuân, B.; et al. Biodegradation of polyhydroxyalkanoates (PHAs) in tropical coastal waters and identification of PHA-degrading bacteria. *Polym. Degrad. Stab.* **2010**, *95*, 2350–2359. [CrossRef]
5. Heimowska, A.; Morawska, M.; Bocho-Janiszewska, A. Biodegradation of poly(ε-caprolactone) in natural water environments. *Polish J. Chem. Technol.* **2017**, *19*, 120–126. [CrossRef]
6. Tsuji, H.; Suzuyoshi, K. Environmental degradation of biodegradable polyesters 2. Poly(ε-caprolactone), poly[(R)-3-hydroxybutyrate], and poly(L-lactide) films in natural dynamic seawater. *Polym. Degrad. Stab.* **2002**, *75*, 357–365. [CrossRef]
7. Bagheri, A.R.; Laforsch, C.; Greiner, A.; Agarwal, S. Fate of So-Called Biodegradable Polymers in Seawater and Freshwater. *Glob. Challenges* **2017**, *1*, 1700048. [CrossRef]
8. Munn, C.B. *Marine Microbiology: Ecology & Applications*, 3rd ed.; CRC Press: Boca Raton, FL, USA, 2019; pp. 1–27.
9. Mackenzie, F.T.; Duxbury, A.C.; Byrne, R.H. Seawater. Available online: https://www.britannica.com/science/seawater (accessed on 12 April 2023).

10. Bher, A.; Mayekar, P.C.; Auras, R.A.; Schvezov, C.E. Biodegradation of Biodegradable Polymers in Mesophilic Aerobic Environments. *Int. J. Mol. Sci.* **2022**, *23*, 12165. [CrossRef]
11. García-Campo, M.J.; Boronat, T.; Quiles-Carrillo, L.; Balart, R.; Montanes, N. Manufacturing and characterization of toughened poly(lactic acid) (PLA) formulations by ternary blends with biopolyesters. *Polymers* **2018**, *10*, 3. [CrossRef]
12. Engler, L.G.; Crespo, J.S.; Gately, N.M.; Major, I.; Devine, D.M. Process Optimization for the 3D Printing of PLA and HNT Composites with Arburg Plastic Freeforming. *J. Compos. Sci.* **2022**, *6*, 309. [CrossRef]
13. Da Silva, D.; Kaduri, M.; Poley, M.; Adir, O.; Krinsky, N.; Shainsky-Roitman, J.; Schroeder, A. Biocompatibility, biodegradation and excretion of polylactic acid (PLA) in medical implants and theranostic systems. *Chem. Eng. J.* **2018**, *340*, 9–14. [CrossRef]
14. Lyu, S.P.; Untereker, D. Degradability of polymers for implantable biomedical devices. *Int. J. Mol. Sci.* **2009**, *10*, 4033–4065. [CrossRef]
15. Narancic, T.; Verstichel, S.; Reddy, C.S.; Morales-Gamez, L.; Kenny, S.T.; De Wilde, B.; Babu, P.R.; O'Connor, K.E. Biodegradable Plastic Blends Create New Possibilities for End-of-Life Management of Plastics but They Are Not a Panacea for Plastic Pollution. *Environ. Sci. Technol.* **2018**, *52*, 10441–10452. [CrossRef] [PubMed]
16. Dalton, B.; Bhagabati, P.; De Micco, J.; Padamati, R.B.; O'Connor, K. A Review on Biological Synthesis of the Biodegradable Polymers Polyhydroxyalkanoates and the Development of Multiple Applications. *Catalysts* **2022**, *12*, 319. [CrossRef]
17. Bhasney, S.M.; Bhagabati, P.; Kumar, A.; Katiyar, V. Morphology and crystalline characteristics of polylactic acid [PLA]/linear low density polyethylene [LLDPE]/microcrystalline cellulose [MCC] fiber composite. *Compos. Sci. Technol.* **2019**, *171*, 54–61. [CrossRef]
18. Siracusa, V.; Karpova, S.; Olkhov, A.; Zhulkina, A.; Kosenko, R.; Iordanskii, A. Gas transport phenomena and polymer dynamics in PHB/PLA blend films as potential packaging materials. *Polymers* **2020**, *12*, 647. [CrossRef] [PubMed]
19. Simão, J.A.; Bellani, C.F.; Branciforti, M.C. Thermal properties and crystallinity of PCL/PBSA/cellulose nanocrystals grafted with PCL chains. *J. Appl. Polym. Sci.* **2017**, *134*, 44493. [CrossRef]
20. Brannigan, R.P.; Dove, A.P. Synthesis, properties and biomedical applications of hydrolytically degradable materials based on aliphatic polyesters and polycarbonates. *Biomater. Sci.* **2017**, *5*, 9–21. [CrossRef]
21. Chen, Y.; Hung, S.T.; Chou, E.; Wu, H.S.; Devine, D.M. *Bioresorbable Polymers and their Biomedical Applications*; Smithers Rapra: Shrewsbury, UK, 2017; Volume 15, pp. 1–26.
22. Meereboer, K.W.; Misra, M.; Mohanty, A.K. Review of recent advances in the biodegradability of polyhydroxyalkanoate (PHA) bioplastics and their composites. *Green Chem.* **2020**, *22*, 5519–5558. [CrossRef]
23. Agarwal, M.; Koelling, K.W.; Chalmers, J.J. Characterization of the degradation of polylactic acid polymer in a solid substrate environment. *Biotechnol. Prog.* **1998**, *14*, 517–526. [CrossRef]
24. Tabasi, R.Y.; Ajji, A. Selective degradation of biodegradable blends in simulated laboratory composting. *Polym. Degrad. Stab.* **2015**, *120*, 435–442. [CrossRef]
25. Limsukon, W.; Auras, R.; Selke, S. Hydrolytic degradation and lifetime prediction of poly(lactic acid) modified with a multifunctional epoxy-based chain extender. *Polym. Test.* **2019**, *80*, 106108. [CrossRef]
26. Hoüglund, A.; Odelius, K.; Albertsson, A.C. Crucial differences in the hydrolytic degradation between industrial polylactide and laboratory-scale poly(L-lactide). *ACS Appl. Mater. Interfaces* **2012**, *4*, 2788–2793. [CrossRef]
27. Kolstad, J.J.; Vink, E.T.H.; De Wilde, B.; Debeer, L. Assessment of anaerobic degradation of Ingeo$^{TM}$ polylactides under accelerated landfill conditions. *Polym. Degrad. Stab.* **2012**, *97*, 1131–1141. [CrossRef]
28. Elhami, V.; Hempenius, M.A.; Schuur, B. Crotonic Acid Production by Pyrolysis and Vapor Fractionation of Mixed Microbial Culture-Based Poly(3-hydroxybutyrate-co-3-hydroxyvalerate). *Ind. Eng. Chem. Res.* **2023**, *62*, 916–923. [CrossRef]
29. Kasuya, K.I.; Takagi, K.I.; Ishiwatari, S.I.; Yoshida, Y.; Doi, Y. Biodegradabilities of various aliphatic polyesters in natural waters. *Polym. Degrad. Stab.* **1998**, *59*, 327–332. [CrossRef]
30. Nair, L.S.; Laurencin, C.T. Biodegradable polymers as biomaterials. *Prog. Polym. Sci.* **2007**, *32*, 762–798. [CrossRef]
31. Woodard, L.N.; Grunlan, M.A. Hydrolytic Degradation and Erosion of Polyester Biomaterials. *ACS Macro Lett.* **2018**, *7*, 976–982. [CrossRef]
32. Artsis, M.I.; Bonartsev, A.P.; Iordanskii, A.L.; Bonartseva, G.A.; Zaikov, G.E. Biodegradation and medical application of microbial poly(3-Hydroxybutyrate). *Mol. Cryst. Liq. Cryst.* **2012**, *555*, 232–262. [CrossRef]
33. Pfau, M.R.; McKinzey, K.G.; Roth, A.A.; Grunlan, M.A. PCL-Based Shape Memory Polymer Semi-IPNs: The Role of Miscibility in Tuning the Degradation Rate. *Biomacromolecules* **2020**, *21*, 2493–2501. [CrossRef]
34. Wang, S.; Kempen, D.H.R.; Ruiter, G.C.W.; Cai, L.; Spinner, R.J.; Windebank, A.J.; Yaszemski, M.J.; Lu, L. Molecularly engineered biodegradable polymer networks with a wide range of stiffness for bone and peripheral nerve regeneration. *Adv. Funct. Mater.* **2015**, *25*, 2715–2724. [CrossRef]
35. Bartczak, Z.; Galeski, A.; Kowalczuk, M.; Sobota, M.; Malinowski, R. Tough blends of poly(lactide) and amorphous poly([R,S]-3-hydroxy butyrate)—Morphology and properties. *Eur. Polym. J.* **2013**, *49*, 3630–3641. [CrossRef]
36. Liu, H.; Chen, F.; Liu, B.; Estep, G.; Zhang, J. Super toughened poly(lactic acid) ternary blends by simultaneous dynamic vulcanization and interfacial compatibilization. *Macromolecules* **2010**, *43*, 6058–6066. [CrossRef]
37. Zenkiewicz, M.; Richert, J.; Rytlewski, P.; Moraczewski, K.; Stepczyńska, M.; Karasiewicz, T. Characterisation of multi-extruded poly(lactic acid). *Polym. Test.* **2009**, *28*, 412–418. [CrossRef]

38. Taubner, V.; Shishoo, R. Influence of processing parameters on the degradation of poly(L-lactide) during extrusion. *J. Appl. Polym. Sci.* **2001**, *79*, 2128–2135. [CrossRef]
39. Chen, C.C.; Chueh, J.Y.; Tseng, H.; Huang, H.M.; Lee, S.Y. Preparation and characterization of biodegradable PLA polymeric blends. *Biomaterials* **2003**, *24*, 1167–1173. [CrossRef]
40. Elsawy, M.A.; Kim, K.H.; Park, J.W.; Deep, A. Hydrolytic degradation of polylactic acid (PLA) and its composites. *Renew. Sustain. Energy Rev.* **2017**, *79*, 1346–1352. [CrossRef]
41. Urquijo, J.; Guerrica-Echevarría, G.; Eguiazábal, J.I. Melt processed PLA/PCL blends: Effect of processing method on phase structure, morphology, and mechanical properties. *J. Appl. Polym. Sci.* **2015**, *132*, 42641. [CrossRef]
42. Ferri, J.M.; Fenollar, O.; Jorda-Vilaplana, A.; García-Sanoguera, D.; Balart, R. Effect of miscibility on mechanical and thermal properties of poly(lactic acid)/polycaprolactone blends. *Polym. Int.* **2016**, *65*, 453–463. [CrossRef]
43. Jamshidian, M.; Tehrany, E.A.; Imran, M.; Jacquot, M.; Desobry, S. Poly-Lactic Acid: Production, applications, nanocomposites, and release studies. *Compr. Rev. Food Sci. Food Saf.* **2010**, *9*, 552–571. [CrossRef]
44. Bao, Q.; Wong, W.; Liu, S.; Tao, X. Accelerated Degradation of Poly(lactide acid)/Poly(hydroxybutyrate) (PLA/PHB) Yarns/Fabrics by UV and $O_2$ Exposure in South China Seawater. *Polymers* **2022**, *14*, 1216. [CrossRef]
45. Vogel, C.; Siesler, H.W. Thermal degradation of poly(ε-caprolactone), poly(L-lactic acid) and their blends with poly(3-hydroxy-butyrate) studied by TGA/FT-IR spectroscopy. *Macromol. Symp.* **2008**, *265*, 183–194. [CrossRef]
46. Herrera-Kao, W.A.; Loría-Bastarrachea, M.I.; Pérez-Padilla, Y.; Cauich-Rodríguez, J.V.; Vázquez-Torres, H.; Cervantes-Uc, J.M. Thermal degradation of poly(caprolactone), poly(lactic acid), and poly(hydroxybutyrate) studied by TGA/FTIR and other analytical techniques. *Polym. Bull.* **2018**, *75*, 4191–4205. [CrossRef]
47. Li, S.; He, J.; Yu, P.H.; Cheung, M.K. Thermal Degradation of Poly(3-hydroxybutyrate) and Poly(3-hydroxybutyrate- co -3-hydroxyvalerate) as Studied by TG, TG–FTIR, and Py–GC/MS. *J. Appl. Pol. Sci.* **2002**, *89*, 1530–1536. [CrossRef]
48. Braun, B.; Dorgan, J.R.; Dec, S.F. Infrared spectroscopic determination of lactide concentration in polylactide: An improved methodology. *Macromolecules* **2006**, *39*, 9302–9310. [CrossRef]
49. Mofokeng, J.P.; Luyt, A.S. Morphology and thermal degradation studies of melt-mixed poly(lactic acid) (PLA)/poly(ε-caprolactone) (PCL) biodegradable polymer blend nanocomposites with $TiO_2$ as filler. *Polym. Test.* **2015**, *45*, 93–100. [CrossRef]
50. Patrício, T.; Bártolo, P. Thermal stability of PCL/PLA blends produced by physical blending process. *Procedia Eng.* **2013**, *59*, 292–297. [CrossRef]
51. Arrieta, M.P.; López, J.; López, D.; Kenny, J.M.; Peponi, L. Development of flexible materials based on plasticized electrospun PLA-PHB blends: Structural, thermal, mechanical and disintegration properties. *Eur. Polym. J.* **2015**, *73*, 433–446. [CrossRef]
52. Yeo, J.C.C.; Muiruri, J.K.; Thitsartarn, W.; Li, Z.; He, C. Recent advances in the development of biodegradable PHB-based toughening materials: Approaches, advantages and applications. *Mater. Sci. Eng. C* **2018**, *92*, 1092–1116. [CrossRef]
53. Farias, N.C.; Major, I.; Devine, D.; Fournet, M.B.; Pezzoli, R.; Taghinezhad, S.F.; Hesabi, M. Multiple recycling of a PLA/PHB biopolymer blend for sustainable packaging applications: Rheology-morphology, thermal, and mechanical performance analysis. *Polym. Eng. Sci.* **2022**, *62*, 1764–1774. [CrossRef]
54. Ramezani Dana, H.; Ebrahimi, F. Synthesis, properties, and applications of polylactic acid-based polymers. *Polym. Eng. Sci.* **2022**, *63*, 22–43. [CrossRef]
55. Navarro-Baena, I.; Sessini, V.; Dominici, F.; Torre, L.; Kenny, J.M.; Peponi, L. Design of biodegradable blends based on PLA and PCL: From morphological, thermal and mechanical studies to shape memory behavior. *Polym. Degrad. Stab.* **2016**, *132*, 97–108. [CrossRef]
56. D'Amico, D.A.; Iglesias Montes, M.L.; Manfredi, L.B.; Cyras, V.P. Fully bio-based and biodegradable polylactic acid/poly(3-hydroxybutirate) blends: Use of a common plasticizer as performance improvement strategy. *Polym. Test.* **2016**, *49*, 22–28. [CrossRef]
57. Ma, P.; Spoelstra, A.B.; Schmit, P.; Lemstra, P.J. Toughening of poly (lactic acid) by poly (β-hydroxybutyrate-co-β- hydroxyvalerate) with high β-hydroxyvalerate content. *Eur. Polym. J.* **2013**, *49*, 1523–1531. [CrossRef]
58. Zhang, K.; Mohanty, A.K.; Misra, M. Fully biodegradable and biorenewable ternary blends from polylactide, poly(3-hydroxybutyrate-co-hydroxyvalerate) and poly(butylene succinate) with balanced properties. *ACS Appl. Mater. Interfaces* **2012**, *4*, 3091–3101. [CrossRef] [PubMed]
59. Gerard, T.; Budtova, T.; Podshivalov, A.; Bronnikov, S. Polylactide/poly(hydroxybutyrate-co-hydroxyvalerate) blends: Morphology and mechanical properties. *Express Polym. Lett.* **2014**, *8*, 609–617. [CrossRef]
60. Wei, L.; Stark, N.M.; McDonald, A.G. Interfacial improvements in biocomposites based on poly(3-hydroxybutyrate) and poly(3-hydroxybutyrate-co-3-hydroxyvalerate) bioplastics reinforced and grafted with α-cellulose fibers. *Green Chem.* **2015**, *17*, 4800–4814. [CrossRef]

polymers

MDPI

*Article*

# One-Step Method for Direct Acrylation of Vegetable Oils: A Biobased Material for 3D Printing

**Cristian Mendes-Felipe [1,2,*], Igor Isusi [3], Olga Gómez-Jiménez-Aberasturi [4], Soraya Prieto-Fernandez [4], Leire Ruiz-Rubio [1,3,*], Marco Sangermano [2] and José Luis Vilas-Vilela [1,3]**

1 BCMaterials, Basque Center for Materials, Applications and Nanostructures, UPV/EHU Science Park, 48940 Leioa, Spain; joseluis.vilas@ehu.eus
2 Department of Applied Science and Technology (DISAT), Politecnico di Torino, 10129 Torino, Italy; marco.sangermano@polito.it
3 Macromolecular Chemistry Group (LABQUIMAC), Department of Physical Chemistry, Faculty of Science and Technology, University of the Basque Country (UPV/EHU), 48940 Leioa, Spain
4 TECNALIA, Basque Research and Technology Alliance (BRTA), Parque Tecnológico de Álava, Leonardo Da Vinci 11, 01510 Minano, Spain; olga.gomez@tecnalia.com (O.G.-J.-A.); soraya.prieto@tecnalia.com (S.P.-F.)
* Correspondence: cristian.mendes@bcmaterials.net (C.M.-F.); leire.ruiz@ehu.eus (L.R.-R.)

**Abstract:** The substitution of fossil resources by alternatives derived from biomass is a reality that is taking on a growing relevance in the chemical and energy industries. In this sense, fats, oils, and their derived products have become indispensable inputs due to their broad functional attributes, stable price and sustainable character. Acrylated vegetable oils are considered to be very versatile materials for very broad applications (such as in adhesives, coatings or inks) since, in the presence of photoinitiators, they can be polymerized by means of UV-initiated free radical polymerizations. The usual process for the synthesis of acrylate vegetable oils consists in reacting epoxidized oils derivatives with acrylic acid. Here, the influence of different catalysts on the activity and selectivity of the process of acrylation of epoxidized soybean oil is studied. In addition, a novel one-step method for direct acrylation of vegetable oils is also explored. This new approach advantageously uses the original vegetable resource and eliminates intermediate reactions, thus being more environmentally efficient. This study offers a simple and low-cost option for synthesizing a biomass-derived monomer and studies the potential for the 3D printing of complex structures via digital light processing (DLP) 3D printing of the thus-obtained novel sustainable formulations.

**Keywords:** soybean oil; one-step reaction; biobased polymers; UV-curable; 3D printing

## 1. Introduction

The modern world is constantly changing in order to be able to satisfy all requirements at both technological and industrial levels. In this sense, the 4th industrial revolution, also called Industry 4.0, began some years ago as a revolution that utilizes information technology in the manufacturing sector to enable smart manufacturing [1], highlighting additive manufacturing (AM) as being among the most useful process. AM is defined as a process that "makes 'objects' from a digital 'model' by depositing the constituent material/s in a layer-by-layer manner using digitally controlled and operated material laying tools" and includes eight different processes: binder jetting (BJ), directed energy deposition (DED), material extrusion (ME), material jetting (MJ), powder bed fusion, sheet lamination (SL) and vat photopolymerization (VP). Among these methods, VP needs to be highlighted as it allows the production of large parts with excellent accuracy, surface finish and details [2]. Particularly, digital light processing (DLP) 3D printing is among the most useful techniques of VP due to its basis in the layer-by-layer curing of liquid resin formulations using digital micro-mirror devices (DMD) that induce an instantaneous entire layer polymerization, compared to stereolithography (SLA) methods that cure liquid formulations using single-laser point technology [3]. In addition, as a UV light-induced polymerization technique, DLP presents advantages because it is a fast, economic and nontoxic method that works at room temperature, with space and energy efficiency, and produces high-resolution patterns compared to the other AM processes [4].

Conversely, the current situation of the planet due to the growth of the greenhouse effect, climate change, and fossil fuel resource depletion, requires the adoption of a circular economy (CE) strategy in which the production process relies on the principles of sustainability, such as green purchasing, reusing, recycling, and remanufacturing [5]. Furthermore, the COVID-19 pandemic situation has drastically increased the consumption of disposable polymer-based materials [6]. Thus, the use of materials obtained from sustainable sources becomes more important year by year. Among the different possible options of renewable materials, terpenes, carbohydrates, and vegetable oils are the most used monomers for the plastic production [7]. In particular, oils obtained from vegetables, such as cardanol [8], are important renewable feedstocks for the synthesis of UV-curable biobased materials because of their low cost in comparison to other biomass material, as well as their availability and biodegradability [9].

Here, soybean oil (SO) stands out as one of the oils presenting the largest global production volumes (together with palm and rapeseed oil) with an economical price for large-scale uses [10]. Additionally, it presents low toxicity, biocompatibility, inherent biodegradability, high thermal stability, and good mechanical properties, and can be easily combined with fillers to make functional materials [11].

However, AM processes, particularly DLP printing, are not optimized for the use of materials obtained from renewable sources. The most promising ones for use as DLP resins are vegetable oils (such as the aforementioned soybean oil) because they have liquid or quasi-liquid states, similar to those of the existing petroleum-based resins, and as they present functional groups along their chains, which are similar to those needed for UV curing, such as the carbon–carbon double bonds of the triglyceride [10,12]. Despite this, it was demonstrated that the reactivity of those groups against UV light in the presence of a photoinitiator was very low [13]. For this reason, several methods have been developed to functionalize vegetable oils, including epoxidation and/or acrylation. Epoxidation is one of the most used methods when photocurable materials for film and coating applications are needed, while acrylation is most appropriately used to obtain 3D printable resins. The epoxidation reaction of vegetable oils has been tested for several oils such as peanut, rapeseed, rose hip, safflower, or camelina oils, among others. This epoxidation consists in reacting internal alkenes of vegetable oils using oxidizing systems to form the epoxy groups. This is typically carried out using one of these four epoxidation methods: in situ with peracids in the presence of an inorganic soluble or supported catalyst; with organic or inorganic peroxides, using transition metals or enzymatic species as catalysts; via halohydrins formation; or using molecular oxygen [14–16]. On the other hand, acrylation of vegetable oils has been generally carried out as a two-step process consisting in an initial epoxidation reaction (as explained above), followed by acrylation using acrylic acid. This is the common way of obtaining acrylated epoxidized vegetable oils, such as the ones made of soybean, palm, castor, linseed, or Jatropha seed [17–19]. However, in terms of green chemistry (atom economy) and cost-effectiveness, it would be desirable to synthesize acrylated vegetable oils via the direct addition of acrylic acids to the carbon–carbon double bond of the triglyceride as it is prone to capturing electrophiles [20]. In this sense, the direct acrylation reaction has been studied for use in soybean oil [21,22].

In this way, the main objective of this work is to synthesize and characterize acrylated epoxidized soybean oil (AESO) and direct acrylated soybean oil (ASO), and to study their application as suitable formulations for DLP 3D printing, comparing the final mechanical and viscoelastic properties of UV-cured materials.

## 2. Materials and Methods

### 2.1. Materials

Epoxidized soybean oil (ESO) EPOVINSTAB H-800D, with a viscosity of 550–650 cP at 20 °C, was obtained from Hebron SA (La Llagosta, Spain) while soybean oil (SO) was obtained from KM Elite products (Petwoth, UK). The acrylic acid anhydrous (AA), presenting a purity of 99%, was purchased on Fluka (Buchs, Switzerland) and hydroquinone with a purity of 99.5% was acquired from Acros Organics (Hampton, NH, USA). Five different materials were used as catalysts of ESO acrylation: triethylamine (TEA) with a purity of 99% from Fisher Chemical, triphenylphosphine (TPP) with a purity of 99% from Acros Organics, and three different chromium salts, namely, chromium (III) 2-ethylhexanoate at 50% purity in 2-ethylhexanoic acid ($Cr(EH)_3$-EHXA) from Alfa Aesar (Kandel, Germany), chromium (III) 2,4-pentanedionate ($Cr(acac)_3$) with a purity of 97% from Alfa Aesar, and chromium (III) 2-ethylhexanoate ($Cr(EH)_3$-MO) at 70% in mineral oil from VladaChem (Malsch, Germany). In the case of acrylation of SO, boron trifluoride ($BF_3$) at 48% purity in diethyl ether ($Et_2O$), obtained from Acros Organics, was used as a catalyst. Ethyl acetate with a purity of 99.8% was obtained from Fisher Scientific (Hampton, NH, USA) and sodium sulphate anhydrous with a purity of 99.0% was purchased from Sigma Aldrich (Darmstadt, Germany). Deuterated chloroform, presenting a purity of 99.8%, was acquired from Sigma Aldrich and used as a solvent in nuclear magnetic resonance experiments.

Phenylbis(2,4,6-trimethylbenzoyl) phosphine oxide with a purity of 98% was purchased from Sigma Aldrich and used as a photoinitiator. Ethanol (EtOH) purchased from Merck (Darmstadt, Germany) with 96% of purity was used as a sample cleaner. Poly(ethyleneglycol) di-acrylate (PEGDA) of a molecular weight of 575 g/mol from Sigma Aldrich (Darmstadt, Germany) was used as a sacrificial material during the 3D printing process.

### 2.2. Synthesis of Acrylated Epoxidized Soybean Oil (AESO)

AESO was obtained by opening the oxirane groups present in the fatty chains of commercial epoxidized oils with acrylic acid. Reactions were carried out in a batch mode using mechanically stirred glass reactors with a 1.1 molar excess from acrylic/epoxide groups. Temperature was maintained at 80 °C for 6 h. In total, 1% ($w/w$) hydroquinone was employed as a polymerization inhibitor. The efficiency of different nucleophilic catalysts was evaluated using the same operational conditions in order to select the most appropriate catalysts for use in the process. TEA and TPP were selected as two of the most common catalysts used in acrylation reactions [18], together with three salts of chromium (III).

Products were dissolved with ethyl acetate and washed several times with brine until the complete removal of the rest of the catalyst, hydroquinone, and acrylic acid. This process was finished with the determination of a neutral pH for the aqueous phase. On the other hand, the organic phase was collected, dried over sodium sulphate and filtered. Finally, the remaining organic solvent was eliminated via vacuum distillation. The evolution of the acrylation of epoxidized oils was followed by epoxide index and acidity measurements.

### 2.3. Synthesis of Acrylated Soybean Oil (ASO)

Soybean oil (SO) and acrylic acid (AA) were premixed in specific proportions in a glass reactor equipped with a condenser, thermometer, and magnetic stirring. When the temperature achieved 80 °C, $BF_3 \cdot Et_2O$ was added as a catalyst to the system. The acrylation reactions were maintained for 10 h. The purification of products was carried out following the same protocol as is employed in the acrylation of epoxidized oils.

### 2.4. Characterization of Acrylated Oils by Epoxy Index (E.I.), Acidity Index and Iodine Value

The disappearance of the oxirane groups present in the ESO gives an idea about the evolution of the acrylation reaction and its corresponding conversion. This factor was assessed from the epoxy oxygen content determination, which was performed with the standard AOCS analysis method Cd 9–57.

Conversely, the presence of free acidic species in the reaction mixture was determined via the acidity index according to the norm (UNE-EN-ISO 660 [23]). We determined the free acrylic acid that remained after the reaction process. Finally, the iodine value was measured via the Wijs method [24]. This analysis permitted us to establish the unsaturations present in the soybean oil.

### 2.5. Fourier Transformed Infrared Spectroscopy (FTIR)

Fourier transformed infrared spectroscopy (FTIR) was used to analyze the main chemical groups of the molecular structures of the different oils synthesized. The five different acrylated epoxidized soybean oils (AESO) obtained depending on the catalyst used, as well as the acrylated soybean oil (ASO), were analyzed. In addition, the synthesis reaction of AESO using the chromium (III) 2-ethylhexanoate at 70% in mineral oil ($Cr(EH)_3$-MO) catalyst was followed at different times via FTIR, and epoxidized soybean oil (ESO) and soybean oil (SO) were studied in order to compare the initial and final products obtained. A Bruker Alpha Platinum ATR-FTIR spectrophotometer (Billerica, MA, USA) was employed for experimental measures. Spectra were collected at a resolution of 4 $cm^{-1}$, in transmittance mode, within a range of 400 to 4000 $cm^{-1}$. A total of 24 scans were acquired for each analysis.

Furthermore, attenuated total reflectance-Fourier transform infrared spectroscopy (ATR-FTIR) equipment was used to monitor the photopolymerization reaction of AESO and ASO formulations (AESO-L and ASO-L prepared as explained in Section 2.7) by assessing the disappearance of the acrylic double bond peak at 1630 $cm^{-1}$. Samples were measured first in a liquid state (pre-cured) and then a solid state (post-cured) after 90 s irradiation at room temperature and under nitrogen atmospheric conditions using a Dymax ECE 5000 UV lamp (130 $mW/cm^2$ of irradiance) (Frankfurt, Germany). For ATR-FTIR, we used a Thermo Scientific Nicolet iS50 FTIR Spectrometer (Waltham, MA, USA) instrument equipped with a diamond crystal ATR accessory. For each measurement, 32 spectra were obtained in the spectral range from 4000 to 600 $cm^{-1}$ at a resolution of 4 $cm^{-1}$. The averaged value after 3 measurements was indicated.

### 2.6. Nuclear Magnetic Resonance ($^1H$-NMR)

The chemical structure and the degree of acrylation (AD) of AESO and ASO were determined via NMR spectroscopy. Briefly, $^1H$-NMR spectra were taken in $CDCl_3$ on a Bruker Avance 300 MHz spectrometer at 25 °C, employing a polymer concentration of 33 mg/mL for all oils measured. For the quantification of the modification of epoxidized soybean oil (ESO) and soybean oil (SO), the relative area of the hydrogen atoms of a group that does not change during the reaction, i.e., hydrogen atoms of the carbon next to the carbonyl group, was used as a reference for the determination of protons at an acrylate group.

### 2.7. Preparation of 3D Printable Formulations and Solid Samples

Liquid formulations for use as 3D printable materials were prepared by mixing the corresponding oil (AESO for the AESO-L sample and ASO for the ASO-L sample) with a 1 wt.% of photoinitiator (sufficient quantity to ensure photopolymerization in relatively short periods of time [10]). A homogeneous solution is obtained after 30 min of magnetic stirring at 50 °C for AESO material and at room temperature for ASO material. Then, in order to de-gas the liquid formulations, samples were sonicated for 1 h in an ultrasound bath at a controlling temperature above 40 °C.

AESO-L and ASO-L formulations were used to obtain the corresponding solid samples AESO-S and ASO-S, respectively. For solid samples, liquid formulations were added to a silicone mold and cured via illumination with a Dymax ECE 5000-UV lamp for 60 s at 130 $mW/cm^2$ of irradiance under a nitrogen atmosphere.

In the case of 3D printing, samples were prepared using the 3D printer Asiga MAX X27 DLP printer (Erfurt, Germany) presenting a diose source emitting at 405 nm, a building volume of 51.8 × 2.92 × 75 mm, and a XY pixel resolution of 27 μm. During the printing process, a sacrificial PEGA base (around 50 μm thickness) was used to easily remove the printed samples. The layer thickness and light intensity were fixed at 50 μm and 40 mW/cm$^2$, respectively. Exposure time was selected for each formulation attending to the photorheology results, while the temperature was controlled to be constant at 35 °C. After printing, samples were washed in an ethanol bath under ultrasonication conditions for 5 min and post-cured 3 min under UV light in a robot factory medium-pressure mercury lamp at 12 mW/cm$^2$ of irradiance.

### 2.8. Characterization of 3D Printable Formulations

The samples AESO-L and ASO-L were characterized in terms of viscosity and photocuring. Viscosity measurements were performed at 25 °C in a cone–plate geometry CP50-1 (diameter of 25 mm and cone angle of 1°) with a fixed gap of 1 mm using an Anton Paar MCR302 instrument (Ganz, Austria). The viscosity was recorded by varying the shear rate ($\dot{\gamma}$) of the samples from 1000 to 0.01 s$^{-1}$.

In the case of photocuring, both the total carbon–carbon double bond conversion and the reactivity of the liquid formulations were evaluated. For the reactivity, photorheology tests were carried out using an Anton PAAR Modular Compact Rheometer (Physica MCR 302, Graz, Austria) in a parallel plate configuration with a plate distance of 0.3 mm (upper disk diameter of 25 mm and quartz as a bottom disk). The measurements were performed at a constant frequency of 1 Hz and temperature of 25 °C. When the signal stabilized after 60 s, samples were irradiated from the bottom with a UV light of Hamamatsu LC8 lamp at 30 mW/cm$^2$ of irradiance. The evolution of both the storage and loss moduli (G′ and G″, respectively) in time was recorded, and the averaged results of each sample after three measurements were presented. For the total C=C double bond conversion, ATR-FTIR technique was used as explained in Section 2.5.

### 2.9. Characterization of Solid Samples

The characterization of AESO-S and ASO-S was performed on samples prepared using a Dymax ECE 5000-UV lamp. Dynamic mechanical thermal analysis (DMTA) was performed on rectangular samples of 20 mm × 7.5 mm × 1 mm using a Tritec 2000 DMA equipment from Triton technology Ltd. (Leicester, UK). Samples were tested by setting a temperature ramp of 2 °C/min at a fixed frequency of 1 Hz with a total displacement of 20 μm in the temperature range of −60 to 120 °C. The apparent crosslinking density ($\nu_c$) of both AESO-S and ASO-S was calculated applying the Equation (1):

$$\nu_c = \frac{E'_R}{3RT} \tag{1}$$

where $E'_R$ is the storage modulus at the rubbery plateau (obtained from the graph at a temperature of 50 °C above the glass transition temperature), R is the gas constant, and T the absolute temperature.

To evaluate the mechanical properties of the prepared solid samples (AESO-S and ASO-S), stress–strain tests were performed in specimens that had dimensions of about 5 mm × 50 mm × 1.10 mm with a Metrotec FTM-50 (20 N load cell) (Lezo, Spain) at room temperature and at a deformation rate of 1 mm/min. The Young modulus (E) was obtained by calculating the slope of the linear region and the strain ($\varepsilon_b$) and stress ($\sigma_b$) at break were provided in a way that takes five replicates into consideration.

Moreover, the 3D-printed samples were also characterized in terms of the quality of the printed materials. Three-dimensional scanning of the printed objects was performed with a E4 3D scanner from 3Shape and a heat map of the difference between the digital file and the digitalization of the real object created by the 3D scanner was reported.

## 3. Results

### 3.1. Synthesis and Characterization of Acrylated Epoxidized Soybean Oil (AESO), Effect of the Catalyst Nature

Acrylated epoxidized soybean oils are commonly synthesized through two consecutive steps. First, the vegetable oil is epoxidized to transform the double bonds of the fatty chains into epoxide functionalities. In the second step, these epoxide groups react with the carboxylic groups of acrylic acid to give the desired product.

Epoxide groups are tensioned and stressed structures which are very prone to reacting with nucleophiles that interact with the electrophile carbon of the C-O bond, making the ring opening easy [25]. This C-O bond is also polarized by electrophilic compounds, such as protic acids, or Lewis acids that can interact with the electron pair of the oxygen. As a result, carboxylic acid reacts with the epoxide group, giving an ester and hydroxyl functionality. As the epoxidation of vegetable oils is a mature industrial process, commercial epoxidized oil was directly reacted to acrylic acid in order to study the second step of the process. Five different catalysts were considered to speed up the acrylation reaction: TEA and TPP had been previously referenced in the bibliography [26]. On the other hand, a commercial chromium (III) complex salt named AMC-2® accelerator had been demonstrated to be effective in the acrylation of epoxides with organic acids [27]. Here, three alternatives to AMC-2® were compared: chromium (III) 2,4-pentanedionate at 97%, chromium (III) 2-ethylhexanoate at 50% in 2-ethyl hexanoic acid, and chromium (III) 2-ethyl hexanoate in 70% in mineral oil. All catalysts were tested in equivalent conditions: 80 °C; a small molar excess of acrylic acid to oxirane oxygen of 1.1:1; 1% hydroquinone to prevent homopolymerization; a molar ratio (MR) of catalyst to epoxidized oil of 1% b (*w*/*w*); and a reaction time of 6 h.

Making a comparison between the different catalysts employed, it was found that the chromium (III) 2-ethyl hexanoate in 70% of mineral oil presented the highest epoxide conversion (98%), followed by the chromium (III) 2-ethylhexanoate 50% in 2-ethyl hexanoic acid. Next, the chromium (III) 2,4-pentanedionate 97% offered a conversion of 69%, and finally, the least active catalysts were triphenylphosphine (68%) and triethylamine (65%).

As depicted in Scheme 1, the role of the catalyst is related to its acidic/basic properties, as well as its ways of activating the reactant molecules to produce the acrylated oil. In this way, the best catalyst is the one that possesses dual behavior. Ethyl hexanoate and petanedionate anions present basic properties. They abstract a proton from the acrylic acid to form an acrylate anion. On the other hand, Cr (III) is a Lewis acid. It polarizes the C-O bond in the epoxide ring (STEP 1), making the carbon atom more electrophilic, thus providing the attack of the nucleophile. It results in easier epoxide ring opening (STEP 2). Finally, the bases transfer the proton to the previously formed anion, giving acrylated oil (STEP 3). TPP and TEA are exclusively basic catalysts, meaning that their efficiency in the reaction is related to their basicity degree.

Although the conversions are quite similar using chromium (III) 2-ethyl hexanoate diluted in either mineral oil or in 2-ethyl hexanoic acid, it is necessary to consider that in the second case, the solvent can compete with acrylic acid in terms of opening the epoxide rings, thus lowering the selectivity and yield of the acrylated oil.

The use of chromium (III) 2-ethyl hexanoate in 70% in mineral oil allowed the completion of the acrylation reaction in equivalent operational conditions to those of the AMC-2® catalyst. The epoxide conversion of the reaction, which was quantified via the disappearance of the oxirane groups of the vegetable oil at the beginning and the end of the reaction, indicated it was concluded after 5 h. On the other hand, the acrylic acid content in the reaction media was quantified via the acidity index in the product. It was stable at a value corresponding to 32 mg KOH/g after 5 h of reaction, indicating that all the reactant needed had already been consumed, as can be observed in Figure 1.

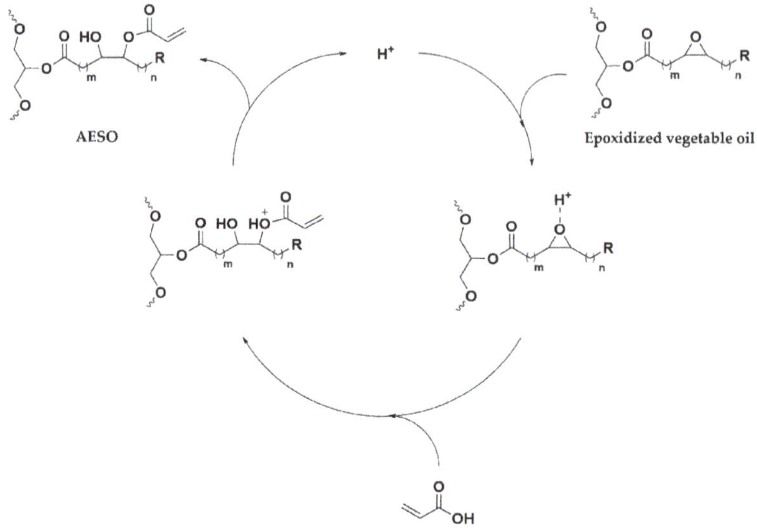


**Scheme 1.** Reaction mechanism of the acrylated epoxidized oil.

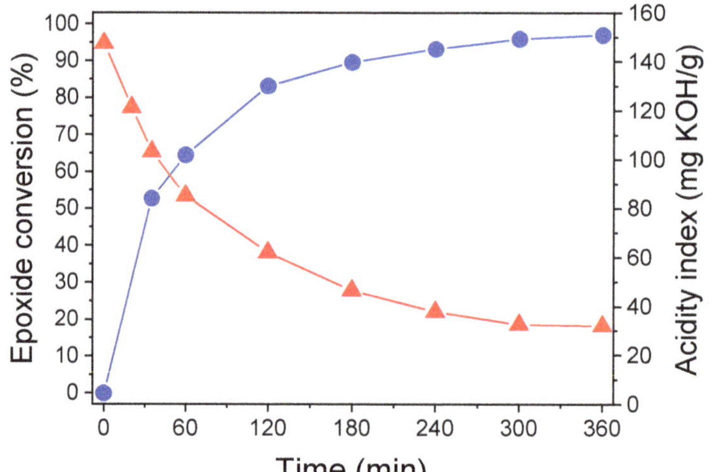


**Figure 1.** Evolution of the acrylation reaction using chromium (III) 2-ethyl hexanoate in 70% in mineral oil. (• epoxide conversion %; ▲ acidity index (mg KOH/g)). Operational conditions: temperature = 80 °C; molar ratio AA/epox. = 1.1:1; 1% (*w*/*w*) hydroquinone.

According to the FTIR spectra (Figure 2), the product obtained was AESO. The complete disappearance of the peak at 823–830 cm$^{-1}$ indicates that epoxide groups react to form other functionalities. Instead, there is an absorption band of 3400–3500 cm$^{-1}$, associated with the band of –OH. The vinyl group (=C–H) from the acrylate group appears at 3100 cm$^{-1}$. The signal of double bonds (C=C) from the acrylate pendant groups is found at 1650, 1617, and 1406 cm$^{-1}$. It can be also observed a peak at 810 cm$^{-1}$, a phenemenon which is related to the acrylate out-of-plane deformation [28]. Finally, other peaks, such as the one at 750 cm$^{-1}$, correspond to the band of several continuous methylenes (-$CH_2CH_2CH_2CH_2$-) from fatty acid chains.

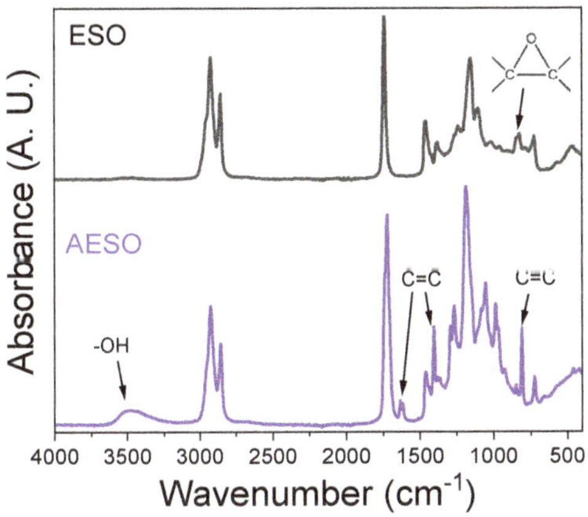


**Figure 2.** FTIR spectra of ESO and AESO.

Figure 3 shows the $^1$H-NMR spectra for ESO and AESO and gives their proton attributions. The epoxidation of the vegetable oils was evidenced by the disappearance of the peaks at 5.2–5.5 ppm, which indicated that the vinylic unsaturated hydrogens of the SO were transformed into epoxy groups. These signals appeared again in the AESO sample due to the presence of new acrylic unsaturations. The signals a ($\delta$ = 5.25 ppm), b ($\delta$ = 4.1 and 4.2 ppm) and c ($\delta$ = 2.4 ppm) appeared in both spectra ESO and AESO since they corresponded to the protons of the tri-ester group, which did not suffer modifications during the acrylation reaction. On the other hand, a peak that corresponded to the epoxy functionalities appeared at $\delta$ = 2.9 ppm (d) and disappeared in the AESO sample. In the same way, the peak at $\delta$ = 1.45 ppm that corresponded to -CH hydrogens adjacent to epoxy groups disappeared in the acrylated sample. The signals at $\delta$ = 6.5, 6.2 and 5.9 ppm (e, f, g) corresponded to the cis, trans and gemminal hydrogens with respect to acrylate ester groups. They confirmed the presence of acrylic functionalities in the AESO product.

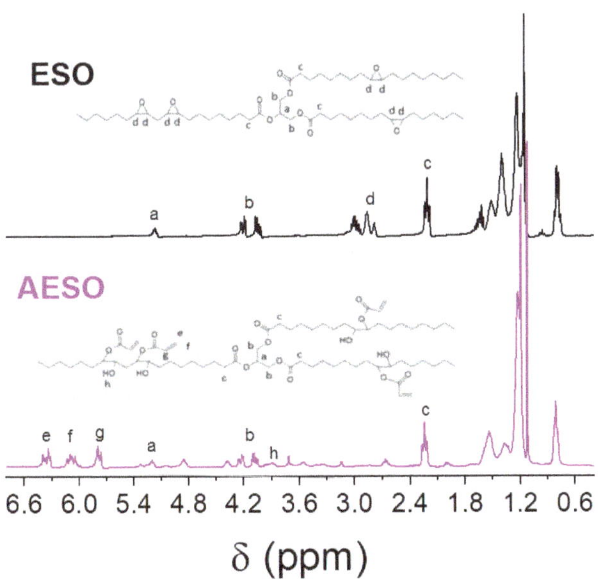


**Figure 3.** $^1$H-NMR spectra of ESO and AESO.

### 3.2. *Synthesis and Characterization of Directly Acrylated Soybean Oil (ASO)*

The double bonds present in the vegetable oils can directly react with acrylic acid through a one-step process without their intermediate transformation into epoxide rings. However, this reaction requires a strongly acidic catalyst [21] and different operational conditions to activate double bonds of the fatty chains with low activity. Additionally, the polymerization of acrylic acid should be avoided at the same time. Zhang et al. [20] propose a plausible mechanism that explains the direct acrylation of SO with AA, as can be seen in Scheme 2.

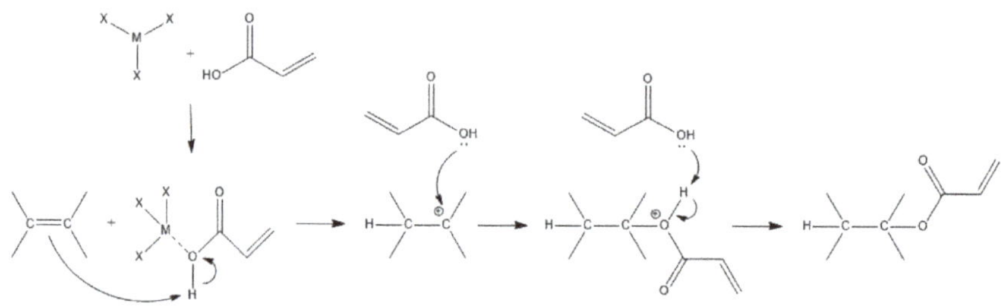

**Scheme 2.** Reaction mechanism of the direct acrylation of double bonds.

First, the SO unsaturation degree was determined via the iodine index analysis (Wij's method). Then, $BF_3OEt_2$ was employed as an efficient Lewis acid capable of inserting the acrylic acid molecules into the double bonds of the fatty chains. Overall, 0.1 equiv of this catalyst were used since higher amounts promoted oligomerization reactions between acrylic acid and SO. A molar ratio between the double bonds and acrylic acid of 1:6 was employed to increase the contact between the reactive groups. It was observed that higher AA amounts were not conducted in higher conversions. However, this was supposedly an important cost of obtaining a valuable reactant and had more drawbacks in the purification stage. The temperature was maintained at 80 °C for 10 h. Increasing the temperature value permitted us to achieve higher conversions faster. However, secondary reactions also appeared. Hydroquinone was used to prevent the polymerization of acrylic acid. However, in this case, the concentration was lowered to 0.25% (*w*/*w*) because hydroquinone acrylate can formed in the presence of this strongly acidic catalyst. The FTIR spectra of ASO indicated some differences from the corresponding one of AESO, as can be seen in Figure 4.

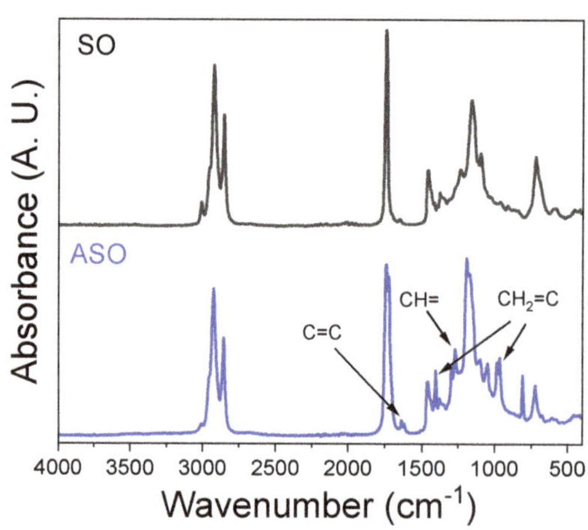


**Figure 4.** FTIR spectra of SO and ASO.

The C=C vibration of the acrylate groups in ASO appeared at 1637 and 1619 $cm^{-1}$ and was considerably more intense than that seen at the corresponding peak at 1652 $cm^{-1}$. This could be attributed to the C=C vibration in SO. Another two peaks appeared at 1400 and 960 $cm^{-1}$, respectively, corresponding to the scissoring vibration and the rocking vibration of $CH_2$ in $CH_2$=C of the acrylate group. Finally, the two peaks at 1296 and 1272 $cm^{-1}$ corresponded to the vibration of CH in acrylate CH= [28]. The most relevant difference between AESO and ASO was evidenced by the absence of the peak 3400–3500 $cm^{-1}$ that corresponded to the -OH groups of AESO since no hydroxyl functionalities were generated by the direct insertion of acrylic acid into SO.

Regarding the $^1$H-NMR results (Figure 5), whereas in the synthesis of AESO the signals corresponding to the epoxide groups of ESO disappeared (δ = 2.9 ppm), the direct acrylation of SO to give ASO showed a partial reaction conversion since the signals of the double bonds of the triglyceride chains decreased (δ = 5.4 ppm) but remained in the product. Referencing Zhang et al. [20], an average rate of 1.44 acrylate molecules per triglyceride was inserted, which was inferior to the percentage achieved in the acrylation of epoxidized oils.

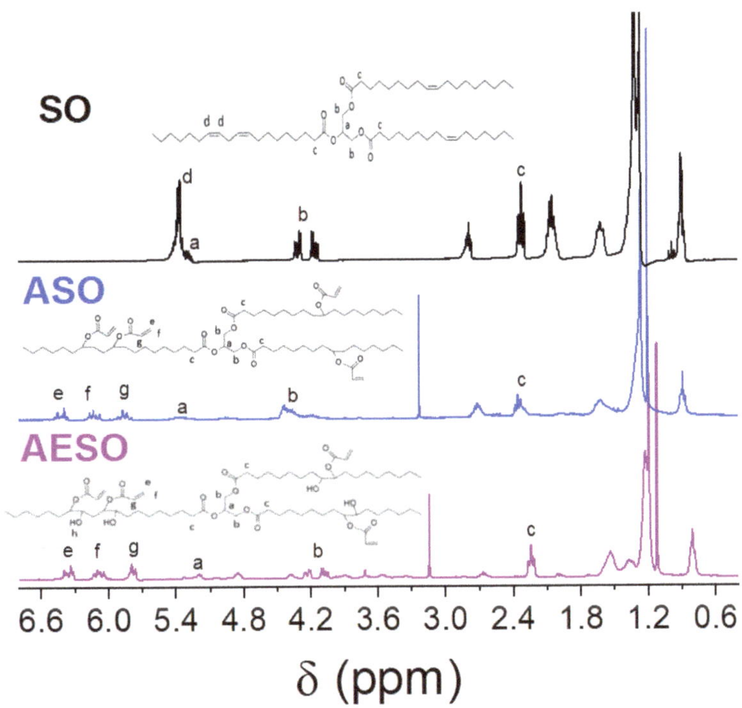


**Figure 5.** $^1$H-NMR spectra of SO, ASO and AESO.

### *3.3. Characterization of 3D Printable Formulations*

Once the starting polymeric precursors AESO and ASO have been prepared and characterized, they are mixed with a 1 wt.% of photoinitiator in order to obtain the photocurable formulations. Mainly, resins for use in DLP printing need to be characterized in a way that attends to photoreactivity and viscosity [29]. The photopolymerization processability of AESO-L and ASO-L formulations was evaluated via real-time photorheology, measuring the evolution of both storage and loss moduli (G′ and G″, respectively), while samples were irradiated with UV-light at 405 nm. Thus, G′ values as a function of time (t) are reported in Figure 6a, while values of gel time ($t_{gel}$) calculated as the time at which the crossover of both moduli and the curing rate ($\Delta G'/\Delta t$) occurs, measured as the slope of the G′ curves in the initial irradiation times, are indicated in Table 1.

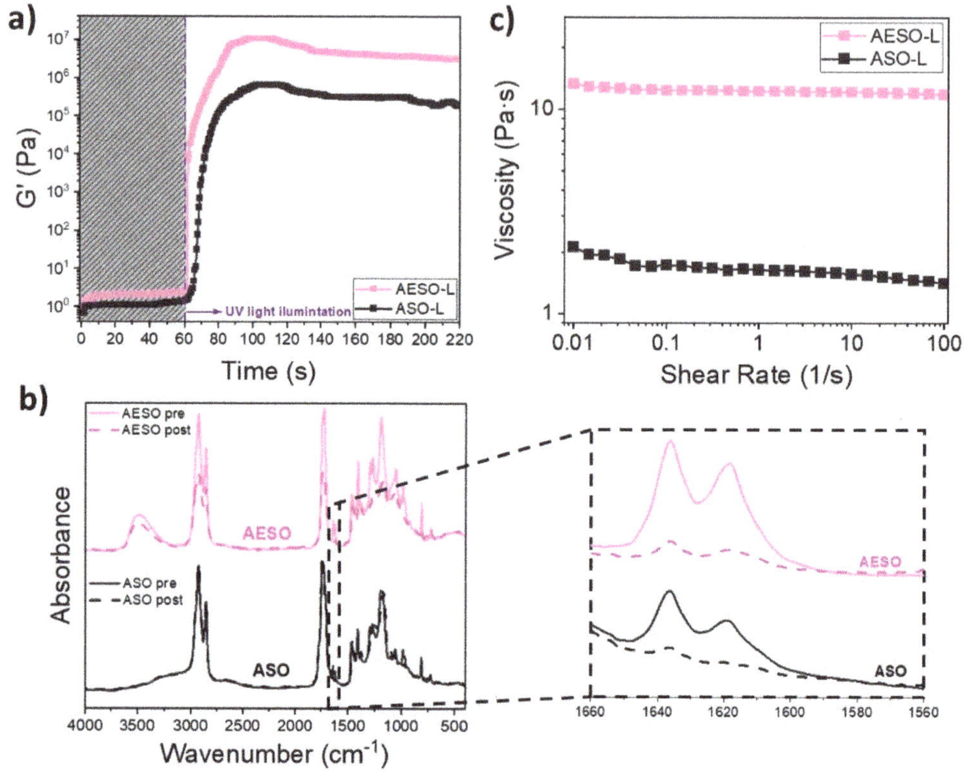


**Figure 6.** Photorheology curves (**a**), FTIR spectra before (pre-) and after (post-) irradiation with UV-light showing the double bond conversion (**b**) and viscosity as a function of shear rate at 25 °C (**c**) of AESO and ASO formulations.

**Table 1.** Main properties for AESO and ASO liquid and solid materials.

| Property | AESO | ASO |
|---|---|---|
| $t_{gel}$ (s) | 1.3 ± 0.2 | 7.0 ± 1.4 |
| $\Delta G'/\Delta t$ (kPa/s) | 72 ± 6 | 8.6 ± 0.2 |
| Conversion (%) | 79 ± 3 | 93 ± 2 |
| Viscosity at $\dot{\gamma} = 1\ s^{-1}$ (Pa·s) | 11.8 ± 0.6 | 1.66 ± 0.04 |
| $E'_R$ (MPa) | 40 | 2.6 |
| $T_g$ (°C) | 48 | −6 |
| $\nu_c$ (mmol·cm$^{-3}$) | 2.58 | 0.26 |
| ΔV(%) | 5.0 ± 0.2 | 4.3 ± 0.4 |
| E (MPa) | 1.433 ± 0.370 | 0.085 ± 0.007 |
| $\sigma_b$ (MPa) | 3.46 ± 0.25 | 0.43 ± 0.07 |
| $\varepsilon_b$ (%) | 3.44 ± 0.46 | 11.04 ± 1.31 |

The photorheology curves show that both acrylated oils present similar behavior and offer a fast cure by UV light, as can be identified by the large slopes of both curves and the short time that they take to arrive at the plateau. When formulations are irradiated, the photolysis of the initiator generates radicals that rapidly react with the double bonds of the acrylated oils, thus generating a crosslinked network rapidly [30]. The photocuring process of AESO is faster than that of ASO, as revealed by an onset of polymerization of at less than 2 s compared to the 5 s required for the ASO polymeric precursor. The slower kinetic for ASO-L than AESO-L is also revealed by the decrease in the slope at the initial part of the G′ curve, the curing rate being 72 ± 6 kPa/s for AESO-L and 8.6 ± 0.2 kPa/s for ASO L, and by the higher $t_{gel}$ of ASO-L (7.0 ± 1.4 s) than AESO-L (1.3 ± 0.2 s). A possible explanation for this is related to the multifunctionality of the synthetized oils and the autoacceleration phenomenon [31]. As $^1$H-NMR reveals, the functionality of the AESO polymeric precursor is significantly higher than that that of the ASO polymeric precursor (2.46 acrylic groups per molecule in the case of AESO versus 1.44 for the ASO) due to the higher effect of the autoacceration for AESO than ASO and, therefore, the higher amount of kinetic energy for the acrylated epoxidized oil than the directly acrylated one.

In addition, G′ values after the photopolymerization plateau are directly related to the mechanical stability of crosslinked materials [32]. Here, the results show that the G′ of the AESO-cured polymer is larger than that of ASO, indicating higher mechanical properties for AESO- than ASO-cured materials, as will be explained in the following section.

Photoreactivity was also characterized by measuring the acrylic double bond (C=C) conversion via FTIR. The average percentages of conversion were calculated following the decrease in the peak at 1630 cm$^{-1}$, corresponding to the mentioned C=C group at the pre- and post-cured materials. The obtained values of conversion (in %), respectively, for the initial C=C content are reported in Table 1, while the FTIR spectra can be seen in Figure 6b.

The FTIR spectra of AESO and ASO formulations display very similar peaks to those characteristic of the main chemical groups presented in the soybean oil-based polymer [11]. Similar to the FTIR spectra of the synthetized polymeric precursors, the stretching vibrations of the -OH, C=O, and C-O-C groups are shown at 3450, 1725 and 1160 cm$^{-1}$, respectively. In addition, the characteristic bands of the asymmetric stretching vibrations and deformations of the C–H bond in the $-CH_2-$ and $-CH_3-$ groups are identified at 2920, 2850 and 1450 cm$^{-1}$. Further, the characteristic band of the stretching of C=C group is presented in both formulations at the mentioned 1630 cm$^{-1}$. When samples are irradiated, the success of the C=C conversion can be confirmed by the decrease in the corresponding peak and the difference between pre- and post-irradiated samples is easily observed when spectra are zoomed in on. The conversion degrees are 79 ± 3% for the AESO formulation, while the ASO one presents a significantly higher conversion degree of 93 ± 2%. This can be ascribed to the difference in the functionalization of each oil due to the number of double bonds being different per molecule of AESO or ASO, being 2.46 in the case of AESO and 1.44 for ASO. As has been reported by several authors, the increased number of functionalities presented in the photocurable formulations results in a decrease in C=C conversion due to the vitrification of the highly crosslinked network and the concomitant restricted mobility of the remaining acrylates [33]. Anyhow, high conversion degrees above 75% are confirmed for both formulations.

Once photoreactivity has been assessed, the viscosity of both formulations is measured. In stereolithography or other layer-by-layer printing techniques (as DLP), the resin viscosity is a crucial parameter, with low viscosities desired to enable an appropriate coating of the last cured layer and the resin tank with the liquid resin before beginning the illumination process of the subsequent layer [34]. Moreover, in the preparation of UV-curable formulations, low viscosity oligomers are preferred due to their reduced demand for reactive diluents [35]. However, there is not a fixed range of appropriated viscosities for DLP printing, and some authors have highlighted different values depending on the material used, such as 0.50–0.80 Pa·s for biorenewable formulations [29], a maximum of 10 Pa·s for polymeric materials in general [36], or less than 20 Pa·s at a shear rate of 10 to 100 s$^{-1}$ for polymeric composites [37]. In this sense, viscosities of AESO-L and ASO-L formulations were characterized at room temperature and in the range of 1000 to 0.01 $s^{-1}$ of shear rate. Figure 6c depicts the viscosity curves thus obtained, and Table 1 includes the viscosity values of both materials at a shear rate of 1 $s^{-1}$.

The acrylated soybean oil-based formulations were found to be Newtonian. This can be ascribed to the fact that near-constant viscosity values were showed in the shear rate range tested, as occurs for other natural oils [38]. Particularly, the viscosity of AESO-L is significantly higher (one order of magnitude) than the one of ASO-L, showing values of 11.8 ± 0.6 Pa·s and 1.66 ± 0.04 Pa·s, respectively, at shear rate of 1 $s^{-1}$. In general, there are two major factors that affect the viscosity of acrylated natural oils: molecular weight and H-bonding through OH groups. The viscosity will be higher when the molecular weight increases and when more -OH groups are presented in the formulation [35]. Thus, the higher viscosity of AESO than ASO is justified due to the fact that AESO polymeric precursor presented a higher acrylic functionalization than ASO, which indicates a high molecular weight [35,39,40]. Further, as it is expected, the AESO presents -OH groups in the structure due to the reaction mechanism that consists in the opening of the epoxy rings of ESO and the formation of pendant -OH groups along the double-bond-reacted triglyceride, while ASO is formed via the direct acrylation of SO and OH groups are not formed (also revealed in the FTIR spectra of the polymeric precursors included in the previous section).

### 3.4. Characterization of Solid Samples

The characterization of solid samples (AESO-S and ASO-S) started with Dynamic Mechanical Thermal Analysis (DMTA), which was employed to fully analyze their thermal and viscoelastic properties. The storage (E′) and loss (E″) moduli were obtained, as well as the ratio between of them named tan $\delta$ (tan δ = E″/E′). Thus, both E′ and tan δ curves in front of temperature are plotted in Figure 7a, and the storage modulus at the rubbery plateau ($E'_R$), the glass transition temperature of each material ($T_g$), calculated as the peak on the tan δ curve, and the apparent crosslinking density ($\nu_c$), obtained as explained in "Section 2", are detailed in Table 1.

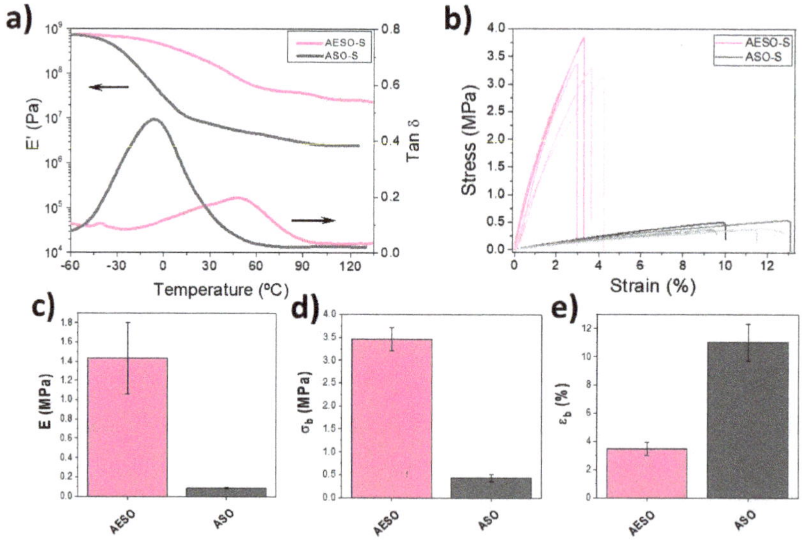


**Figure 7.** E′ and tan δ curves from dynamic mechanical thermal analysis, arrows indicate the axes of each data (**a**), stress–strain curves, (**b**), Young modulus or E (**c**), strain at break or $\varepsilon_b$ (**d**) and stress at break or $\sigma_b$ (**e**) of AESO-S and ASO-S samples.

It is possible to observe that both curves present similar shapes. They are characterized by a decrease in the storage modulus when temperature increases and the appearance of a peak in the tan δ plot that corresponds to the decay of the E′. This single E′ step decrease changes from 680 MPa to 40 MPa in AESO-S sample and to 2.6 MPa in ASO-S, corresponding to the transition between the glassy and rubbery states (glass transition) of both materials [41]. Attending to the temperature at which this glass transition occurs, that is, observing the temperature at which the tan δ peak appears, a significantly difference between the acrylated epoxidized and the directly acrylated materials is presented. The $T_g$ of AESO-S is located at a higher temperature (around 48 °C) than the one of ASO-S, which appears at −6 °C. Further, the peak is particularly broader for AESO-S than ASO-S, suggesting the existence of a more heterogeneous network for directly acrylated than acrylated epoxidized material [42].

As it is described in the experimental part, the modulus at a rubbery sate ($E'_R$) can be related to the crosslinking density ($\nu_c$) of the polymer networks. Thus, the calculated $\nu_c$ of AESO-S is significantly higher (2.58 mmol·cm$^{-3}$) than the one of ASO-S (0.26 mmol·cm$^{-3}$). Attending to the results of photopolymerization process, a higher double bond conversion for ASO-S than AESO-S was obtained. This, in theory, should indicate a higher crosslinking degree of directly acrylated material than that for the epoxidized acrylated one [43]. However, the functionality of AESO was estimated as 2.46 acrylic groups per molecule, while for ASO polymeric precursor it was 1.44. In this sense, a conversion of around 80% for AESO supposes a higher number of double bonds converted than a conversion of around 90% of ASO. Therefore, the number of crosslinking points, and hence, the crosslinking density, is higher for AESO than ASO. This effect of crosslinking density also explains the difference observed in the $T_g$ of both materials, as occurs in other soybean oil-based materials [44].

The mechanical characterization of AESO-S and ASO-S samples was carried out via stress–strain measurements. Figure 7b reports the obtained curves and the main parameters, that is, Young's modulus (E) and, the strain ($\varepsilon_b$), and the stress ($\sigma_b$) at break are plotted in Figure 7c–e. These values are summarized in Table 1.

As can be clearly seen, AESO-S presents a typical stress–strain curves of thermoset brittle materials, characterized by a continued linear increase in the stress with the increase in the strain and the sample break before reaching the yield point at a low value of strain [45]. This indicates the rigid characteristics of the AESO-S material. In the case of ASO-S, a similar behaviour is obtained, even if it shows a larger strain at break and a lower Young modulus. The AESO-S sample presents an E value of around 1.433 ± 0.370 MPa. This is significantly higher (around 20 times more) than the E value obtained for ASO-S, which is 0.085 ± 0.007 MPa. In the case of parameters at break, $\sigma_b$ varies from 3.46 ± 0.25 MPa to 0.43 ± 0.07 MPa and $\varepsilon_b$ changes from 3.44 ± 0.46% to 11.04 ± 1.31% when soybean oil is directly acrylated.

Typically, the mechanical properties of natural-based oil-cured films are dependent on the inherent chemical composition and the crosslinking density of the oil presented. When photocured polymers are considered, the increase in grafting number is an essential parameter to be considered as the functionality of the oil directly depends on it [46]. Thus, when functionality increases, the number of crosslinking points increases helping this to the increase in the rigidity of the cured films and enhancing the mechanical properties of photocrosslinked materials [47]. In this case, as NMR results reveal, the functionality of AESO is higher than that of ASO, which the subsequent increase in the crosslinking degree as DMA results indicate. Therefore, AESO-S presents a more rigid structure than ASO-S. This is characterized by higher E and $\sigma_b$ values, as well as by a lower $\varepsilon_b$.

Having confirmed the good reactivity and the appropriate viscosities of both soybean oil-based materials for DLP printing, the formulations are successfully 3D-printed. To evaluate the resolution and the capabilities of both AESO-L and ASO-L materials in DLP, a hexagonal honeycomb 3D form is printed. The 3D-printed structures are characterized via a 3D scanner; Figure 8 shows the real photos of structures printed using both soybean oil-based materials, as well as an analysis of them obtained using the scanner.

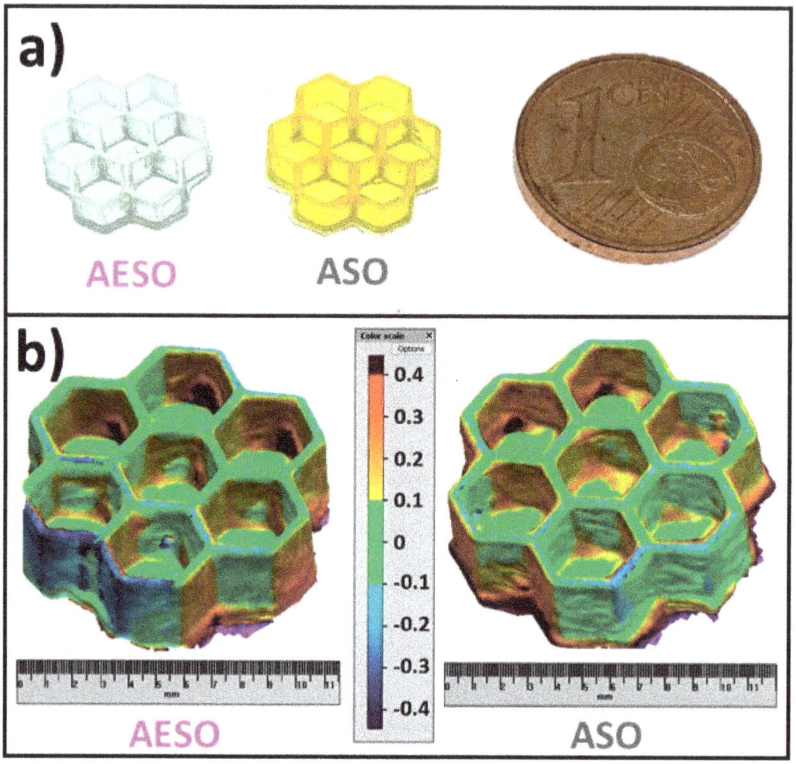


**Figure 8.** Photographs of honeycomb printed structures for AESO and ASO materials (**a**) and 3D image obtained with 3D scanner with the evaluation of the displacement between the CAD project and the real device for same printed structures (**b**).

In previous probes, the different printing parameters were properly adjusted depending on the material employed. As the AESO-based formulation presents a higher viscosity compared to ASO-based one, the working temperature for 3D printer slightly increases from room temperature to 40 °C in order to facilitate the movement of the formulation during the layer-by-layer process. Further, different parameters such as exposure time, platform approximation velocity, or light intensity, were adjusted to obtain 3D structures with a consistent form (self-supporting structures) in short periods of time. The methods did not prioritize prevention over curing and had the highest possible quality. Thus, in a honeycomb structure, great results are obtained for both AESO and ASO materials, as Figure 8a depicts. In the case of an AESO-printed structure, an exposure time of 2 s per layer is enough to properly obtain the whole structure. Meantime, for ASO material, an exposure time of 4 s is needed (light intensity of 35 mW/cm$^2$ for both materials). This is in concordance with the photorheology results that demonstrate a lowest reactivity for the ASO formulation than the AESO one. In terms of the platform approximation velocity, the lowest viscosity of ASO material than AESO allows for the use of a higher velocity, reducing the total time needed for printing. In this sense, the exposure time of both parameters, and the platform approximation velocity present the opposite effect, and the total times required to print the honeycomb structures are similar for both materials. Similarly, other photocurable resins, printed using the same printer as this study, also use the same printing parameters as the ones employed here [48], indicating the good ability of these materials to be used as DLP resins.

Attending to the resolution and the quality of the printed structures, both materials present a good similar appearance. Both AESO-S and ASO-S 3D-printed structures are investigated via a 3D scanner in order to evaluate their CAD fidelity. Figure 8b presents the comparison between the 3D image and the CAD virtual project. As can be seen, both printed structures are colored in green, indicating the great resolution and the fidelity of the project for both of them. The green color represents a displacement at about $\pm$ 0.1 mm and yellow and light blue colors are about $\pm$0.2 mm. These results are to those obtained on other soybean oil-based materials printed via DLP [49].

## 4. Conclusions

In summary, acrylated epoxidized soybean oil (AESO) and acrylated soybean oil (ASO) have been successfully obtained from epoxidized soybean oil (ESO) and soybean oil (SO), respectively. In the case of AESO, five catalysts have been studied. In particular, we examined chromium (III) 2-ethyl hexanoate (Cr(EH)3-MO) in 70% in mineral oil, the one with higher reaction yield according to the epoxide conversion and acidity index. For ASO, the reaction has been performed by employing the boron trifluoride at 48% in diethyl ether (BF3OEt2) as an efficient Lewis acid, obtaining the product with a lower rate of acrylate molecules per triglyceride than AESO. Both AESO and ASO structures have been confirmed via the 1H-NMR technique. Additionally, oils have been mixed with an appropriate quantity of BAPO photoinitiator and the resulting formulations have been analyzed in terms of 3D printability via photorheology and viscosity measurements. Both AESO and ASO have been demonstrated to be suitable materials for DLP printing and their printability has been studied by analyzing the quality of 3D-printed part by means of a 3D scanner that indicates an accuracy of $\pm$0.1 mm for both oils. The AESO photocured materials showed higher glass transition temperature, crosslinking density and Young's modulus than ASO, which can be ascribed to the lower acrylate group rate presented. In conclusion, this work demonstrates the possibility of using biobased materials as photocurable inks for 3D printing and the possibility of obtaining them via one-step reactions from the original vegetable resource.

**Author Contributions:** Conceptualization, J.L.V.-V. and S.P.-F.; methodology, M.S., C.M.-F. and L.R.-R.; investigation, C.M.-F., L.R.-R. and I.I.; data curation, M.S., C.M.-F. and L.R.-R.; writing—original draft preparation, C.M.-F. and L.R.-R.; writing—review and editing, M.S., L.R.-R., J.L.V.-V. and S.P.-F.; supervision, J.L.V.-V., O.G.-J.-A. and M.S.; funding acquisition, J.L.V.-V. and L.R.-R. All authors have read and agreed to the published version of the manuscript.

**Funding:** Financial support from the Basque Government postdoctoral grant (POS-E_2021_2_0001), Grupos Consolidados (IT1756-22) and Elkartek program (KK-2021/00082) grants are acknowledged. Technical and human support provided by SGIker (UPV/EHU, MICINN, GV/EJ, EGEF, and ESF) is gratefully acknowledged.

**Institutional Review Board Statement:** Not applicable.

**Data Availability Statement:** The raw/processed data required to reproduce these findings cannot be shared at this time as the data also forms part of an ongoing study.

**Conflicts of Interest:** The authors declare that the research was conducted in the absence of any commercial or financial relationships that could be construed as a potential conflict of interest.

## References

1. Ghobakhloo, M. Industry 4.0, Digitization, and Opportunities for Sustainability. *J. Clean. Prod.* **2020**, *252*, 119869. [CrossRef]
2. Tofail, S.A.M.; Koumoulos, E.P.; Bandyopadhyay, A.; Bose, S.; O'Donoghue, L.; Charitidis, C. Additive Manufacturing: Scientific and Technological Challenges, Market Uptake and Opportunities. *Mater. Today* **2018**, *21*, 22–37. [CrossRef]
3. Al Rashid, A.; Ahmed, W.; Khalid, M.Y.; Koç, M. Vat Photopolymerization of Polymers and Polymer Composites: Processes and Applications. *Addit. Manuf.* **2021**, *47*, 102279. [CrossRef]
4. Mendes-Felipe, C.; Oliveira, J.; Etxebarria, I.; Vilas-Vilela, J.L.; Lanceros-Mendez, S. State-of-the-Art and Future Challenges of UV Curable Polymer-Based Smart Materials for Printing Technologies. *Adv. Mater. Technol.* **2019**, *4*, 1800618. [CrossRef]

5. Khan, S.A.R.; Razzaq, A.; Yu, Z.; Miller, S. Industry 4.0 and Circular Economy Practices: A New Era Business Strategies for Environmental Sustainability. *Bus. Strateg. Environ.* **2021**, *30*, 4001–4014. [CrossRef]
6. Benson, N.U.; Bassey, D.E.; Palanisami, T. COVID Pollution: Impact of COVID-19 Pandemic on Global Plastic Waste Footprint. *Heliyon* **2021**, *7*, e06343. [CrossRef] [PubMed]
7. Sousa, A.F.; Silvestre, A.J.D. Plastics from Renewable Sources as Green and Sustainable Alternatives. *Curr. Opin. Green Sustain. Chem.* **2022**, *33*, 100557. [CrossRef]
8. Kalita, D.J.; Tarnavchyk, I.; Kalita, H.; Chisholm, B.J.; Webster, D.C. Bio-Based Coating Resins Derived from Cardanol Using Carbocationic Polymerization and Their Evaluation as One-Component Alkyd-Type Coatings. *Prog. Org. Coat.* **2023**, *174*, 107252. [CrossRef]
9. Fertier, L.; Koleilat, H.; Stemmelen, M.; Giani, O.; Joly-Duhamel, C.; Lapinte, V.; Robin, J.J. The Use of Renewable Feedstock in UV-Curable Materials-A New Age for Polymers and Green Chemistry. *Prog. Polym. Sci.* **2013**, *38*, 932–962. [CrossRef]
10. Zhang, C.; Garrison, T.F.; Madbouly, S.A.; Kessler, M.R. Recent Advances in Vegetable Oil-Based Polymers and Their Composites. *Prog. Polym. Sci.* **2017**, *71*, 91–143. [CrossRef]
11. Mendes-Felipe, C.; Costa, P.; Roppolo, I.; Sangermano, M.; Lanceros-Mendez, S. Bio-Based Piezo- and Thermoresistive Photocurable Sensing Materials from Acrylated Epoxidized Soybean Oil. *Macromol. Mater. Eng.* **2022**, *307*, 2100934. [CrossRef]
12. Voet, V.S.D.; Guit, J.; Loos, K. Sustainable Photopolymers in 3D Printing: A Review on Biobased, Biodegradable, and Recyclable Alternatives. *Macromol. Rapid Commun.* **2021**, *42*, 2000475. [CrossRef]
13. Gan, Y.; Jiang, X. Photo-Cured Materials from Vegetable Oils. In *Green Materials from Plant Oils*; Royal Society of Chemistry: London, UK, 2014.
14. Malburet, S.; Di Mauro, C.; Noè, C.; Mija, A.; Sangermano, M.; Graillot, A. Sustainable Access to Fully Biobased Epoxidized Vegetable Oil Thermoset Materials Prepared by Thermal or UV-Cationic Processes. *RSC Adv.* **2020**, *10*, 41954–41966. [CrossRef]
15. Noè, C.; Hakkarainen, M.; Sangermano, M. Cationic UV-Curing of Epoxidized Biobased Resins. *Polymers* **2020**, *13*, 89. [CrossRef]
16. Danov, S.M.; Kazantsev, O.A.; Esipovich, A.L.; Belousov, A.S.; Rogozhin, A.E.; Kanakov, E.A. Recent Advances in the Field of Selective Epoxidation of Vegetable Oils and Their Derivatives: A Review and Perspective. *Catal. Sci. Technol.* **2017**, *7*, 3659–3675. [CrossRef]
17. Khot, S.N.; Lascala, J.J.; Can, E.; Morye, S.S.; Williams, G.I.; Palmese, G.R.; Kusefoglu, S.H.; Wool, R.P. Development and Application of Triglyceride-Based Polymers and Composites. *J. Appl. Polym. Sci.* **2001**, *82*, 703–723. [CrossRef]
18. Ho, Y.H.; Parthiban, A.; Thian, M.C.; Ban, Z.H.; Siwayanan, P. Acrylated Biopolymers Derived via Epoxidation and Subsequent Acrylation of Vegetable Oils. *Int. J. Polym. Sci.* **2022**, *2022*, 6210128. [CrossRef]
19. Briede, S.; Barkane, A.; Jurinovs, M.; Thakur, V.K.; Gaidukovs, S. Acrylation of Biomass: A Review of Synthesis Process: Know-How and Future Application Directions. *Curr. Opin. Green Sustain. Chem.* **2022**, *35*, 100626. [CrossRef]
20. Zhang, P.; Xin, J.; Zhang, J. Effects of Catalyst Type and Reaction Parameters on One-Step Acrylation of Soybean Oil. *ACS Sustain. Chem. Eng.* **2014**, *2*, 181–187. [CrossRef]
21. Zhang, P.; Zhang, J. One-Step Acrylation of Soybean Oil (SO) for the Preparation of SO-Based Macromonomers. *Green Chem.* **2013**, *15*, 641. [CrossRef]
22. Briede, S.; Jurinovs, M.; Nechausov, S.; Platnieks, O.; Gaidukovs, S. State-of-the-Art UV-Assisted 3D Printing via a Rapid Syringe-Extrusion Approach for Photoactive Vegetable Oil Acrylates Produced in One-Step Synthesis. *Mol. Syst. Des. Eng.* **2022**, *7*, 1434–1448. [CrossRef]
23. *UNE-EN-ISO 660*; Animal and Vegetable Fats and Oils—Determination of Acid Value and Acidity. Asociación Española de Normalización, UNE: Madrid, Spain, 2021.
24. McCutcheon, J.W. Wijs Iodine Method. *Ind. Eng. Chem. Anal. Ed.* **1940**, *12*, 465. [CrossRef]
25. Bukowska, A.; Bukowski, W. Reactivity of Some Carboxylic Acids in Reactions with Some Epoxides in the Presence Chromium (III) Ethanoate. *Org. Process Res. Dev.* **2002**, *6*, 234–237. [CrossRef]
26. Baghban, S.A.; Ebrahimi, M.; Khorasani, M. A Facile Method to Synthesis of a Highly Acrylated Epoxidized Soybean Oil with Low Viscosity: Combined Experimental and Computational Approach. *Polym. Test.* **2022**, *115*, 107727. [CrossRef]
27. Li, Y.; Sun, X.S. Synthesis and Characterization of Acrylic Polyols and Polymers from Soybean Oils for Pressure-Sensitive Adhesives. *RSC Adv.* **2015**, *5*, 44009–44017. [CrossRef]
28. Fu, L.; Yang, L.; Dai, C.; Zhao, C.; Ma, L. Thermal and Mechanical Properties of Acrylated Expoxidized-Soybean Oil-Based Thermosets. *J. Appl. Polym. Sci.* **2010**, *117*, 2220–2225. [CrossRef]
29. Lebedevaite, M.; Talacka, V.; Ostrauskaite, J. High Biorenewable Content Acrylate Photocurable Resins for DLP 3D Printing. *J. Appl. Polym. Sci.* **2021**, *138*, 50233. [CrossRef]
30. Cosola, A.; Conti, R.; Grützmacher, H.; Sangermano, M.; Roppolo, I.; Pirri, C.F.; Chiappone, A. Multiacrylated Cyclodextrin: A Bio-Derived Photocurable Macromer for VAT 3D Printing. *Macromol. Mater. Eng.* **2020**, *305*, 2000350. [CrossRef]
31. Anseth, K.S.; Wang, C.M.; Bowman, C.N. Reaction Behaviour and Kinetic Constants for Photopolymerizations of Multi(Meth)Acrylate Monomers. *Polymer* **1994**, *35*, 3243–3250. [CrossRef]
32. Barkane, A.; Platnieks, O.; Jurinovs, M.; Kasetaite, S.; Ostrauskaite, J.; Gaidukovs, S.; Habibi, Y. UV-Light Curing of 3D Printing Inks from Vegetable Oils for Stereolithography. *Polymers* **2021**, *13*, 1195. [CrossRef]

33. Arslan, A.; Steiger, W.; Roose, P.; Van den Bergen, H.; Gruber, P.; Zerobin, E.; Gantner, F.; Guillaume, O.; Ovsianikov, A.; Van Vlierberghe, S.; et al. Polymer Architecture as Key to Unprecedented High-Resolution 3D-Printing Performance: The Case of Biodegradable Hexa-Functional Telechelic Urethane-Based Poly-ε-Caprolactone. *Mater. Today* **2021**, *44*, 25–39. [CrossRef]
34. Weng, Z.; Zhou, Y.; Lin, W.; Senthil, T.; Wu, L. Structure-Property Relationship of Nano Enhanced Stereolithography Resin for Desktop SLA 3D Printer. *Compos. Part A Appl. Sci. Manuf.* **2016**, *88*, 234–242. [CrossRef]
35. Rengasamy, S.; Mannari, V. Development of Soy-Based UV-Curable Acrylate Oligomers and Study of Their Film Properties. *Prog. Org. Coat.* **2013**, *76*, 78–85. [CrossRef]
36. Kuhnt, T.; Morgan, F.L.C.; Baker, M.B.; Moroni, L. An Efficient and Easily Adjustable Heating Stage for Digital Light Processing Set-Ups. *Addit. Manuf.* **2021**, *46*, 102102. [CrossRef]
37. Chen, Z.; Li, J.; Liu, C.; Liu, Y.; Zhu, J.; Lao, C. Preparation of High Solid Loading and Low Viscosity Ceramic Slurries for Photopolymerization-Based 3D Printing. *Ceram. Int.* **2019**, *45*, 11549–11557. [CrossRef]
38. Noè, C.; Iannucci, L.; Malburet, S.; Graillot, A.; Sangermano, M.; Grassini, S. New UV-Curable Anticorrosion Coatings from Vegetable Oils. *Macromol. Mater. Eng.* **2021**, *306*, 2100029. [CrossRef]
39. Liu, P.; Zhang, X.; Liu, R.; Liu, X.; Liu, J. Highly Functional Bio-Based Acrylates with a Hard Core and Soft Arms: From Synthesis to Enhancement of an Acrylated Epoxidized Soybean Oil-Based UV-Curable Coating. *Prog. Org. Coat.* **2019**, *134*, 342–348. [CrossRef]
40. Sahoo, S.K.; Khandelwal, V.; Manik, G. Synthesis and Characterization of Low Viscous and Highly Acrylated Epoxidized Methyl Ester Based Green Adhesives Derived from Linseed Oil. *Int. J. Adhes. Adhes.* **2019**, *89*, 174–177. [CrossRef]
41. Landel, R.F.; Nielsen, L.E. *Mechanical Properties of Polymers and Composites*; CRC Press: Boca Raton, FL, USA, 1993.
42. Bandzierz, K.; Reuvekamp, L.; Dryzek, J.; Dierkes, W.; Blume, A.; Bielinski, D. Influence of Network Structure on Glass Transition Temperature of Elastomers. *Materials* **2016**, *9*, 607. [CrossRef]
43. Wu, N.; Wu, B.; Xu, Y.; Tang, X. Relationships of the Degree of C=C Double Bond Conversion with the Dielectric Properties for $SiO_2$/1,2-PB/SBS/EPDM Composites Cured by Organic Peroxide. *ChemistrySelect* **2022**, *7*, e202104078. [CrossRef]
44. Sahoo, S.K.; Mohanty, S.; Nayak, S.K. Study of Thermal Stability and Thermo-Mechanical Behavior of Functionalized Soybean Oil Modified Toughened Epoxy/Organo Clay Nanocomposite. *Prog. Org. Coat.* **2015**, *88*, 263–271. [CrossRef]
45. Chen, D.; Li, J.; Yuan, Y.; Gao, C.; Cui, Y.; Li, S.; Liu, X.; Wang, H.; Peng, C.; Wu, Z. A Review of the Polymer for Cryogenic Application: Methods, Mechanisms and Perspectives. *Polymers* **2021**, *13*, 320. [CrossRef]
46. Su, Y.; Lin, H.; Zhang, S.; Yang, Z.; Yuan, T. One-Step Synthesis of Novel Renewable Vegetable Oil-Based Acrylate Prepolymers and Their Application in UV-Curable Coatings. *Polymers* **2020**, *12*, 1165. [CrossRef] [PubMed]
47. Huang, J.; Yuan, T.; Yang, Z.; Man, L.; Hu, Y.; Yang, Z. UV/Thermal Dual Curing of Tung Oil-Based Polymers Induced by Cationic Photoinitiator. *Prog. Org. Coat.* **2019**, *126*, 8–17. [CrossRef]
48. Noè, C.; Cosola, A.; Tonda-Turo, C.; Sesana, R.; Delprete, C.; Chiappone, A.; Hakkarainen, M.; Sangermano, M. DLP-Printable Fully Biobased Soybean Oil Composites. *Polymer* **2022**, *247*, 124779. [CrossRef]
49. Gastaldi, M.; Roppolo, I.; Chiappone, A.; Garino, C.; Fin, A.; Manachino, M.; Sirianni, P.; Viscardi, G.; Scaltrito, L.; Zanetti, M.; et al. Thermochromic Photoluminescent 3D Printed Polymeric Devices Based on Copper-Iodide Clusters. *Addit. Manuf.* **2022**, *49*, 102504. [CrossRef]

polymers

MDPI

*Article*

# Particle Shedding from Cotton and Cotton-Polyester Fabrics in the Dry State and in Washes

**Tanja Pušić [1], Branka Vojnović [1], Sandra Flinčec Grgac [1,*], Mirjana Čurlin [2] and Rajna Malinar [1]**

[1] Faculty of Textile Technology, University of Zagreb, 10000 Zagreb, Croatia; tanja.pusic@ttf.unizg.hr (T.P.); branka.vojnovic@ttf.unizg.hr (B.V.); rajna.malinar@ttf.unizg.hr (R.M.)
[2] Faculty of Food Technology and Biotechnology, University of Zagreb, 10000 Zagreb, Croatia; mcurlin@pbf.hr
* Correspondence: sflincec@ttf.unizg.hr

**Abstract:** The influence of 3, 10 and 50 washing cycles on the properties of cotton fabric and cotton-polyester blend in plain weave, was investigated in this study. In addition to the analysis of tensile properties in weft and warp directions and thickness, the number of particles produced in the dry state was also measured after 3, 10 and 50 washes. After washing, the entire effluent was analysed by determining the total suspended solids (TSS), the total solids (TS), the pH value and the conductivity. To determine the similarity of the observed wash cycles and properties of all processed samples, hierarchical cluster analysis (HCA) was performed. The fabric changes indicated by total wear in the warp direction after 50 washing cycles compared to unwashed ones amounting to 41.2% for cotton and 30.9% for cotton-polyester blend, may be attributed to the synergy of washing factors and raw material composition. Cotton fabric produced significantly more particles than cotton-polyester fabric in the dry state after the examined washing cycles in all size categories. A smaller number of released particles are in the larger size category >25 μm. The obtained TSS values confirm the degree of loading of the effluent with particulate matter from the analysed fabrics, since the detergent consists of water-soluble components. The HCA dendrograms confirmed that the release of particles during the first washing cycles is mainly determined by the structural properties of fabrics, while in the subsequent cycles the synergistic effect of chemical, mechanical and thermal effects in the interaction with the material prevailed.

**Keywords:** cotton; cotton-polyester; washing process; particles; release; dry and wet state

## 1. Introduction

Environmental protection and methods to analyse the impact of textiles on quality and human health have been mandatory for many years and are now legal norms that need to be clarified through extensive product monitoring, such as circular economy principles. The choice of fibre is significant for the appearance of garments [1] and the environmental impact in the dry state and after washes. Previous research described textile dust released into the atmosphere as a result of the production and use of various types of textiles. Exposure to textile dust occurs during the production process, finishing, use, textile care and textile recycling. A major problem with the presence of textile dust is the possibility of various diseases arising from daily exposure. Despite the focus on cotton dust, the same challenges exist in the processing, finishing, and application of a variety of natural and synthetic materials. Particles shed from synthetic textiles represent a bigger problem, as they are deposited and cannot be broken down for the most part, which further pollutes the environment and people [2].

During the washing process, textiles are exposed to chemicals and mechanical agitation, which can create changes after repeated cycles. The extent of changes depends on textile characteristics (construction parameters, polymerisation degree, swelling capacity) as well as Sinner factors (chemicals, temperature, time, mechanical agitation) [3].

Detergent formulations are often alkaline and contain bleach as oxidising agents, which can trigger the chemical damage of cellulose when combined with other elements. The loss of tensile properties in cotton textiles is strongly linked to the depolymerisation of cellulose [4,5]. Lower washing temperatures, high efficiency detergents, and low bath ratios are required for the washing process to be sustainable, resulting in external fibrillation and pigment particle migration [6]. Cotton textile fibrillation may be associated with hydrophilicity, swelling, and construction features, namely the presence of shorter fibres in the yarn. This can be due to abrasion of the textile material in a wet (washing baths) or dry environment (tumble drying) [7,8]. Friction and deformation are two of the most common causes of clothing damage caused by washing machines [9,10]. The drying process at high temperature may affect the formation of cracks, and increased fibrillation.

Polyester fibres are characterised by high strength, crease resistance and fast drying, while cotton fibres are characterised by high strength, hydrophilicity and comfort [11]. These two fibre groups account for more than 75% of global production, and thus the largest share of textile waste [12]. Polyester fibres are semi-crystalline [13], hydrophobic, do not swell in water and are not prone to certain degradation and fibrillation. The influence of abrasion under wet and dry conditions is related to the generation of pilling [14,15]. Cotton and polyester textiles have different dimensional stability in wet and dry environments, in addition to significant variances in characteristics.

Cotton fibres in the alkaline medium can swell radially and longitudinally during finishing, causing shrinkage and increasing the take-up and yarn cross-section. In fibres where the orientation of the cellulose chains is in the direction of the fibres, swelling is greater in the transverse direction than in the longitudinal direction [16]. Increasing the crimp of washed cotton fabrics increases the elongation [17].

Polyester materials may shrink or become damaged if exposed to a temperature higher than the temperature of thermal transition. It was found that recycled polyester sheds almost 2.3 times more microplastics in washing compared to virgin polyester, as the strength of the fibre is reduced due to thermal exposure and shear degradation during the recycling process [18].

Cotton and polyester blends can meet special requirements for functional products, such as feel, appearance, dimensional stability, easy care and sufficient comfort [14]. The blends offer a number of advantages over pure cotton materials, including the ability to wash at lower temperatures, to reduce deposit content and are less damaged after multiple wash cycles.

On the other hand, the presence of microplastic particles (MP) in natural and wastewater, sediment, soil, aquatic organisms, and air has been linked to textile sources [19–24].

According to [25], "microplastic means particles containing solid polymer, to which additives or other substances may have been added, and where 1% $w/w$ of particles have all dimensions 0.1 nm–5 mm, or a length of 0.3 mm–15 mm and length to diameter ratio of >3".

The fragmentation [26], degradation [27], ageing [28], washing [29], and drying [30] of synthetic textile products are all possible sources of MP particles in environment [24]. Such products used for wet and dry cleaning of surfaces (mops), isolated synthetic fibres from vacuuming and drying, and fibres that come loose from clothing during home washing are the most common sources of MP in the environment [31,32]. Washing is thought to be responsible for around 35% of synthetic fibres in the environment [15,33].

The qualitative and quantitative determination of MP emissions and other released substances is difficult due to the variability and complexity of sources, as well as the fact that particulate matter is a vector for the dispersion of other emissions with varying degrees of risk in a real system. Numerous methods for determining particles were applied, either directly for the characterisation of dispersed systems [20] or after separation methods [34]. The choice of one or more methods of particle identification and/or separation in wet and dry environments is determined by knowledge of the system in which the particles originate. Total suspended solids (TSS) were found to be one of the best parameters for assessing the degree of particle loading in effluents and their separation [31]. Various filters with different pore sizes can be used to determine this parameter, such as glass fibre filters [35], polyethersulphonic filters [36], cellulosic filters [37,38], polyamide, polycarbonate, metallic and aluminosilicate filters [39,40].

The purpose of this research was to examine cotton fabrics and cotton-polyester blends after three, ten, and fifty cycles by measuring the released particles in the dry state and in washes. Particles in the dry state were released by cyclic bending, so the particle sizes of the washed fabrics were measured separately. The distribution of particles released from cotton fabrics and fabrics made of a blend of cotton and polyester was observed both individually and as a combined system. The effluents of the washing process of the aforementioned fabrics were analysed by determining the physico-chemical parameters, TSS, TS (total solids), pH and conductivity to determine the degree of particle load.

At the same time, the effect of 3, 10 and 50 washing cycles on the thickness and strength, which are the structural properties of the investigated fabrics, were monitored. It is well known that Fourier transform infrared spectroscopy (FTIR) provides information about changes in the chemical structure and environment of polymeric materials, such as: the presence or absence of certain functional groups: shifts in the frequency of absorption bands and changes in the relative intensity of the bands; the appearance of new peaks due to modifications; and the monitoring of changes during the life cycle of the material [41]. In addition, attenuated total reflectance Fourier' transform infrared spectroscopy (FTIR-ATR) is a non-destructive analytical method that does not require lengthy sample preparation and has an exceptional wave number accuracy of 0.01 $cm^{-1}$, which allows for the determination of low concentrations of individual groups of compounds [42]. For this reason, FTIR-ATR was applied in this study to evaluate the influence of 3, 10, and 50 washing cycles on the physicochemical changes of the cellulose polymer and polyethylene terephthalate that make up the cotton and cotton-polyester fabrics.

The originality of the research carried out can be seen in the connection between the amount of particles released from the fabrics in the dry state and and after the washing process.

## 2. Materials and Methods

### 2.1. Materials

Fabrics of 100% cotton and cotton-polyester (50:50) fabric in plain weave were produced on a Picanol OMNIplus 800 loom (air-jet loom, width 190 cm) at the Čateks d.o.o. textile mill, Čakovec, Croatia. The fabrics prepared in this way were scoured and, bleached according to the factory's recipes and purchased for research purposes.The mass per unit area of cotton-polyester fabric is 158.6 $gm^{-2}$, while the cotton fabric has a mass per unit area of 160.8 $gm^{-2}$. Cotton and cotton-polyester fabrics have a warp and weft density of 20 picks per cm, and the fineness of two-ply yarn 14.2 tex.

The washing of the 3.6 kg fabrics, which contain 2.7 kg of cotton and 0.9 kg of polyester (3:1), was carried out in accordance with HRN EN ISO 15797:2002 using a standard detergent with fluorescent whitening agent, the composition of which is shown in Table 1, to which 2 g/L peracetic acid (PAA) was added as a bleach in the washing process.

### 2.2. Washing Process

The cotton and cotton-polyester blend fabrics were washed in the laboratory washing machine Wascator FOM71 CLS by Electrolux at 75 °C using programme 2 with 3, 10, and 50 washing cycles. All process parameters comply with the standard, but due to the importance of the influence of the process conditions, a detailed description of programme 2 is included in Table 2.

**Table 1.** Composition of reference detergent according to HRN EN ISO 15797.

| Composition | Percentage |
|---|---|
| ABS-Na (C-12 chain) | 0.425 |
| Nonionic surfactant (C13/15 7EO or C12/14 7EO) | 6.0 |
| Sodium citrate dihydrate | 5.0 |
| Hydroxyethanediphosporic acid Na salt (HEDP) | 1.0 |
| Metasilicate anhydrous | 42.3 |
| Polymer (polymaleic acid) | 2.0 |
| Foam inhibitor (phosphoric acid ester) | 3.0 |
| Sodium carbonate | 39.5 |
| Fluorescent whitening agent | 0.3 |
| Water | 0.475 |

**Table 2.** Washing procedure.

| | |
|---|---|
| load ratio | 1:17 |
| agitation during heating, washing and rinsing | normal |
| **washing** | |
| liquor ratio | 1:4 |
| detergent<br>additive | 4 g/L detergent<br>2 g/L PAA |
| temperature | 75 ± 2 °C |
| time | 20 min |
| cool down | yes |
| drain | 1 min |
| interspin | No |
| **rinse 1** | |
| liquor ratio | 1:5 |
| time | 3 min |
| drain | 1 min |
| interspin | 1 min |
| **rinse 2** | |
| liquor ratio | 1:5 |
| time | 3 min |
| drain | 1 min |
| interspin | 1 min |
| **rinse 3** | |
| liquor ratio | 1:5 |
| time | 3 min |
| drain | 1 min |
| final extraction | 6 min |
| residual moisture | 35–40% |

After washing, the samples were air-dried. Table 3 shows the designations applied to samples before and after washing.

**Table 3.** Designations of samples.

| Designation | Sample |
|---|---|
| CO-0w | Cotton fabric before washing |
| CO-3w | Cotton fabric after 3 washing cycles |
| CO-10w | Cotton fabric after 10 washing cycles |
| CO-50w | Cotton fabric after 50 washing cycles |
| CO/PES-0w | Cotton-polyester fabric after 3 washing cycles |
| CO/PES-3w | Cotton-polyester fabric after 10 washing cycles |
| CO/PES-10w | Cotton-polyester fabric after 50 washing cycles |
| CO/PES-50w | Cotton-polyester fabric after 3 washing cycles |

### 2.3. Methods

In order to monitor the influence of washing on the structural properties of fabrics before and after the washing cycle, the thickness was tested according to HRN EN ISO 5084:2003 Textiles -- Determination of thickness of textiles and textile products at ten different places using a thickness gauge DM 2000—Wolf, Germany with a precision of 0.001 mm. The thickness gauge is made up of two parts: a support to hold the material in place and a device called a pressure plate, which is a 25 $cm^2$ circular plate that presses down on the material at a specific pressure (preload of 0.5 kPa). Tensile properties were measured on the samples in the weft direction before and after the washing cycles according to EN ISO 13934-1:1999 Textiles—Tensile properties of fabrics—Part 1: Determination of maximum force and elongation at maximum force using the strip method on a TensoLab Strength Tester (Mesdan S.p.A., Puegnago del Garda, Italy), distance between clamps 100 mm, bursting speed 100 mm/min and pretension 2 N.

Determination of overall decrease in breaking strength (total wear) was calculated according to ISO 4312:1989: Surface active agents—Evaluation of certain effects of laundering—Methods of analysis and test for unsoiled cotton control cloth.

The total wear, $U_t$ of the fabrics was calculated according to Equation (1):

$$U_t = \frac{F_0 - F}{F_0} \cdot 100 \tag{1}$$

where $U_t$ is total wear (%), $F_0$ is breaking force of unwashed fabric (N), and $F$ is breaking force of washed fabric (N).

Cotton and cotton-polyester fabric before and after 3, 10 and 50 washing cycles were analysed using Fourier transform infrared spectroscopy (FTIR, PerkinElmer, Spectrum 100, Shelton, CT, USA) with the attenuated total reflection (ATR) measurement technique obtained spectral curves were processed in Spectrum 100. Four scans were performed for each sample with a resolution of 4 $cm^{-1}$ between 4000 $cm^{-1}$ and 380 $cm^{-1}$

The generation lint and other particles was measured on LasAir III (Particle Measuring Systems) laser particle counter connected to a particle generator in a laminar airflow booth. Samples were prepared according to EN ISO 9073-10 and mounted on the particle generator and subjected to controlled bending. The number of particles released during the test was measured in the following size categories: 0.3 μm for particle sizes of 0.3–0.5 μm; 0.5 μm for particle sizes of 0.5–1 μm; 1 μm for particle sizes of 1–5 μm; 5 μm for particle sizes of 5–10 μm; 10 μm for particle sizes of 10–25 μm; 25 μm for particle sizes larger than 25 μm. For each sample, the measurements were performed on 5 test tubes. This method was adopted from EN ISO 9073-10 Textiles—Test methods for nonwovens—Part 10: Lint and other particles generation in the dry state, with the test time adjusted to 30 min [2].

The entire effluent (hence referred to as effluent) collected from the washing procedure and three rinse cycles was analysed as part of the research. Methods for characterising effluent after 3, 10 and 50 washing cycles include selected physicochemical parameters such as TSS, TS, pH, and conductivity.

The total suspended solids (TSS) of effluents were determined by the standard gravimetric method. After membrane filtration of effluent using 0.7 μm fibre glass filter (GF), mass of GF with filter cake as residue after drying at 100 °C was determined.

The total solids (TS) of effluents were determined by the standard gravimetric method in evaporating dish at 105 °C until constant mass was achieved.

pH and conductivity of effluents were determined using pH meter, Schott, ProLab 3000 and conductometer, CG 853, Schott, respectively.

To determine the similarity of the observed wash cycles and similar properties of all processed samples, hierarchical cluster analysis (HCA) was performed using a Minitab software, and graphs in the form of Ward's dendrograms showed the homogeneous groups or clusters whose variables are connected by a certain similarity [43].

## 3. Results

The effect of the cyclic washing procedure on the qualities of cotton fabrics and cotton-polyester blends was studied by monitoring structural features, thickness, and tensile properties. These fabrics were washed under the same conditions, and it is predicted that the percentage of cotton component will have an effect on thickness variations during washing cycles.

### 3.1. Structural Parameters

The average values and coefficients of variation of the measured fabric thicknesses before and after 3, 10, and 50 washing cycles are presented in Table 4.

**Table 4.** Average value and standard deviation of the fabrics before and after the washing cycles.

| Sample | Thickness (mm) | CV (%) |
|---|---|---|
| CO-0w | 0.440 | 2.83 |
| CO-3w | 0.489 | 6.35 |
| CO-10w | 0.516 | 6.66 |
| CO-50w | 0.553 | 8.61 |
| CO/PES-0w | 0.384 | 2.20 |
| CO/PES-3w | 0.443 | 1.52 |
| CO/PES-10w | 0.443 | 1.52 |
| CO/PES-50w | 0.461 | 2.97 |

The data in Table 4 show that the cotton fabric has a larger thickness before washing than the unwashed fabric consisting of a cotton-polyester blend. Due to the influence of the parameters of the washing process, the thickness of the cotton fabric increases. Statistical results indicate that cotton textiles shrink unevenly when washed. The cotton-polyester blend showed the highest difference in fabric thickness after three cycles as compared to unwashed material. Further washing cycles had a significantly lower influence on the thickness change, indicating no further scatter in relation to the structural features of the sample in terms of the coefficient of variation. The swelling of cotton fibres in the alkaline medium of the washing bath causes shrinkage, which modifies the cross-section of the yarn and results in differences in fabric thickness. The hydrophobic polyester component of the blend is more resistant to all washing parameters and guarantees dimensional stability of the fabric.

Table 5 groups the average values of breaking strength and elongation of cotton and cotton-polyester samples before and after the third, tenth, and fifth washing cycles.

**Table 5.** Breaking strength (F) and elongation (Ɛ) of cotton and cotton-polyester fabrics in warp and weft directions.

| Samples | | F (N) | Ɛ (%) |
|---|---|---|---|
| CO-0w | weft | 483.8 | 6.46 |
| | warp | 1028.8 | 8.94 |
| CO-3w | weft | 501.8 | 6.82 |
| | warp | 706.0 | 15.16 |
| CO-10w | weft | 582.2 | 7.84 |
| | warp | 729.8 | 18.00 |
| CO-50w | weft | 518.8 | 8.42 |
| | warp | 605.0 | 20.14 |
| CO/PES-0w | weft | 480.2 | 14.64 |
| | warp | 1074.6 | 16.85 |
| CO/PES-3w | weft | 541.6 | 14.25 |
| | warp | 1002.8 | 21.01 |
| CO/PES-10w | weft | 511.8 | 14.05 |
| | warp | 983.8 | 22.20 |
| CO/PES-50w | weft | 442.6 | 13.10 |
| | warp | 741.6 | 22.60 |

The breaking forces of unwashed cotton fabrics and cotton-polyester blend fabrics differed in the warp and weft directions, and the value in the warp direction was twice that of the weft direction. The breaking force of the examined fabrics (Table 5) in the warp direction exceeds 1000 N, indicating a high degree of structural integration. In contrast to the consolidation of the fabric in the warp direction, the breaking force in the weft direction is modest and is roughly 480 N for both fabrics.

Total wear (Ut) was calculated in the direction of the warp and weft throughout the washing process, which included tensile changes caused by mechanical and chemical degradation of the washed sample in relation to the unwashed sample, Table 6.

**Table 6.** Total wear (Ut) of fabrics in warp and weft directions.

| | Ut (%) | | | | | |
|---|---|---|---|---|---|---|
| | CO | | | CO/PES | | |
| | 3w | 10w | 50w | 3w | 10w | 50w |
| weft | −3.721 | −20.339 | −7.234 | −12.786 | −6.581 | 7.830 |
| warp | 31.376 | 29.063 | 41.194 | 6.6816 | 8.450 | 30.988 |

The results for the breaking forces and calculated Ut (Tables 5 and 6) show that there is a decrease in breaking strength, i.e., an increase in damage (total wear) after 3, 10, and 50 washing cycles in the direction of the warp threads for both fabrics tested (cotton and cotton-polyester blend). The increase in strength of the cotton-polyester fabric after 3 and 10 washing cycles was recorded in the weft direction. Structural properties, fibrillation and shrinking of materials during the washing process are the contributing factors to this relationship. The total wear of the cotton fabric (Table 6) after 50 washing cycles in the warp direction was 41.19%, which was higher than the 30.98% of the cotton-polyester blend fabric. The resulting damage can be attributed to a synergistic effect, including the raw material composition, the degree of soiling of the fabrics, and all components of the Sinner's circle. It is in line with the study of oxidation and alkali promoted reactions on cotton cellulose [4]. Standard alkaline detergent with peracetic acid (PAA) as an oxidising agent at 75 °C acts on the fibrillation and depolymerisation of cotton cellulose, as well as the high alkalinity of the polyester component of the blend. However, total wear in the weft direction of almost all washed fabrics has a negative sign, indicating that no damage has been caused by the wash. Given the predicted equivalent influence of significant washing process parameters on the fabrics in the direction of the weft threads, these values may be explained by fabric shrinkage during the washing process. Based on the results presented, the impact of the washing cycle on the cotton component is more pronounced under the applied conditions.

Standard detergent does not include the enzyme cellulase, which eliminates cellulose fibrils during the washing process [44]. When textiles are dried in a stationary ambient atmosphere, the migrating fibrils remain on the surface [45] and tend to loosen in both dry and wet conditions.

*3.2. Physicochemical Analysis of Samples with FTIR-ATR*

The spectral curves of the cotton and cotton-polyester fabrics before and after 3, 10, and 50 wash cycles are showen in Figure 1.

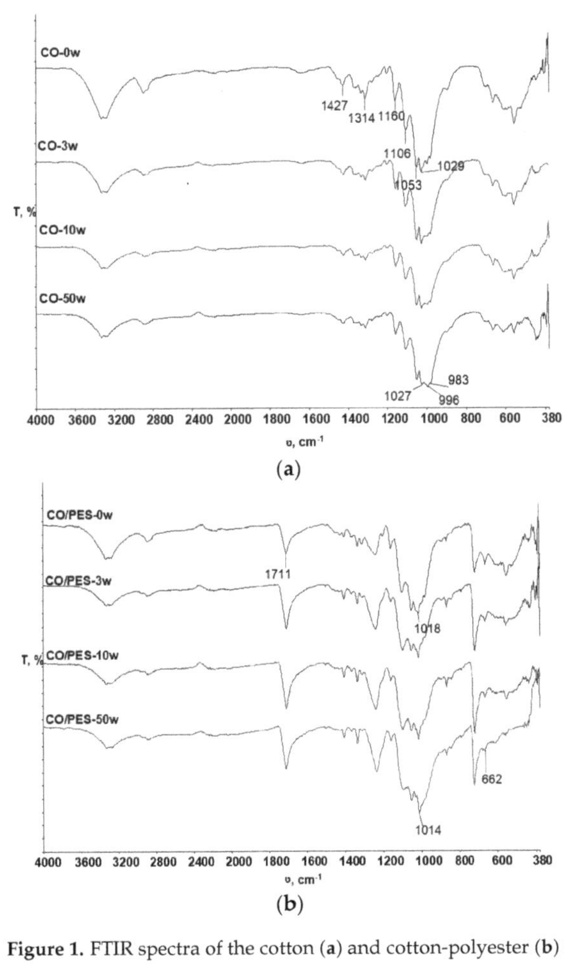


**Figure 1.** FTIR spectra of the cotton (**a**) and cotton-polyester (**b**) samples before and after 3, 10, and 50 wash cycles.

From Figure 1a,b, no significant changes were observed in the spectral bands obtained for the cotton and cotton-polyester fabric samples (CO-0w and CO /PES-0w) before and after 3 and 10 washing cyclesn (CO-3w, CO-10w and CO/PES-3w, CO/PES-10w), respectively. The spectral band of the sample CO-50w shows changes in the shape and intensity of the peaks in the range of 1027 $cm^{-1}$, where the peak is due to the vibration of the C-OH bond of primary alcohols, and in the ranges of wavenumbers 996 $cm^{-1}$ and 983 $cm^{-1}$, where the peaks indicate vibrations within the—CH—bond, indicating that minor changes occurred within the polymer as a result of the washing process under the influence of chemistry as described. Minor changes can also be seen in the CO/PES-50w sample, specifically by the increase in the intensity of the peak at wavenumber 1014 $cm^{-1}$, indicating an enhancement of the vibrations in the ester group (O=C-O-C) of the polyester component, which also indicates a higher surface wear of the cotton component within the CO/PES blend, which was also confirmed by the analytical methods in the previous research [46,47].

### 3.3. Particle Shedding from Fabrics in Dry State

The number of particles thus produced in the dry state was determined by applying cyclic bending. In view of the material differences mentioned above, the test was carried out for cotton and cotton-polyester fabrics. The obtained results refer to the release of lint and other particles from the material due to the deposits formed during the washing process.

Based on the number of particles released in Table 7, a minor proportion of the particles released are in the larger size category >25 μm. During the washing cycles analysed, significantly more particles in all size categories were released from the cotton fabric compared to the cotton-polyester fabric.

**Table 7.** Average number of particles released in each category.

| | | Number of Released Particles | | | | | |
|---|---|---|---|---|---|---|---|
| **Fabric** | **Cycles** | **0.3–0.5 μm** | **0.5–1 μm** | **1–5 μm** | **5–10 μm** | **10–25 μm** | **≥25 μm** |
| CO | 3w | 1,972,940.2 | 1,642,870.4 | 808,332.4 | 16,146.0 | 3305.8 | 745.8 |
| | 10w | 2,782,562.8 | 2,133,212.3 | 1,030,349.3 | 28,086.5 | 4983.0 | 628.8 |
| | 50w | 4,347,945.4 | 3,452,533.2 | 1,655,430.8 | 65,939.8 | 12,165.2 | 744.8 |
| CO/PES | 3w | 51,192.0 | 28,188.6 | 12,296.4 | 347.0 | 120.6 | 80.4 |
| | 10w | 45,924.0 | 20,549.2 | 7887.8 | 233.0 | 82.0 | 47.4 |
| | 50w | 63,962.8 | 33,833.4 | 13,122.0 | 428.8 | 116.2 | 60.4 |

The relative values of the number of particles for each material were determined in proportion to the largest number of released particles in order to compare the results based on the washing cycles. The results are shown in Figures 2 and 3 with the following size categories: 0.3 μm for particle sizes of 0.3–0.5 μm (1); 0.5 μm for particle sizes of 0.5–1 μm (2); 1 μm for particle sizes of 1–5 μm (3); 5 μm for particle sizes of 5–10 μm (4); 10 μm for particle sizes of 10–25 μm (5); 25 μm for particle sizes larger than 25 μm (6).

According to the graph shown in Figures 2 and 3, the smaller particle size category in both materials fluctuates more depending on the wash cycle. The number of wash cycles has a different effect on the number of particles released from different materials. For cotton fabrics, the correlation between the number of washing cycles and particle size in the first four categories is higher as the number of washing cycles increases, but the number of washing cycles has a different effect on cotton-polyester fabrics. There are more particles between 0.3 and 5 μm in size than particles larger than 5 μm. These results also reflect the thermal, chemical, and mechanical effects of the washing cycles on the fibre. Standard program 2 was used to wash cotton fabrics and cotton-polyester blended fabrics. This program required the addition of a chemical bleach, PAA, in addition to a standard detergent. Cotton textiles become more fibrillated as the number of washing cycles rises [7], but in this system, due to the lack of soiling, peracetic acid in alkaline at 75 °C may affect cotton cellulose depolymerisation and cellulose fibril formation.

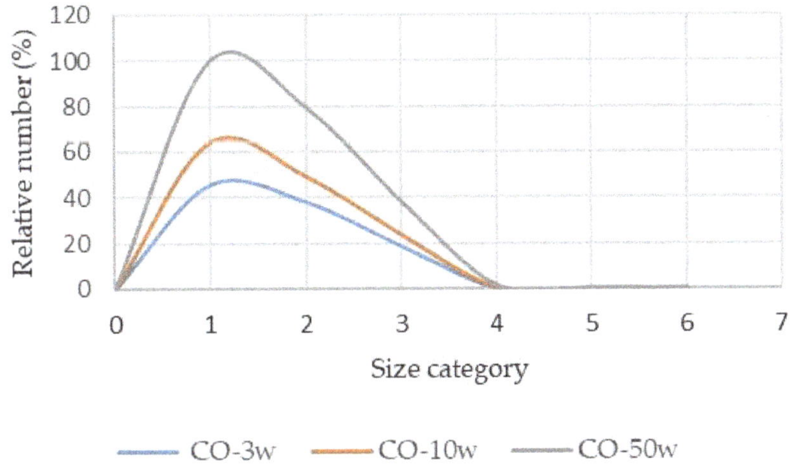


**Figure 2.** Relative number of particles released from cotton fabric according to size category (1–6).

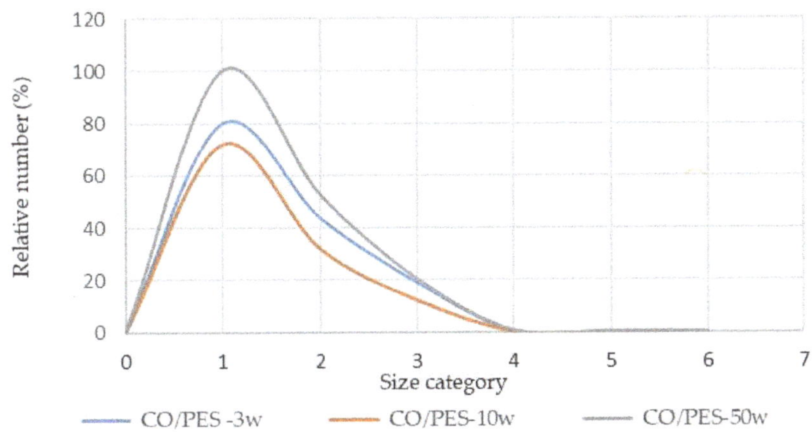


**Figure 3.** Relative number of particles released from cotton-polyester fabric according to size category (1–6).

### *3.4. Particle Shedding from Fabrics in Wash Cycles*

The effluent after washing can be contaminated by soil from textiles, lint, dyes, finishing agents, and detergents [48]. Some detergents contain insoluble components that pollute the effluent and at the same time obscure the filter along with the MP particles, making microscopic analysis of the released MP particles difficult. Therefore, washing synthetic textiles in water is carried out in some studies [49].

Since cotton and cotton-polyester fabrics in the study have not been finished, dyed and soiled, the effluent can be contaminated with particles of textile origin and detergent. The components of the used standard detergent are soluble, so contamination by particles originating from the detergent is not expected. It can therefore be assumed that the contamination of the wash effluent contains particles from cotton and cotton-polyester fabrics. Effluent characteristics after 3, 10 and 50 washes of cotton and cotton-polyester fabrics are presented in Table 8.

**Table 8.** Physicochemical properties of effluent from washing cotton and cotton-polyester fabrics.

| | Washing Effluent | | |
|---|---|---|---|
| **Parameter** | **3w** | **10w** | **50w** |
| pH | 9.94 | 9.99 | 9.89 |
| Conductivity (μS/cm) | 1859 | 1927 | 1782 |
| TSS (mg/L) | 117 | 77 | 81 |
| TS (mg/L) | 3394 | 3452 | 3298 |

The high pH value of the studied effluents is nearly consistent during all washing cycles and is caused by the alkaline components in the detergent. This characteristic is not expected to change despite the dilution that occurs during the three-step washing process. In this procedure for washing textiles without soiling [50], the alkalis could not be used to saponify grease and neutralise acidic soiling, but they functioned synergistically with anionic surfactants and bleach (PAA). A high alkali concentration can also be hazardous to particular textiles, such as damaged cotton and PES textiles, which get hydrolysed [51] as a result, causing wear and tear and shortening the product's life.

The conductivity of the effluents under consideration given in Table 8 reflects the quantity of detergent components that are water soluble. Again, the washing process was performed on fabrics with no soiling, implying that the detergent ingredients were not focused on their removal but on potential interactions with the fabrics, causing structural changes and a loss of fabric integrity in the direction of the warp threads.

Based on this, it is clear that the composition of the reference detergent with the addition of PAA and other parameters of the Sinner cycle (temperature, mechanics, time) influence the polymeric structures of cotton cellulose and cotton-polyester during the washing cycles. The TSS values obtained, as reported in Table 8, confirm the degree of particulate load in the effluent. Given the composition of the detergent in Table 1, which consists exclusively of water-soluble components, it is probable that the TSS consists of particles released by the studied textile materials. Although the relationship and type of the particles were not determined in this study, earlier studies on the effect of washing on the release of particulate matter from cotton-polyester blend fabrics have shown that the filter cake contains fibrillary particulate matter of both polymer substances [52].

### *3.5. MVA-Similarities and Disimilarities of Observed Systems in Dry State and Washes*

The different systems (wash-wet and dry) to which the cotton and cotton-polyester fabrics studied were exposed and the different physicochemical properties in each system required the use of advanced statistical methods. Multivariant analysis was used to determine system similarities in terms of the number of particles released from the examined fabrics in procedures with varying numbers of washing cycles. The results using the Ward dendrogram are shown in Figures 4–6.

The different distribution of clustering with regard to the number of washing cycles indicates that cotton contributes to the increasing release of particles with a higher number of washing cycles, while the contribution of polyester is greater in a lower number of washing cycles.

The results include the formation of clusters, where a data set was selected for particle release from cotton and cotton-polyester fabrics in washes, Figure 4 and dry state, Figure 5 for different washing cycles. The clusters formed are very similar, but a separate cluster can be identified that includes effluents from 10 and 50 wash cycles, while the effluent from 3 washing cycles was separated.

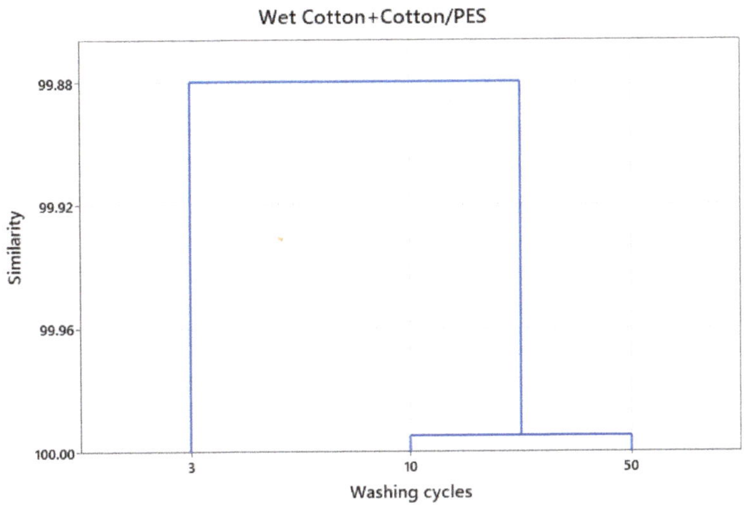


**Figure 4.** Dendrogram of the HCA according to Ward for similarities/dissimilarities of washing cycles for cotton and cotton-polyester in washes.

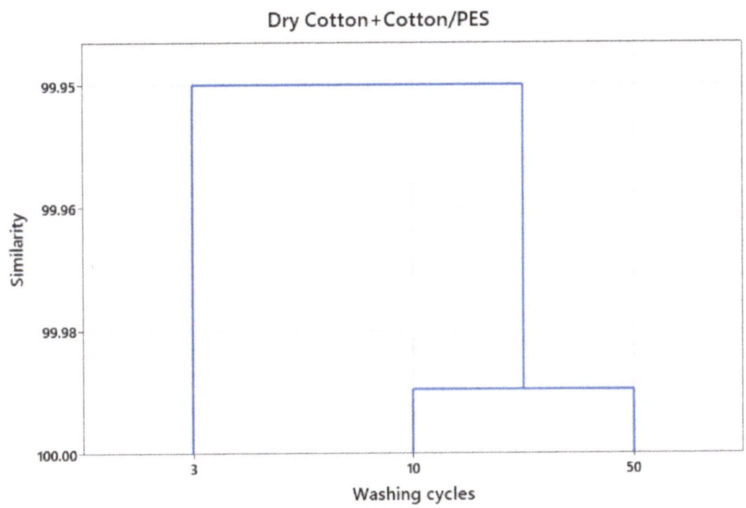


**Figure 5.** Dendrogram of the HCA according to Ward for similarities/dissimilarities of washed cotton and cotton-polyester fabrics in dry state.

The dendrograms of clustering also show a high degree of similarity, but the cluster distribution is different with respect to the number of washing cycles. The washing conditions, especially the high alkali content, affect the damage of the textiles, which contributes to the release of particles.

Tensile properties, thickness, and total wear in the weft and warp directions were examined to establish the importance of textile composition and washing cycles to particle release, Figure 6.

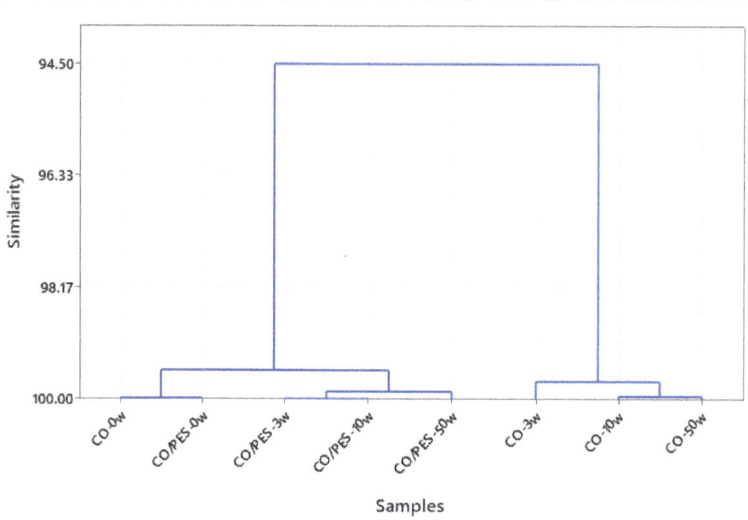


**Figure 6.** Dendrogram of the HCA according to Ward for similarities/dissimilarities of samples based on data set of tensile properties, total wear and thickness.

In the initial washing cycles, the structural properties of the material play a significant role in the release of particulate matter; however, in subsequent cycles, the interaction between the chemical, mechanical and thermal factors and of the fabrics becomes increasingly prominent.

## 4. Conclusions

This research examines the chemical, mechanical, and thermal impacts of 3 washing cycles, 10 washing cycles, and 50 washing cycles on 100% cotton and a 50/50 cotton-polyester blend in plain weave with no soiling in the dry state and after washes utilising fabric and effluent characteristics. Tensile properties, thickness and generated particles in the dry state depend on the number of washing cycles. Increased strength of cotton-polyesteru fabric after 3 and 10 washing cycles in the weft direction is the result of fibrillation and shrinkage. Changes in fabric properties, expressed as total wear in the warp direction after 50 washing cycles compared to unwashed amounting to 41.2% for cotton and 30.9% for cotton-polyester blend, can be attributed to the synergy of process parameters, fabric structure and raw material composition.

The number of particles released in the dry state > 25 µm is significantly lower than the number of particles released in the size of 0.3 to 5 µm. In all size categories, the quantity of particles released in the dry state is much larger from washed cotton fabric than from washed cotton-polyester fabric. Because the detergent contains water-soluble components, the TSS values obtained confirm the degree of contamination of the effluent with particles from the tested textile materials.

From the spectral bands of CO and CO /PES samples before and after 3, 10 and 50 washing cycles, no significant changes at the physicochemical level within the polymers of the tested samples were identified.

According to HCA dendrograms, particle release in early wash cycles is mostly regulated by material structure. Future wash cycles are influenced by chemical, mechanical, and thermal interactions. The findings of the research highlight the need for the use of an analytical approach to categorise and quantify particles released from textiles in both the dry state and washes in order to decrease the potential for detrimental impacts on humans and the environment.

**Author Contributions:** Conceptualisation, T.P. and B.V.; methodology, R.M. and B.V., formal analysis, R.M., B.V., M.Č. and S.F.G.; investigation, T.P. and R.M.; writing—original draft preparation, T.P., B.V., S.F.G. and M.Č. All authors have read and agreed to the published version of the manuscript.

**Funding:** This research was funded by Croatian Science Foundation, grant numbers HRZZ-IP-2020-02-7575 and HRZZ-UIP-2017-05-8780.

**Institutional Review Board Statement:** Not applicable.

**Data Availability Statement:** Not applicable.

**Acknowledgments:** The part of the research was performed on equipment purchased by K.K.01.1.1.02.0024 project "Modernization of Textile Science Research Centre Infrastructure" (MI-TSRC).

**Conflicts of Interest:** The authors declare no conflict of interest.

## References

1. Sigaard, A.S.; Laitala, K. Natural and Sustainable? Consumers' Textile Fiber Preferences. *Fibers* **2023**, *11*, 12. [CrossRef]
2. Malinar, R.; Flinčec Grgac, S.; Katović, D. Textile particle generation: Test method for nonwovens modified for use on woven materials. *Text. Res. J.* **2020**, *90*, 19–20. [CrossRef]
3. Sinner, H. *Über das Waschen mit Haushaltswaschmaschinen*, 2nd ed.; Auflage Haus und Heim Verlag: Hamburg, Germany, 1960; pp. 9–10.
4. Palme, A.; Idström, A.; Nordstierna, L.; Brelid, H. Chemical and ultrastructural changes in cotton cellulose induced by laundering and textile use. *Cellulose* **2014**, *21*, 4681–4691. [CrossRef]
5. Fijan, S.; Šostar Turk, S.; Neral, B.; Pušić, T. The Influence of Industrial Laundering of Hospital Textiles on the Properties of Cotton Fabrics. *Text. Res. J.* **2007**, *77*, 247–255. [CrossRef]
6. Gotoh, K.; Harayama, K.; Handa, K. Combination effect of ultrasound and shake as a mechanical action for textile cleaning. *Ultrason. Sonochemistry* **2015**, *22*, 412–421. [CrossRef]
7. Rogina-Car, B.; Pušić, T.; Dekanić, T. Impact of washing and sterilization on properties of fabrics used for medical applications. *Indian J. Fibre Text. Res.* **2016**, *41*, 426–431.
8. Buisson, Y.L.; Rajasekaran, K.; French, A.D.; Conrad, D.C.; Roy, P.S. Qualitative and Quantitative Evaluation of Cotton Fabric Damage by Tumble Drying. *Text. Res. J.* **2000**, *70*, 739–743. [CrossRef]
9. Hazlehurst, A.; Tiffin, L.; Sumner, M.; Taylor, M. Quantification of microfibre release from textiles during domestic laundering. *Environ. Sci. Pollut. Res.* **2023**, *30*, 43932–43949. [CrossRef]
10. Tiffin, L.; Hazlehurst, A.; Sumner, M.; Taylor, M. Reliable quantification of microplastic release from the domestic laundry of textile fabrics. *J. Text. Inst.* **2022**, *113*, 558–566. [CrossRef]
11. Mitic, J.; Amin, G.; Kodric, M.; Šmelcerović, M.; Đorđević, D. Polyester fibres structure modification using some organic solutions. *Tekstil* **2016**, *65*, 196–200.
12. Wang, S.; Salmon, S. Progress toward Circularity of Polyester and Cotton Textiles. *Sustain. Chem.* **2022**, *3*, 376–403. [CrossRef]
13. East, A.J. The structure of polyester fibres. In *Handbook of Textile Fibre Structure: Volume 1: Fundamentals and Manufactured Polymer Fibres*; Eichorn, S.J., Hearl, J.W.S., Jaffe, M.K., Eds.; T. Woodhead Publishing Limited: Oxford, UK, 2009; Volume 185, pp. 181–225.
14. Kurz, J. Laundering in the prevention of Skin Infections. In *Textiles and the Skin*; Elsner, P., Hatch, K.K., Eds.; Karger: Basel, Switzerland, 2003; Volume 5, pp. 64–81.
15. Palacios-Mateo, C.; van der Meer, Y.; Seide, G. Analysis of the polyester clothing value chain to identify key intervention points for sustainability. *Environ. Sci. Eur.* **2021**, *33*, 1–25. [CrossRef]
16. Zhang, S.; Wang, W.-C.; Li, F.-X.; Yu, J.-Y. Swelling and dissolution of cellulose in NaOH aqueous solvent systems. *Cellul. Chem. Technol.* **2013**, *47*, 671–679.
17. Toshikj, E.; Jordanov, I.; Demboski, G.; Mangovska, B. Influence of Multiple Laundering on Cotton Shirts Properties. *Tekst. Ve Konfeksiyon* **2016**, *26*, 393–399.
18. Özkan, İ.; Gündoğdu, S. Investigation on the microfiber release under controlled washings from the knitted fabrics produced by recycled and virgin polyester yarns. *J. Text. Inst.* **2021**, *112*, 264–272. [CrossRef]
19. Volgare, M.; Avolio, R.; Castaldo, R.; Errico, M.E.; El Khiar, H.; Gentile, G.; Sinjur, A.; Susnik, D.; Znidarsic, A.; Cocca, M. Microfiber Contamination in Potable Water: Detection and Mitigation Using a Filtering Device. *Microplastics* **2022**, *1*, 322–333. [CrossRef]
20. Mariano, S.; Tacconi, S.; Fidaleo, M.; Rossi, M.; Dini, L. Micro and Nanoplastics Identification: Classic Methods and Innovative Detection Techniques. *Front. Toxicol.* **2021**, *3*, 636640. [CrossRef]
21. Magnusson, K.; Eliasson, K.; Fråne, A.; Haikonen, K.; Hultén, J.; Olshammar, M.; Stadmark, J.; Voisin, A. *Swedish Sources and Pathways for Microplastics to the Marine Environment—Report C 183*; IVL Swedish Environmental Research Institute Ltd.: Stockholm, Sweden, 2016.
22. Lim, S.J.; Park, Y.-K.; Kim, H.; Kwon, J.; Moon, H.M.; Lee, Y.; Watanabe, A.; Teramae, N.; Ohtani, H.; Kim, Y.-M. Selective solvent extraction and quantification of synthetic microfibers in textile laundry wastewater using pyrolysis-gas chromatography/mass spectrometry. *Chem. Eng. J.* **2022**, *434*, 134653. [CrossRef]

23. Hendrickson, E.; Minor, E.C.; Schreiner, K. Microplastic Abundance and Composition in Western Lake Superior as Determined via Microscopy, Pyr-GC/MS, and FTIR. *Environ. Sci. Technol.* **2018**, *52*, 1787–1796. [CrossRef]
24. Piribauer, B.; Laminger, T.; Ipsmiller, W.; Koch, D.; Bartl, A. Assessment of Microplastics in the Environment—Fibres: The Disregarded Twin? *Detritus* **2020**, *9*, 201–212.
25. Committee for Risk Assessment and Committee for Socio-economic Analysis. *Background Document on Intentionally Added Microplastics*; ECHA: Helsinki, Finland, 2020; p. 18.
26. Šaravanja, A.; Pušić, T.; Dekanić, T. Microplastics in Wastewater by Washing Polyester Fabrics. *Materials* **2022**, *15*, 2683. [CrossRef] [PubMed]
27. Singh, B.; Sharma, N. Mechanistic implications of plastic degradation. *Polym. Degrad. Stab.* **2008**, *93*, 561–584. [CrossRef]
28. Silva, R.R.A.; Marques, C.S.; Arruda, T.R.; Teixeira, S.C.; de Oliveira, T.V. Biodegradation of Polymers: Stages, Measurement, Standards and Prospects. *Macromolecules* **2023**, *3*, 371–399. [CrossRef]
29. Hernandez, E.; Nowack, B.; Mitrano, D.M. Polyester Textiles as a Source of Microplastics from Households: A Mechanistic Study to Understand Microfiber Release During Washing. *Environ. Sci. Technol.* **2017**, *51*, 7036–7046. [CrossRef]
30. Choi, S.; Kwon, M.; Park, M.J.; Kim, J. Analysis of Microplastics Released from Plain Woven Classified by Yarn Types during Washing and Drying. *Polymers* **2021**, *13*, 2988. [CrossRef]
31. Roth, J.; Zerger, B.; De Geeter, D.; Benavides, J.G.; Roudier, S. *Best Available Techniques (BAT) Reference Document for the Textile Industry*; Publications Office of the European Union: Luxembourg, 2023; p. 164.
32. Raja Balasaraswathi, S.; Rathinamoorthy, R. Effect of fabric properties on microfiber shedding from synthetic textiles. *J. Text. Inst.* **2021**, *113*, 789–809. [CrossRef]
33. Zhang, Y.; Lykaki, M.; Alrajoula, M.T.; Markiewicz, M.; Kraas, C.S.; Kolbe Klinkhammer, K.; Rabe, M.; Klauer, R.; Bendtd, E.; Stolte, S. Microplastics from textile origin—Emission and reduction measures. *Green Chem.* **2021**, *23*, 5247–5271. [CrossRef]
34. Gaylarde, C.; Baptista-Neto, J.A.; da Fonseca, E.M. Plastic microfibre pollution: How important is clothes' laundering? *Heliyon* **2021**, *25*, e07105. [CrossRef]
35. Carney Almroth, B.M.; Åström, L.; Roslund, S.; Petersson, H.; Johansson, M.; Persson, N.K. Quantifying shedding of synthetic fibers from textiles; a source of microplastics released into the environment. *Environ. Sci. Pollut. Res.* **2018**, *25*, 1191–1199. [CrossRef]
36. Abreu-Silva, J.; Ribeirinho-Soares, S.; Oliveira-Inocêncio, I.; Pedrosa, M.; Silva, A.M.T.; Nunes, C.O.; Manaia, C.M. Performance of polycarbonate, cellulose nitrate and polyethersulfone filtering membranes for culture-independent microbiota analysis of clean waters. *J. Environ. Chem. Eng.* **2023**, *11*, 109132. [CrossRef]
37. Sillanpää, M.; Sainio, P. Release of polyester and cotton fibers from textiles in machine washings. *Environ. Sci. Pollut. Res.* **2017**, *24*, 19313–19321. [CrossRef]
38. Cotton, L.; Hayward, A.S.; Lant, N.J.; Blackburn, R.S. Improved garment longevity and reduced microfibre release are important sustainability benefits of laundering in colder and quicker washing machine cycles. *Dyes Pigment.* **2020**, *177*, 108120. [CrossRef]
39. Cai, H.; Chen, M.; Chen, Q.; Du, H.; Liu, J.; Shi, H. Microplastic quantification affected by structure and pore size of filters. *Chemosphere* **2020**, *257*, 127198. [CrossRef]
40. Shen, M.; Hu, T.; Huang, W.; Song, B.; Zeng, G.; Zhang, Y. Removal of microplastics from wastewater with aluminosilicate filter media and their surfactant-modified products: Performance, mechanism and utilization. *Chem. Eng. J.* **2021**, *421*, 129918. Available online: https://www.sciencedirect.com/science/article/pii/S1385894721015023 (accessed on 15 May 2023).
41. Zupan, J. *Kemometrija in Obdelava Experimentalnih Podatkov*; Inštitut Nove revije, Zavod za humanistiko and National Institute of Chemistry: Ljubljana, Slovenia, 2009; pp. 154–167.
42. Maiken, U.; Johanna, M.H.; Shari, L.F.; Barbara, H.S. Degradation patterns of natural and synthetic textiles on a soil surface during summer and winter seasons studied using ATR-FTIR spectroscopy. *Spectrochim. Acta Part A Mol. Biomol. Spectrosc.* **2017**, *185*, 69–76.
43. Physical_and_Theoretical_Chemistry_Textbook_Maps. Available online: https://chem.libretexts.org/Bookshelves/Physical_and_Theoretical_Chemistry_Textbook_Maps/Supplemental_Modules_(Physical_and_Theoretical_Chemistry)/Spectroscopy/Vibrational_Spectroscopy/Infrared_Spectroscopy/How_an_FTIR_Spectrometer_Operates (accessed on 15 June 2023).
44. Gürkök, S. Microbial Enzymes in Detergents: A Review. *Int. J. Sci. Eng. Res.* **2019**, *10*, 75–81.
45. Soljačić, I.; Pušić, T. *Njega Tekstila, Sveučilište u Zagrebu Tekstilno-tehnološki fakultet*; University of Zagreb: Zagreb, Croatia, 2005; p. 123.
46. Smulders, E.P. *Smulders*; Wiley-VCH: Weincheim, Germany, 2002; p. 165.
47. Gore, A.; Venkataraman, A. Identification of Polyester/Cellulosic Blends Using FT-IR Spectrometer. *Indian J. Fibre Text. Res.* **1998**, *23*, 165–169.
48. Sangeetha, V.H.; Varghese, T.O.; Nayak, S.K. Isolation and characterisation of nanofibrillated cellulose from waste cotton: Effects on thermo-mechanical properties of polylactic acid/MA-g-SEBS blends. *Iran. Polym. J.* **2019**, *28*, 673–683. [CrossRef]
49. Jönsson, C.; Levenstam Arturin, O.; Hanning, A.-C.; Landin, R.; Holmström, E.; Roos, S. Microplastics Shedding from Textiles—Developing Analytical Method for Measurement of Shed Material Representing Release during Domestic Washing. *Sustainability* **2018**, *10*, 2457. [CrossRef]
50. Nayak, R.; Ratnapandian, S. *Care and Maintenance of Textile Products Including Apparel and Protective Clothing*; CRC Press: Boca Raton, FL, USA, 2018; p. 54.

51. Bishop, P.D. Physical and chemical effects of domestic laundering processes. In *Chemistry of the Textiles Industry*; Carr, C.M., Ed.; Springer: Dordrecht, The Netherlands, 1995; pp. 125–172.
52. Čurlin, M.; Pušić, T.; Vojnović, B.; Dimitrov, N. Particle Characterization of Washing Process Effluents by Laser Diffraction Technique. *Materials* **2021**, *14*, 7781. [CrossRef]

*polymers* MDPI

*Article*

# Utilization of Torrefied and Non-Torrefied Short Rotation Willow in Wood–Plastic Composites

**Jaka Gašper Pečnik [1,2,*], Mariem Zouari [1,2], Matthew Schwarzkopf [1,2] and David B. DeVallance [3,4]**

1 InnoRenew CoE, Livade 6a, 6310 Izola, Slovenia; mariem.zouari@innorenew.eu (M.Z.); matthew.schwarzkopf@innorenew.eu (M.S.)
2 Faculty of Mathematics, Natural Sciences and Information Technologies, University of Primorska, Titov Trg 4, 6000 Koper, Slovenia
3 College of Science and Technology, Commonwealth University of Pennsylvania, 401 North Fairview Street, Lock Haven, PA 17745, USA; ddevallance@lockhaven.edu
4 Division of Forestry and Natural Resources, West Virginia University, Morgantown, WV 26506, USA
* Correspondence: jaka.pecnik@innorenew.eu

**Abstract:** The torrefaction process is widely used in the energy field, but the characteristics of the torrefied wood also have positive effects on the production of wood plastic composites. In this study, short-rotation shrub willow was torrefied at 225 and 300 °C and incorporated into polypropylene composites filled with changing levels of weight percent (wt%) of non-torrefied and torrefied (5, 15, 25, and 40 wt%) wood. Nine different formulations were extruded for mechanical, thermal, and water absorption properties. The tensile properties of composites were not affected by any level of torrefaction, while higher flexure properties were in favor of lower wt% of torrefied wood. The slowest rate of thermal degradation was confirmed for the highest wt% of torrefied wood with a torrefaction temperature of 300 °C. In contrast, the presence of torrefied wood in composites did not show a difference in crystallization or melting temperatures. The most noticeable contribution of torrefaction temperature and wt% was found for water-absorbing properties, where the higher torrefaction temperature and largest wt% of torrefied wood in the composite resulted in decreased water uptake.

**Keywords:** torrefaction; wood–plastic composite; short rotation willow

## 1. Introduction

Woody crop biomass represents a significant source of renewable feedstock for bioenergy generation, bioproducts with a positive impact on soil and water conservation, recycled nutrients, and sequestered carbon [1]. Short-rotation wood species grown in plantations have higher potential than traditionally grown wood plantations due to shorter cultivation time intervals and high yields [2]. Agricultural perennial crops such as poplar and willow are considered energy crops grown for bioenergy production [3]. Short-rotation willows, such as Millbrook (*Salix purpurea* × *S. miyabeana*), with high growing yields, are often planted on marginal land from reclaimed mines or agricultural land to produce woody biomass. Willow stems grow rapidly and can be harvested every 3–4 years, up to 15 years. Biomass yield by dry stem mass in two growing seasons after coppice is estimated at around 9.3 T/year/hectare [4].

Willow biomass used in producing wood–plastic composites (WPC) was found to be an economically available source of natural fillers with improved and comparable mechanical performance of composites against classically used soft and hardwood fillers [5]. However, alternative pretreatments are key to adding value to fast-growing species grown primarily as energy crops to enhance the WPC material properties, specifically the hydrophobicity. WPCs are used in processed advanced materials in products for construction in the automotive industry, furniture applications, and consumer goods [6–9]. Wood is an environmentally friendly alternative to mineral fillers in the production of plastic composites [10]. Wood-based fillers have beneficial properties such as a good strength-to-weight ratio, low density, and easy integration into existing product lines [11]. Different types of wood fillers have been investigated to tailor the properties of WPC. Pudlik et al. [5] compared willow (*Salix viminalis*) filler to pine and beech fillers for WPC production and found comparable mechanical properties. Ayrilmis et al. [12] used fast-growing Pawlonia wood (*Paulownia elongata*) to produce WPC. They reported lower water absorption and thickness swelling properties against other fast-growing species such as Poplar spp. or Eucalyptus spp. The results were attributed to the lower capacity of bonding water, which depends on wood species. More recently, torrefied biomass has been gaining attention for non-energy applications. Tripathi et al. [13] indicated many applications where torrefied biomass is likely to outperform raw biomass and biochar. For example, the higher yield, as compared to biochar, and hydrophobic properties, as compared to raw biomass, make torrefied biomass a more attractive filler and additive in composites.

Kim et al. [14] observed small differences in tensile strength using several different wood species and pointed out that many wood species are appropriate for producing WPC. Another study [15] showed that selecting wood species for WPC material may improve its durability due to factors of water absorption and consequently create better conditions for decay growth [15]. Fabiyi et al. [16] found that water absorption in the case of wood species with lower diffusion coefficients also resulted in lower WPC water absorption due to differences in the chemical composition of the wood species. Bouafif et al. [17] reported that water absorption is also related to the particles size and contents in the composite material. Thermally modified wood is typically associated with being a more durable, dimensionally stable, and ecological alternative for construction products with a longer service life [18].

Moreover, studies related to WPC focused on using thermal modification wood to reduce water absorption, improve durability, and enhance the thermal resistance of WPC. For instance, Ayrilmis et al. [19] used cellulose fibers from fast-growing Eucalyptus camaldulensis treated at variable temperatures. They found that better water resistance for WPC panels was obtained when using thermally modified fibers. A review by Samaniego [20] also discussed applications of thermal treatments to improve wood-based composites, with the potential to apply the torrefaction process to enhance biomass properties.

Torrefaction involves heating biomass between 200 °C and 300 °C in an inert atmosphere. The hemicelluloses are removed through this process, while the remaining material retains the original molecular structure of cellulose and lignin. Maintaining this structure provides the opportunity to utilize torrefied biomass in composite applications where cellulose and lignin have been shown to improve properties. During the grinding, the degraded hemicellulose matrix results in a more brittle material, which can impact the sphericity of particles and particle size [21]. By removing hemicellulose, torrefied wood can potentially improve WPC's water-resistance properties. Recent torrefaction development has primarily focused on using biomass to improve caloric energy efficiency [22]. Besides the higher caloric value of torrefied biomass, the process enhances grindability, fungi decay, and water absorption. A few studies investigated using torrefied biomass as filler and reinforcement materials for polymers.

Chiou et al. [23] found that adding torrefied biomass to polypropylene improved the distortion under heat and increased the glass transition temperature and tensile modulus of polypropylene. However, they found that the tensile strength decreased with increasing percentages of biomass. McCaffrey et al. [24] utilized torrefied almond shells when studying recycled polypropylene–polyethylene bio-composites. They reported that tensile and bending moduli could be increased by adding torrefied almond shells at high loading levels (up to 40%). They also reported that the particle size did not impact the mechanical properties. However, using smaller particles resulted in higher heat deflection temperature (HDT) values. Volfson et al. [25] found that torrefied pine filler in a 50% wood-to-polypropylene ratio improved the tensile strength by 38% and reduced the elongation at the break by 12%.

They found a slight reduction in tensile strength when using torrefied birch filler. Vold et al. [26] investigated torrefied flax shives and sunflower hulls as fillers for polyamides. Their results suggested that the tensile elastic modulus of the polyamide was improved by adding torrefied flax filler up to 30%. Furthermore, the flexural and tensile modulus of the neat polyamide 6 (PA6) were found to increase, and the flexural and tensile strength decreased with increasing percentages (up to 30%) of torrefied sunflower hulls. They also reported a decrease in moisture uptake when torrefied fillers were added. This improvement in water absorption was also reported by Lu et al. [27], who found that the use of torrefied sweet sorghum slag fibers improved water absorption when used as fillers for high-density polyethylene (HDPE), as compared to non-torrefied sweet sorghum slag fibers. Berthet et al. [28] reported a 30% decrease in the water vapor permeability (WVP) of torrefied wheat straw fiber/poly(3-hydroxybutyrate-co-3-hydroxyvalerate) PHBV composites when using 20% of torrefied fibers. However, when loading levels were increased to 30%, fiber agglomeration occurred, and WVP increased. However, issues related to water absorption still exist when using willow as an additive for WPCs. On the other hand, torrefaction could be a useful pretreatment for short-rotation willow being used for composite-related applications. Research is necessary to determine how effective the torrefaction process, and therefore torrefied willow, is in improving the mechanical and physical properties of WPCs. Furthermore, while short-rotation willow is generally grown for energy-related applications, finding an alternative non-energy crop could provide growers with an additional value-added stream for their biomass feedstock. Given these opportunities, this study investigated non-torrefied and torrefied short-rotation shrub willow as a filler originating from low-value biomass and used to produce WPC. Combinations of polypropylene (PP), non-torrefied willow (NTW), and torrefied willow (TW) at varying weight percentages (wt%) were used to produce the WPCs. The TW was made using both 225 °C and 300 °C modification temperatures. The composites were evaluated under mechanical, thermal, and physical properties to assess the influence of filler particles as an alternative low-value filler to produce WPC.

## 2. Materials and Methods

### 2.1. Materials

Millbrook (*Salix purpurea* × *S. miyabeana*) short-rotation willow was used as a natural-based filler. The three-year-old willow was harvested at The Pennsylvania State University's Rockview demonstration field site near State College, Pennsylvania, USA. The willow was ground without debarking into 4 mm particle size and air-dried to room equilibrium conditions (approximately 8% moisture content). Polypropylene (PP) pellets were supplied by PolyOne Corporation (Avon Lake, OH, USA), and a lubricant agent, Struktol TPW 113, a blend of complex modified fatty acid ester (Stow, OH, USA), was added to all formulations with a fixed percentage.

### 2.2. Preparation and Characterization of the Filler

The woodchips were oven-dried to reach <5% moisture content for the torrefaction process. Two temperature levels of 225 °C and 300 °C were selected for the process. The torrefaction process occurred in the tube furnace under an inert atmosphere using $N_2$ gas over a 30-min heating regime at a target temperature. The torrefied chips were ground in a Fritsch Pulverisette 25 grinder with a 1 mm mesh sieve size and stored in vacuum bags. The particle size distribution of dry torrefied wood (TW) and non-torrefied wood (NTW) flour was determined by laser scattering technology using a Horiba LA-960A2 analyzer (Horiba, Kyoto, Japan), and results for $D_{10}$, $D_{50}$, $D_{90}$, and mean size are reported.

Fourier transform infrared spectra were obtained using an FTIR spectrometer (Bruker Optik, GmbH, Germany) connected to an ATR (attenuated total reflection) module. The spectra were collected over a wavelength range of 400 to 4000 cm$^{-1}$ and a resolution of 4 cm$^{-1}$. For each material (non-torrefied and torrefied at 225 °C and 300 °C), 64 scans were performed, and three repetitions were conducted to increase the accuracy of the results and reduce the effect of the atmospheric noise. Average spectra were collected using Opus software (https://www.bruker.com/en/products-and-solutions/infrared-and-raman/opus-spectroscopy-software/downloads.html, accessed on 28 September 2023).

### 2.3. Blending and Production of WPC

In total, nine different samples of pellets were compounded using various mixing ratios of TW and NTW. At the same time, the amount of added polypropylene (PP) and lubricant remained constant at 56% and 4%, respectively. A sample of 0% TW was compounded without TW and served as a reference sample. For each torrefaction temperature level, four samples were compounded with variations in TW and NTW amounts (5, 15, 25, and 40% wt). Details about sample nomenclature and mixing composition are summarized in Table 1.

**Table 1.** Formulation of blending mixtures with sample name, the ratio of non-torrefied and torrefied wood, and torrefaction temperature.

| Sample | Sample Description | NTW | TW | Torrefaction Temperature |
|---|---|---|---|---|
| MA | 0%TW | 40% | 0% | No torrefaction |
| MB | 5%TW-225 °C | 35% | 5% | 225 °C |
| MC | 15%TW-225 °C | 25% | 15% | |
| MD | 25%TW-225 °C | 15% | 25% | |
| ME | 40%TW-225 °C | 0% | 40% | |
| MF | 5%TW-300 °C | 35% | 5% | 300 °C |
| MG | 15%TW-300 °C | 25% | 15% | |
| MH | 25%TW-300 °C | 15% | 25% | |
| MI | 40%TW-300 °C | 0% | 40% | |

The compounding of wood-PP pellets was conducted using the lab twin screw extruder (TSE 16TC) with four temperature zones heated at 185 °C, 190 °C, 200 °C, and 210 °C from the material feedthrough to the outlet die to ensure proper melting of the polymers. A constant feeding rate of 40 RPM was provided by an automatic single-screw feeder. For each formulation, the desired amount of different components (PP, filler, and lubricant) was mixed manually and then placed into the hopper.

During the extrusion process, the blend was cooled in a water bath located between the extruder nozzle and the pelletizer. The continuous line of extruded material was connected to the pelletizer with synchronized cutting and feeding speeds. Special care was taken to ensure that only well-blended and extruded material was pelletized and stored, disregarding the initial and end portions of the batch extrusion. This process limited the variability of extruded material in each blend and ensured the mixing process was as consistent as possible in the samples used for testing. Pellets were dried to remove excess water from the surface.

Steel molds were used to manufacture standardized testing specimens for tensile and three-point flexural tests according to ASTM D638 and ASTM D790, respectively. A 50 kN hydraulic hot press (LZT-UK-30-L, Langzauner, Lambrechten, Austria) equipped with water cooling was used to melt and compress compounded pellets into the molds for the final specimen shape. The steel plates were preheated to 200 °C in the hot press. Manufactured pellets were placed on the preheated molds and left for 10 min to soften. Waxed paper was inserted between steel plates to prevent samples from sticking. After softening the pellets, the heated plates were closed for 5 min at 200 °C. The cooling process followed the melting step by cooling the press plates to 80 °C. The mass of pellets per specimen volume was 11 g and 3 g for tensile and flexural mold, respectively, to reach a target density of approximately 1 g/cm$^3$. Each pressing cycle generated five specimens of each shape. Figure 1 shows the process of manufacturing dog-bone specimens.

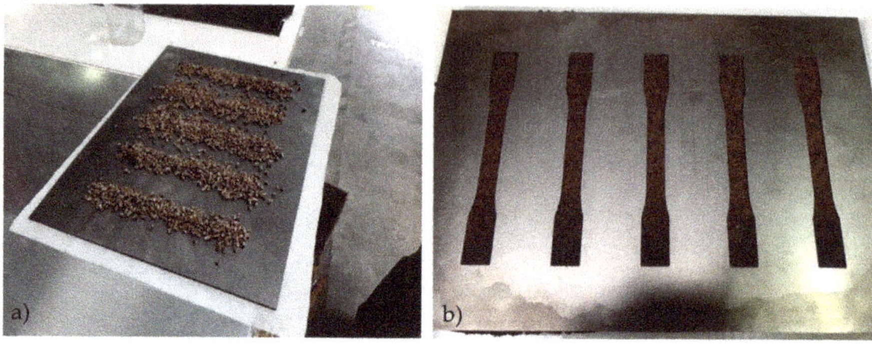


**Figure 1.** Manufacture of tensile specimens before (**a**) and after (**b**) the pressing process in a steel mold plate.

### 2.4. Scanning Electron Microscopy

The morphology of the reference sample and the two samples with the largest TW levels at each temperature (40%TW-225 °C and 40%TW-300 °C) was evaluated by scanning electron microscopy (SEM). Samples were cut with the sharp blade, and images were collected using a scanning electron microscope (JEOL JSM-IT500, Oxford Instruments, Tokyo, Japan) operating at a low vacuum (70–80 Pa), a working distance of 10 mm, and an accelerated voltage of 15 kV.

### 2.5. Mechanical Properties

Tensile mechanical testing was performed on a universal testing machine (Zwick Roell, Ulm, Germany) with a 50 kN load cell equipped with hydraulic grips and extensometer arms. The testing speed was adjusted to 3 mm/min due to the brittle behavior of composites with a gauge length of 50 mm to determine the elongation of the specimen. The three-point bending flexural test was performed on a universal testing machine with a 1 kN load cell to ensure accurate measurements. The span was set to 53 mm, and the loading rate was 1.34 mm/min. Experiments were performed at room temperature (23 ± 2 °C) and 50 ± 10% relative humidity. For the rectangular flexural specimens, density was calculated before the test. Five and ten replicates were tested for tensile and flexural tests, respectively, and the average results were reported.

### 2.6. Water Absorption

After completing the flexural tests, ten rectangular specimens with approximate 24 mm × 13 mm dimensions were cut from the tested specimens and designated for water absorption tests. The specimens were oven-dried at 100 °C for 24 h, then weighed, and the initial mass ($m_0$) was recorded. Specimens were then immersed in distilled water, and the mass after 24 h and 48 h ($m_{24}$ and $m_{48}$, respectively) was determined. Ten replicates were tested for each sample, and mean values were reported.

### 2.7. Thermal Properties

Thermogravimetric analyses (TGA) were conducted using a Discovery TGA (TGA 5500, TA Waters Instruments, New Castle, DE, USA). On average, 51 mg of specimen mass were placed on the aluminum pans and heated from room temperature up to 700 °C at a heating rate of 10 °C/min under nitrogen gas using a sample and balance purge flow of 25 mL/min and 10 mL/min, respectively. Data for $T_{5\%}$, $T_{10\%}$, and $T_{max}$ correspond to temperatures at which 5% and 10% weight loss were achieved, and temperatures at the maximum rate of specimen degradation (i.e., the highest points in the peaks appearing in the derivative curves) were extracted.

Differential scanning calorimetry (DSC) analysis was performed using a Discovery DSC 25 (TA Waters Instruments, New Castle, DE, USA). On average, 3.2 mg of sample in fine powder format were analyzed under a nitrogen gas atmosphere using the heat–cool–heat method as follows:

First heating cycle from 25 to 190 °C at 20 °C/min.
Cooling cycle from 190 to −30 °C at 20 °C/min.
Second heating cycle from −30 to 190 °C at 20 °C/min.

The first heating cycle was conducted to eliminate the thermal history of the composites. Data for crystallization temperature ($T_c$), melting temperature ($T_m$), and crystallization and melting enthalpies were determined from the cooling and second heating curves.

The degrees of crystallinity ($X_c$) of the wood-PP-based composites were calculated using the following Equation (1):

$$X_c = \frac{\Delta H_m}{w \times \Delta H_{m0}} \times 10 \quad (1)$$

where $\Delta H_m$ represents the samples' melting enthalpies, $w$ is the weight fraction of PP, and $\Delta H_{m0}$ is the melting enthalpy of 100% crystalline PP, which is equal to 207 J g$^{-1}$ [29].

### 2.8. Statistical Evaluation

Data analysis was conducted in R (version 4.3.1) using RStudio (version 2023.06.0+421). The results of the measurements were reported as mean values with standard deviations (SD), including the number of analyzed specimens (No). An analysis of variance (ANOVA) was assessed to check for significant differences between samples that differ by TW loading levels and temperature levels. An ANOVA was conducted for each of the following outcomes: tensile modulus of elasticity, tensile strength, and flexural modulus of elasticity and flexural strength. When ANOVA showed a statistically significant result, post hoc pairwise comparisons were performed with Tukey's HSD test for each of the above-mentioned outcomes (adjusted p values for multiple comparisons). The significance level was set at $\alpha = 0.05$.

## 3. Results and Discussion

This section may be divided by subheadings. It should provide a concise and precise description of the experimental results, their interpretation, and the experimental conclusions that can be drawn.

### 3.1. Particle Size of the Raw and Torrefied Wood Fillers

Size distributions of the non-torrefied wood (NTW) and torrefied wood (TW) reinforcement are presented in Table 2. Values for $D_{10}$, $D_{50}$, and $D_{90}$ correspond to the cumulative particle size percentiles and indicate the size below which 10%, 50%, and 90% of all particles are obtained, respectively.

**Table 2.** Representative particle size diameters of the woody fillers.

| Cumulative Particle Size Distribution | Particle Size (μm) | | |
|---|---|---|---|
| | NTW | TW 225 °C | TW 300 °C |
| $D_{10}$ | 163 | 102 | 43 |
| $D_{50}$ | 639 | 530 | 290 |
| $D_{90}$ | 1148 | 1140 | 840 |
| Mean | 661 | 589 | 379 |

The TW samples had a smaller particle size that ranged between <102 μm and 1140 μm and between <43 μm and 840 μm for fillers torrefied at 225 °C and 300 °C, respectively, compared to the NTW sample with a particle size between <163 μm and

1148 μm. The mean size decreased by 11% and 43% after torrefaction at 225 °C and 300 °C, respectively, indicating that the torrefaction process and temperature influenced the particles' diameter. The decrease in particle size can be attributed to the shrinkage of the wood particles upon water evaporation during torrefaction. Moreover, the exposure to heat during torrefaction likely increased the grindability of the wood particles by provoking the decomposition of natural polymers, mainly hemicelluloses, that tend to degrade at temperatures ranging from 220 °C to 315 °C [21,30]. Therefore, pre-degraded particles were less resistant to the mechanical force during grinding after torrefaction. In this context, Moustafa et al. [31] reported that the average particle size of coffee grounds diminished gradually by 24% and 29% after torrefaction for 2 h at 250 °C and 270 °C, respectively. This was attributed to the shrinkage of particles under the heat effect. A similar trend with a decrease in particle size by increasing the modification temperature was also reported for temperatures at 160 and 170 °C for hydrothermally modified birch wood [32].

### 3.2. FTIR

The FTIR spectra of non-torrefied and torrefied wood particles are represented in Figure 2. The obtained spectra revealed some changes in the composition of the surface functional groups of the wood after the torrefaction treatment.

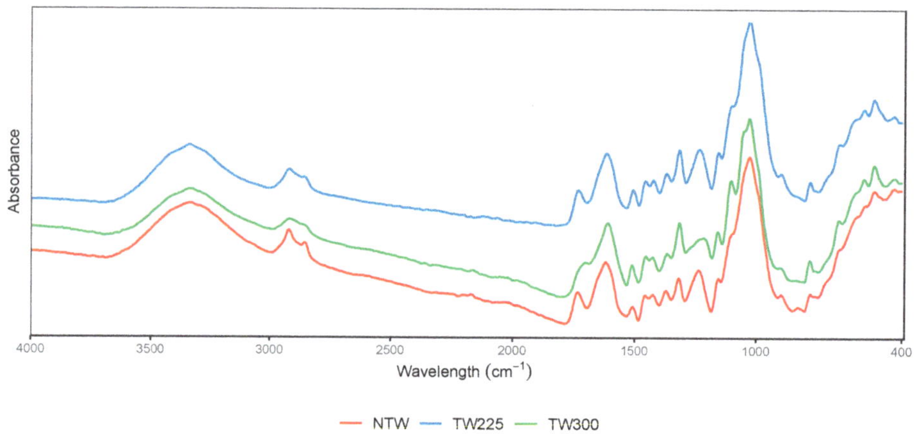


**Figure 2.** FTIR spectra of non-torrefied and torrefied wood biomasses.

The broad peak at 3336 cm$^{-1}$ was observed in all samples and was attributed to the occurrence of polar O-H stretching that corresponds to hydroxyl groups. These groups can correspond to water, alcohol, and carboxylic groups originating from natural polysaccharides and lignin [33]. The peak at 2916 cm$^{-1}$ assigned to asymmetrical and symmetric -CH vibrations was detected in all samples. However, this peak was more visible in the non-torrefied wood material, while it tended to fade progressively in the materials torrefied at 225 °C and 300 °C. The reduction in -CH vibrations was likely associated with the decomposition of hemicellulose, given that they start the thermal degradation at a temperature ranging between 220 °C and 315 °C [30]. Similarly, a peak at 1730 cm$^{-1}$ attributed to C=O functional groups from hemicelluloses was present in all samples. However, the peak intensity decreased after torrefaction at 300 °C [34].

Peaks at 1602 cm$^{-1}$ and 1511 cm$^{-1}$ were ascribed to C=C stretch vibrations and aromatic ring stretching from lignin. Similar peaks were observed previously in the FTIR spectra of Chinese fir and Eucalyptus globulus wood [33,34]. Peaks corresponding to lignin material remained almost constant after the torrefaction treatment, given that lignin has high thermal stability and its degradation occurs at a broad temperature range of 160 °C to 900 °C [30].

Peaks from 1000 to 1100 cm$^{-1}$ attributed to C-O groups from cellulose were soundly prominent in all samples, suggesting that part of the cellulose in the wood materials remained intact after the torrefaction treatment, given that cellulose can degrade in a temperature range from 315 °C to 400 °C [30]. Prior research [35] reported that a similar peak vanished after treatment of the Arundo donax and olive stone biomass at high temperatures up to 600 °C.

### 3.3. SEM

The distribution of wood particles and their interaction with the polymeric matrix were investigated by SEM at different magnifications. Images are presented in Figure 3, where the scale bar represents 100 μm.

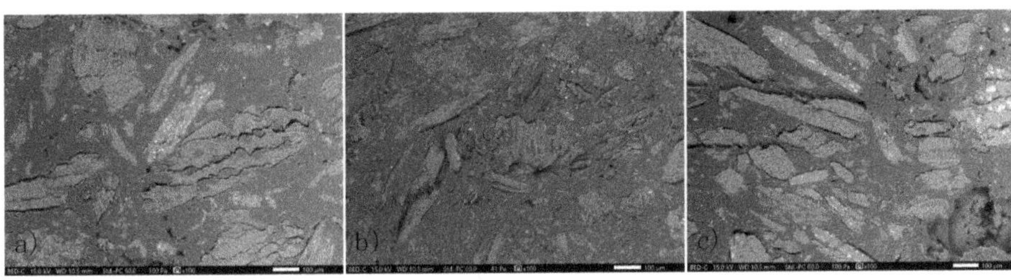

**Figure 3.** SEM images of the (**a**) reference sample (0%TW), (**b**) 40%TW-225 °C, and (**c**) 40%TW-300 °C at 100 μm magnification.

Wooden particles are randomly embedded in the polymer matrix, showing different slenderness. The wood particles were mostly long in shape, suggesting that they were horizontally oriented. However, often, the porous wooden structure can be observed. The difference between the samples (Figure 3) and the difference in the mean particle size (Table 2) is not clear in the images. The incorporation of wooden particles is also often loose from the matrix, indicating poor adhesion between the particles. It can also be observed that the wooden particles often collapse, and cell lumens are compressed together due to the high density of the material. The formation of aggregates was also observed in all samples, likely due to the poor distribution of the filler during the preparation process. Moreover, a void in the polymeric matrix was detected in Figure 3c, which can be attributed to the formation of air bubbles within the polymer during the blending process.

### 3.4. Mechanical Properties

Results from the tensile test are presented in Table 3.

**Table 3.** Mean tensile strength and modulus of elasticity with standard deviation (SD) and number of recorded specimens (No.).

| Sample | Modulus of Elasticity (MPa) | SD | Tensile Strength (MPa) | SD | No. |
|---|---|---|---|---|---|
| MA | 1336 | 84 | 10.8 | 0.7 | 4 |
| MB | 1417 | 86 | 11.6 | 1.1 | 5 |
| MC | 1426 | 47 | 11.3 | 0.6 | 5 |
| MD | 1293 | 80 | 9.9 | 1.0 | 5 |
| ME | 1476 | 48 | 11.0 | 0.2 | 5 |
| MF | 1431 | 118 | 11.6 | 0.8 | 5 |
| MG | 1361 | 93 | 10.5 | 1.8 | 5 |
| MH | 1396 | 26 | 10.5 | 0.5 | 5 |
| MI | 1398 | 119 | 11.0 | 0.7 | 5 |

The values for modulus of elasticity ranged from 1293 MPa to 1476 MPa. Sample ME resulted in the highest modulus of elasticity with 1476 MPa, while the lowest MD showed 12% lower values. The post hoc pairwise comparisons between samples showed that only one pair had a statistically significant difference in average modulus of elasticity (ME-MD: 183 MPa, 95% CI: 8 to 358, $p = 0.034$). The results of the analyses are reported in Table S1 in the supplementary material. There is no clear explanation of these differences, but rather the variability in the material matrix composition.

The results of tensile strength ranged between 9.9 MPa and 11.6 MPa (Table 3). Statistical analyses conducted with ANOVA on the tensile strength results revealed no statistically significant differences between samples. The results of post-hoc pairwise comparisons are presented in Table S2 in the supplementary material. These results indicate that *tw*% of TW loading and the temperature of torrefaction had no noticeable impact on the changes in tensile properties. Experimental results from three-point flexural tests are summarized in Table 4.

**Table 4.** Mean values for flexural modulus of elasticity, strength, density with standard deviation (SD), and number of specimens (No).

| Sample | Modulus of Elasticity | | Flexural Strength | | | Density | | |
|---|---|---|---|---|---|---|---|---|
| | (MPa) | SD | (MPa) | SD | No. | (kg/m$^3$) | SD | No. |
| MA | 1777 | 123 | 26.6 | 2.2 | 10 | 973.8 | 32.5 | 10 |
| MB | 1682 | 75 | 23.0 | 1.6 | 10 | 975.3 | 9.3 | 10 |
| MC | 1604 | 82 | 23.5 | 1.8 | 10 | 992.1 | 8.9 | 10 |
| MD | 1663 | 122 | 24.0 | 3.2 | 10 | 981.7 | 7.5 | 10 |
| ME | 1726 | 97 | 24.5 | 1.6 | 10 | 997.6 | 9.0 | 10 |
| MF | 1603 | 81 | 25.0 | 1.4 | 10 | 984.6 | 8.3 | 10 |
| MG | 1546 | 36 | 23.3 | 1.6 | 10 | 980.5 | 12.5 | 10 |
| MH | 1577 | 175 | 22.0 | 2.8 | 9 | 984.6 | 7.4 | 10 |
| MI | 1622 | 82 | 22.7 | 1.4 | 10 | 979.3 | 7.6 | 10 |

Regarding flexural performance, the reference sample with 0%TW performed with the highest modulus of elasticity. Values for flexural strength showed that the strength values decreased with the increasing level of TW for the higher torrefaction temperature. Around 8% and 15% lower strengths were observed for the samples with the highest level of TW (ME, MI) at 225 °C and 300 °C against the reference, respectively.

Results from the ANOVA suggested at least one group had a statistically significant difference for both flexural properties. In the case of the statistical analyses conducted for mean flexural modulus of elasticity, statistically significant differences were found for pairs MC-MA (−173 MPa, 95% CI: −319 to −27, $p = 0.009$), MF-MA (−174 MPa, 95% CI: −320 to −28, $p = 0.008$), MG-MA (−231 MPa, 95% CI: −377 to −85, $p = 0.000$), MH-MA (−200 MPa, 95% CI: −351 to −50, $p = 0.002$), MI-MA (−155 MPa, 95% CI: −301 to −9, $p = 0.029$), and MG-ME (−180 MPa, 95% CI: −362 to −34, $p = 0.005$). Apart from the MC-MA pair, the remaining statistically significant differences were found in samples with TW torrefied at 300 °C. In contrast, the MC sample had the lowest measured modulus of elasticity from TW at 225 °C. The complete results from post hoc analyses are presented in Table S3 in supplementary material.

Post-hoc comparisons of flexural strength resulted in the following pairs being significantly different: MB-MA (−3.6 MPa, 95% CI: −6.5 to −0.7, $p = 0.005$), MC-MA (−3.1 MPa, 95% CI: −6.0 to −0.2, $p = 0.026$), MG-MA (−3.3 MPa, 95% CI: −6.2 to −0.4, $p = 0.013$), MH-MA (−4.6 MPa, 95% CI: −7.6 to −1.6, $p = 0.000$), MI-MA (−3.9 MPa, 95% CI: −6.8 to −1.0, $p = 0.002$), and pair MH-MF (−3.0 MPa, 95% CI: −6.0 to −0.0, $p = 0.047$). The remaining results from post hoc analyses are presented in Table S4 in supplementary material.

The flexural modulus of elasticity and strength performances of the reference sample containing NTW can be related to the higher average particle size (661 μm) compared to TW-225 °C and TW-300 °C (589 μm and 379 μm, respectively). With the higher torrefication temperature, a higher decrease in flexural properties was observed for 25 and 40% TW loadings. Torrefaction improves grindability and changes the chemical composition of the particles. This change can be attributed to different particle aspect ratios, such as the ratio between cellulose, lignin, and hemicellulose. In addition, smaller wood particles had a higher tendency to agglomerate, preventing uniform dispersion of the filler within the polymeric matrix, which can lead to lower flexibility of the material and poor adhesion of the matrix. Some of these phenomena were observed and described by other studies. Chiou et al. [23] injection molded extruded torrefied almond shells with wood chips and PP. They pointed out that the addition of particles to the matrix did not substantially impact the tensile modulus of elasticity but more on its strength, while with the increasing filler size, the modulus was reduced. However, mid and high particle size levels were subjected to 854 and 1545 um large particles in that prior study. Another study utilized torrefied almond shells (TAS) [24] with three different particle sizes and seven different loading levels in WPC. Increasing wt% of TAS increased the flexural modulus of elasticity, but samples with 50% TAS were already too brittle to evaluate flexural properties. The study found no effect of particle size on flexural properties. Regarding flexural strength, higher loadings of particles led to decreasing strength, while no effect on particle size was found. Chaudemanche et al. [36] evaluated the effect of particle size on extruded HDPE-WPC using three different size ranges and pointed out the importance of particle direction and aspect ratio. Flexural properties were better with larger particles. Also, in the tensile test, larger particles improved properties in the direction of sample extrusion. On the other hand, the finest particles resulted in better performance in the transverse direction. This result was attributed to smaller particles better aligning and compacting during processing into a composite. In our study, this might be true for flexural properties, while no changes due to particle size can be attributed to tensile properties. Pudlik et al. [5] also discussed the impact of chemical composition by comparing soft and hardwood species used in the HDPE-WPC. They emphasized that softwood species are more favorable for producing WPC due to their higher lignin and lower hemicellulose content. However, these properties result in higher strengths, lower flexibility, and poor internal bonding. In the study, the authors confirmed that a higher proportion of cellulose in the willow contributed to the greater strength of WPC. Due to the torrefaction process conducted in this study, these observations likely explain the reduction in flexural properties due to the lower mechanical properties of TW filler particles. Moreover, Patula pine biomass torrefied at 200 °C, 250 °C, and 300 °C showed that both lower torrefaction temperatures did not contribute to significant changes in chemical composition [37], which may also explain why more visible changes were also attributed to TW at 300 °C.

### 3.5. Water Absorption

Results of water absorption (WA) after 24 h and 48 h of immersion in water are presented in Figure 4.

The water uptake of the reference sample (MA) reinforced with NTW (40 wt%) was about 3% and 3.9% after 24 and 48 h immersion, respectively (Figure 4). These findings agree with data reported by Ayrilmis et al. [19] for wood (50 wt%)-PP composites, where water uptake after one day was equal to 6.45%. Prior research [38] reported higher values for water absorption by wood (40 wt%)-PP material, which were about 4.5% and 6.3% after 24 h and 48 h, respectively. These differences in water absorption findings can be related to the different grades of the utilized PP matrix (ratio of crystalline and amorphous regions) or the dimensions of the tested material.

The WA remained steady after incorporating 5, 15, 25, and 40 wt% of wood treated at 225 °C and 5 wt% treated at 300 °C. However, it started to decrease proportionally when the TW-300 °C load was further increased to 15, 25, and 40 wt%, and the decrease in water uptake after 24 h was 2.3, 2.2, and 1.4%, respectively, compared to the reference sample (MA). These findings show that a higher torrefaction temperature (i.e., 300 °C) had a greater effect on the hygroscopic behaviour of the composites. The increase in torrefaction temperature likely contributed to the degradation of most of the present hemicelluloses, which are characterized by their hydrophilic character.

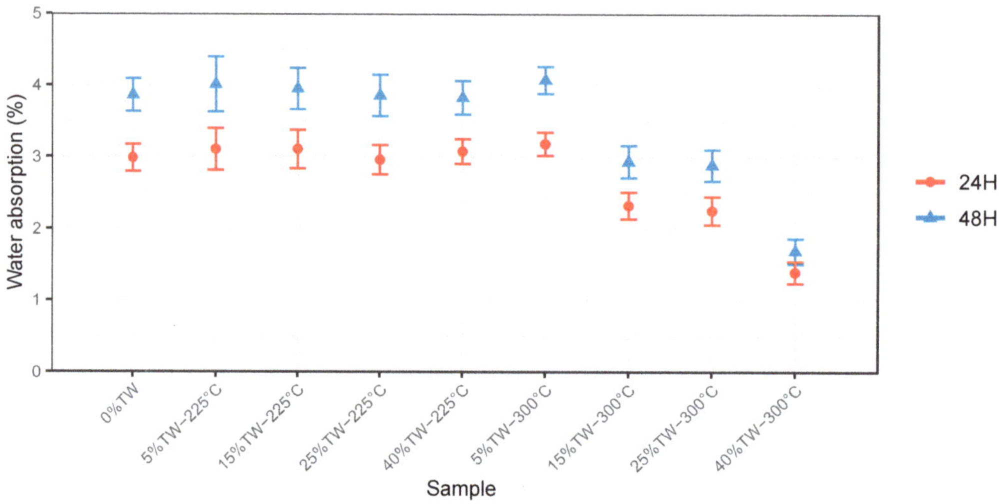


**Figure 4.** Change in water absorption variation in the wood-PP composites after 24 and 48 h.

Prior research [39] investigated the relationship between torrefaction treatment and the moisture uptake of woody biomass. They utilized softwood (*Fir, Abies pectinates*) samples torrefied at variable temperatures (from 200 °C to 230 °C for 1000 min). They reported that the equilibrium moisture content of torrefied wood decreased by 34 and 58% with the increase in torrefaction temperature from 200 to 230 °C, respectively. The authors explained that the hydroxyl groups present in the biomass, mainly hemicelluloses and amorphous regions of celluloses, are responsible for attracting and retaining water molecules via hydrogen bonding. These groups were hydrolyzed during exposure to heat, which increased the hydrophobicity of the material. Similarly, pressed wood plastic panels made of an equal ratio (50:50) of thermally treated wood fibers (*Eucalyptus camaldulensis*) and PP matrix were manufactured by Ayrilmis et al. [19], and the hygroscopic features were explored. They found that the composites water absorption (in 24 h) significantly decreased with increasing the treatment temperature from 120 °C to 180 °C (40 min) by 33.7%, which was linked to the degradation of hydrophilic hemicelluloses.

### 3.6. Thermal Properties

#### 3.6.1. Thermogravimetric Analysis

The TGA mass loss plots obtained for NTW and TW-reinforced PP composites were very similar, and the curves overlapped. Therefore, only curves from the reference sample (0%TW), 40%TW-225 °C, and 40%TW-300 °C were selected and represented in Figure 5. The degradation temperature points $T_{5\%}$, $T_{10\%}$, and $T_{max}$ corresponding to 5%, 10%, and maximum weight loss, respectively, are summarized in Table 5.

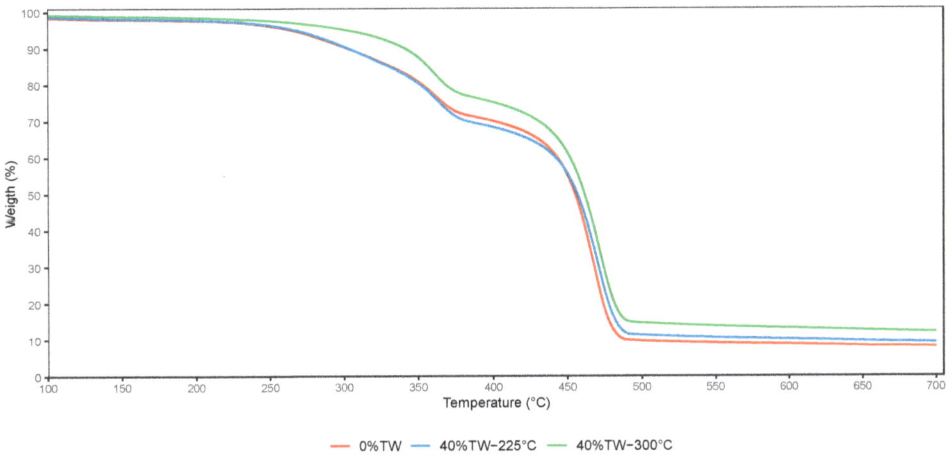


**Figure 5.** TGA thermograms of 0%TW, 40%TW-225 °C, and 40%TW-300 °C.

**Table 5.** Thermal degradation points and final residue of TW and NTW-reinforced PP.

| Sample | Degradation Temperature (°C) | | | Final Residue at 700 °C (%) |
|---|---|---|---|---|
| | $T_{5\%}$ | $T_{10\%}$ | $T_{max}$ | |
| MA | 265.8 | 303.4 | 468.2 | 8.1 |
| MB | 263.2 | 302.2 | 468.3 | 8.9 |
| MC | 271.6 | 307.8 | 466.8 | 8.6 |
| MD | 272.3 | 308.7 | 468.6 | 8.5 |
| ME | 268.5 | 303.9 | 471.0 | 9.4 |
| MF | 271.9 | 309.4 | 471.0 | 9.2 |
| MG | 281.9 | 323.5 | 470.9 | 9.7 |
| MH | 280.0 | 322.6 | 469.4 | 10.9 |
| MI | 303.4 | 342.0 | 472.6 | 12.3 |

The thermal decomposition started earlier for the 0%TW (reference sample MA) compared to the reinforced samples. The incorporation of TW reinforcement had a favorable effect on the material's thermal resistance. Thermal degradation was delayed, especially when TW-300 °C was utilized. When 5% of TW-225 °C was incorporated, $T_{5\%}$ and $T_{10\%}$ decreased slightly, whereas $T_{max}$ remained steady. Thus, no remarkable changes were observed in the thermal degradation compared to the reference sample. However, when the loading of TW-225 °C was further increased, $T_{5\%}$, $T_{10\%}$, and $T_{max}$ increased. The highest increment in $T_{5\%}$ and $T_{10\%}$ was 6.4 °C and 5.3 °C, respectively, for the sample containing 25% of TW (MD), while $T_{max}$ remained almost the same compared to the reference sample. Up to 25% of TW, the increase in TW loading had a negative impact on the thermal stability, and a decrease in the degradation points was noted (Table 5). When TW-300 °C was used as reinforcement, $T_{5\%}$, $T_{10\%}$, and $T_{max}$ increased gradually with the filler loading up to 40%, with a slight difference observed between samples loaded with 15% and 25%. The highest thermal stability was exhibited by sample 40% TW-300 °C (MI) with an elevation in $T_{5\%}$, $T_{10\%}$, and $T_{max}$ of 37.6 °C, 38.5 °C, and 4.4 °C, respectively, compared to the reference sample. The final residue percentage increased with the increase in TW percentage from 5% to 40%, and the increment was more significant in the case of filler treated at 300 °C. The final residue at 700 °C increased by 16% and 52% for samples with 40%TW treated at 225 °C (ME) and 300 °C (MI) compared to the reference sample. These findings suggested that utilizing TW as a reinforcement for PP, particularly TW-300 °C, could improve the thermal stability of wood–polymer composites at high temperatures. The enhancement in thermal resistance can be attributed to torrefied wood's higher stability than raw wood, owing to the pre-degradation of hemicelluloses during the treatment. Hemicellulose is known as the most thermolabile natural polymer, given that it starts to degrade at earlier stages (from 220 °C to 315 °C [30]) compared to other wood components (i.e., cellulose and lignin). TW filler likely contained reduced hemicellulose content, which improved the thermal heat resistance and led to the generation of more residue at the end of the heating cycle. Barajas et al. [40] observed similar trends when they studied the thermal properties of PLA composites reinforced with raw and torrefied (250 °C) coffee husk flour (CHF). They reported that $T_{5\%}$ and $T_{max}$ of PLA-20% torrefied CHF increased by 23 °C and 4 °C, respectively, compared to PLA-20% raw CHF. The authors explained that the torrefaction led to removing the less thermally stable components, specifically hemicellulose and cellulose. However, they found that a further increase in the torrefied CHF loading (up to 50%) negatively affected the composites thermal stability, attributed to the potential occurrence of metals and water in the filler powder that catalyzed and favored the thermal degradation. Moreover, they found that [40] observed similar trends when they studied the thermal properties of PLA composites reinforced with raw and torrefied (250 °C) coffee husk flour. They reported that $T_{5\%}$ and $T_{max}$ decreased gradually from 322.4 ± 1.7 °C to 301.1 ± 1.3 °C and from 354.7 ± 2.2 °C to 341.4 ± 1.6 °C, respectively, when torrefied coffee husk loadings were increased from 20% to 50% wt%. However, they recorded a continuous increase in residual mass from 5.6 ± 0.5 to 10.9 ± 0.8% (from 20 to 50 wt% loading). They also reported that within the same filler loading (20 wt%), $T_{5\%}$ and $T_{max}$ of composites with torrefied filler increased by 23.5 °C and 3.6 °C, respectively, and final residue at 700 °C increased continuously with the increase in torrefied CHF content, which was justified by the ability of torrefied material to generate carbon-rich residue. Similarly, prior research [31] stated that the thermal resistance of polybutylene adipate terephthalate (PBAT) composites was enhanced when the coffee ground was torrefied at 270 °C when incorporated, compared to composites containing raw or coffee ground torrefied at a lower temperature (250 °C). The findings were attributed to the hydrolysis of less heat-stable compounds during the torrefaction, which increased with the severity of the torrefaction temperature (i.e., from 250 to 270 °C).

#### 3.6.2. Differential Scanning Calorimetry

DSC-derived heating and cooling isotherms for the reinforced PP composites were considered to understand the effect of the torrefied filler's incorporation on the PP polymeric matrix's crystallization and melting behaviors. The crystallization temperature ($T_c$), crystallization enthalpy ($\Delta H_c$), melting temperature ($T_m$), melting enthalpy ($\Delta H_m$), and crystallinity degree ($X_c$) of all samples are listed in Table 6.

**Table 6.** Thermal transitions of the wood-PP-based composites.

| Sample | $T_c$ (°C) | $\Delta H_c$ (J g$^{-1}$) | $T_m$ (°C) | $\Delta H_m$ (J g$^{-1}$) | $X_c$ (%) |
|---|---|---|---|---|---|
| MA | 112.7 | 48.0 | 160.5 | 40.3 | 32 |
| MB | 112.6 | 64.7 | 160.9 | 51.6 | 42 |
| MC | 113.2 | 52.5 | 160.4 | 40.5 | 33 |
| MD | 116.7 | 56.8 | 160.6 | 35.7 | 29 |
| ME | 113.2 | 44.8 | 160.1 | 31.2 | 25 |
| MF | 112.9 | 62.0 | 160.6 | 41.2 | 33 |
| MG | 112.4 | 44.3 | 160.8 | 30.7 | 25 |
| MH | 113.1 | 34.2 | 159.2 | 25.5 | 21 |
| MI | 111.4 | 48.3 | 159.6 | 34.2 | 28 |

The crystallization temperature remained steady at 5% and 15% TW loadings, regardless of the torrefaction temperatures. As the amount of TW-225 °C increased to 25%, $T_c$

increased by 4 °C compared to the reference sample. However, the increment in $T_c$ was negligible in the case of 25% TW-300 °C (MH).

When TW levels were further increased to 40%, $T_c$ decreased by 3.5 °C and 1.6 °C in the case of 250 °C and 300 °C torrefied fillers, respectively, compared to the corresponding 25% samples. The $T_c$ values of composites reinforced with TW-250 °C wood filler were higher than those reinforced with TW-300 °C within the same loading levels (Table 6). This result was attributed to the differences in particle size that affected the dispersion of the wood particles within the matrix. TW-225 °C had a larger average particle size (589 µm) compared to TW-300 °C (379 µm). Bigger particles likely prevented the formation of crystals by creating inter-spaces between the PP chains, delaying the $T_c$ from reaching higher values. The degree of crystallization decreased as the amount of TW increased, from 5% to 40%, regardless of the torrefaction temperature. Therefore, with the addition of higher TW loadings, the formation of crystals within the polymeric matrix was harder, which means that the wood particles likely limited the nucleation of the materials by lowering the mobility of PP molecular chains.

$T_m$ values remained almost constant for all samples except for composites containing 25% and 40% of TW-300 °C (MH and MI), where a slight decrease was observed (Table 6). Chiou et al. [41] obtained similar results, incorporating variable loadings (5, 12.5, and 20 wt%) of almond shell torrefied at 280 °C and 300 °C and wood chips torrefied at 280 °C in a PP matrix. They reported that melting temperature values for all the specimens ranged between 166 °C and 167 °C, comparable to the neat PP (167 °C). They concluded that the filler did not significantly influence the melting behaviors of the PP chains. Moustafa et al. [31] reported different findings when they studied the different properties of PBAT composites reinforced with raw and torrefied coffee ground at variable loadings (10, 20, and 30%). They found that torrefied filler had a favorable effect on the thermal resistance of PBAT composites. The $T_m$ of PBAT filled with untreated coffee ground ranged between 96.7 °C and 97.4 °C, whereas samples filled with coffee ground torrefied at 250 °C had higher $T_m$ values that ranged between 109.8 °C and 115.1 °C. However, using coffee ground torrefied at higher temperatures (270 °C) reduced the $T_m$, and values ranged between 95.8 and 105.6 °C.

## 4. Conclusions

Wood plastic composites were prepared from polypropylene reinforced with torrefied and non-torrefied short rotation willows via extrusion and molding processes. The morphology, mechanical, thermal, and water absorption properties of the materials were examined. The mean particle size of wood fillers decreased by 11% and 43% after torrefaction at 225 °C and 300 °C, respectively, suggesting that the torrefied particles had better grindability. Observation of SEM images revealed the random distribution of filler particles and the occurrence of voids within the polymeric matrix. The flexural strength decreased with the increase in torrefied wood particles, and all samples' flexural modulus of elasticity was lower than the reference sample. Changes in flexural behaviors may be attributed to differences in wood particle size and distribution within the polymeric matrix. However, the tensile strength did not change significantly. Samples containing particles torrefied at 300 °C exhibited higher water resistance than other samples. The TGA results showed that the addition of torrefied wood particles delayed the thermal degradation of the composites. The thermolabile components in the wood were partially removed by torrefaction, which enhanced the thermal stability of the composites. The crystallization degree decreased with increased torrefied wood loadings while the melting temperatures remained almost steady.

Torrefied short-rotation willow is a good candidate as a low-value-added filler in polypropylene and wood–polypropylene composites. The optimization of the filler's loadings is crucial to achieving optimal performance for the final composites.

**Supplementary Materials:** The following supporting information can be downloaded at: https://www.mdpi.com/article/10.3390/polym15193997/s1. Table S1: Results of the Tukey HSD comparisons between pairs for tensile modulus of elasticity with difference between the pairs, 95% confident interval and *p*-value. Table S2: Results of the Tukey HSD comparisons between pairs for tensile strength with difference between the pairs, 95% confident interval and *p*-value. Table S3: Results of the Tukey HSD comparisons between pairs for Flexural modulus of elasticity with difference between the pairs, 95% confident interval and *p*-value. Table S4: Results of the Tukey HSD comparisons between pairs for Flexural strength with difference between the pairs, 95% confident interval and *p*-value.

**Author Contributions:** Conceptualization, D.B.D. and J.G.P.; methodology, J.G.P., M.Z. and M.S.; validation, D.B.D.; formal analysis, J.G.P., M.Z. and D.B.D.; investigation, J.G.P. and M.Z.; resources, D.B.D. and M.S.; data curation, J.G.P. and M.Z.; writing—original draft preparation, J.G.P. and M.Z.; writing—review and editing, J.G.P., M.Z., M.S. and D.B.D.; visualization, J.G.P.; supervision, D.B.D.; project administration, D.B.D.; funding acquisition, D.B.D. and M.S. All authors have read and agreed to the published version of the manuscript.

**Funding:** This research was funded by the United States Department of Agriculture (USDA), National Institute of Food and Agriculture (NIFA) grant number 2012-68005-19703, NORTHEAST WOODY/WARM-SEASON BIOMASS CONSORTIUM (NEWBIO). Funding was also provided by the European Commission through the InnoRenew project (Grant Agreement #739574 under the Horizon 2020 WIDESPREAD-2-Teaming program) and the Republic of Slovenia (investment funding from the Republic of Slovenia and the European Regional Development Fund), the ARRS infrastructural program IO-0035 at the University of Primorska, and the SEAFARER (J4-4546) project. The ForestValue Research Program and the Republic of Slovenia's Ministry of Higher Education, Science, and Innovation [BarkBuild: grant agreement No. C3330-21-252003]. Project BarkBuild is supported under the umbrella of ERA-NET Cofund ForestValue by Vinnova (Sweden), Valsts izglītības attīstības aģentūra (Latvia), the Ministry of Higher Education, Science, and Innovation (MVZI) (Slovenia), the Academy of Finland, The Research Council of Norway, and the National Science Centre, Poland. ForestValue has received funding from the European Union's Horizon 2020 research and innovation program under grant agreement No. 773324.

**Institutional Review Board Statement:** Not applicable.

**Data Availability Statement:** The data presented in this study are available on request from the corresponding author.

**Acknowledgments:** The authors would like to acknowledge the Slovenian Research Agency and Grant Agency of the Czech Republic for funding the WEAVE project (22-14942K), which is complementary and has enhanced this study, and Pondelak Andreja (Slovenian National Building and Civil Engineering Institute Dimičeva 12, 1000 Ljubljana) for providing the SEM images.

**Conflicts of Interest:** The authors declare no conflict of interest.

## References

1. Zalesny, R.S.; Cunningham, M.W.; Hall, R.B.; Mirck, J.; Rockwood, D.L.; Stanturf, J.A.; Volk, T.A. Woody Biomass from Short Rotation Energy Crops. In *ACS Symposium Series*; Zhu, J., Zhang, X., Pan, X., Eds.; American Chemical Society: Washington, DC, USA, 2011; Volume 1067, pp. 27–63, ISBN 978-0-8412-2643-2.
2. Pleguezuelo, C.R.R.; Zuazo, V.H.D.; Bielders, C.; Bocanegra, J.A.J.; PereaTorres, F.; Martínez, J.R.F. Bioenergy Farming Using Woody Crops. A Review. *Agron. Sustain. Dev.* **2015**, *35*, 95–119. [CrossRef]
3. European Commission; Joint Research Centre; Camia, A.; Giuntoli, J.; Jonsson, R.; Robert, N.; Cazzaniga, N.; Jasinevičius, G.; Avitabile, V.; Grassi, G.; et al. *The Use of Woody Biomass for Energy Production in the EU*; Publications Office: Luxembourg, 2021.
4. Abrahamson, L.P.; Kopp, R.F.; Smart, L.B.; Volk, T.A. Fast-Growing Willow Shrub Named 'Millbrook'. United States: N. p., 2007. Available online: https://image-ppubs.uspto.gov/dirsearch-public/print/downloadPdf/PP17646 (accessed on 28 September 2023).
5. Barton-Pudlik, J.; Czaja, K. Fast-Growing Willow (*Salix viminalis*) as a Filler in Polyethylene Composites. *Compos. Part B Eng.* **2018**, *143*, 68–74. [CrossRef]
6. Darwish, M.; Ismail, H.; Orz, A.; Ammar, A.; Eldebaby, O.; Solaiman, O.; Nassar, K.; Abouzeid, M.N.; Youssef, P. Manufacturing and Characteristics of Structural Sections from Wood Plastic Composites. In *Proceedings of the Canadian Society of Civil Engineering Annual Conference 2021*; Walbridge, S., Nik-Bakht, M., Ng, K.T.W., Shome, M., Alam, M.S., El Damatty, A., Lovegrove, G., Eds.; Springer Nature: Singapore, 2023; pp. 583–594.

7. Ashori, A. Wood–Plastic Composites as Promising Green-Composites for Automotive Industries! *Bioresour. Technol.* **2008**, *99*, 4661–4667. [CrossRef]
8. Kelleci, O.; Aksu, S.; Aydemir, D.; İStek, A.; Köksal, S.E. Wood Plastic Composite (WPC) Application inIndoor Furniture Sector. In Proceedings of the 2nd International FurnitureCongress, Muğla, Turkey, 13–15 October 2016; pp. 310–314.
9. Partanen, A.; Carus, M. Wood and Natural Fiber Composites Current Trend in Consumer Goods and Automotive Parts. *Reinf. Plast.* **2016**, *60*, 170–173. [CrossRef]
10. Kim, J.K.; Pal, K. *Recent Advances in the Processing of Wood-Plastic Composites*; Engineering Materials; Springer: Berlin, Heidelberg, 2011; Volume 32, ISBN 978-3-642-14876-7.
11. Schwarzkopf, M.J.; Burnard, M.D. Wood-Plastic Composites—Performance and Environmental Impacts. In *Environmental Impacts of Traditional and Innovative Forest-based Bioproducts*; Kutnar, A., Muthu, S.S., Eds.; Environmental Footprints and Eco-design of Products and Processes; Springer: Singapore, 2016; pp. 19–43, ISBN 978-981-10-0653-1.
12. Ayrilmis, N.; Kaymakci, A. Fast Growing Biomass as Reinforcing Filler in Thermoplastic Composites: Paulownia Elongata Wood. *Ind. Crops Prod.* **2013**, *43*, 457–464. [CrossRef]
13. Jaya Tripathi, A.B.; Causer, T.; Ciolkosz, D.E.; DeVallance, D.B.; Nunes, L.J.R. Non-Energetic Application of Carbon-Rich Torrefied Biomass in the Bioeconomy: A Review. *Biofuels* **2023**, 1–17. [CrossRef]
14. Kim, J.-W.; Harper, D.P.; Taylor, A.M. Effect of Wood Species on the Mechanical and Thermal Properties of Wood-Plastic Composites. *J. Appl. Polym. Sci.* **2009**, *112*, 1378–1385. [CrossRef]
15. Kim, J.-W.; Harper, D.P.; Taylor, A.M. Effect of wood species on water sorption and durability of wood–plastic composites. *Wood Fiber Sci.* **2008**, *40*, 519–531.
16. Fabiyi, J.S.; McDonald, A.G.; Morrell, J.J.; Freitag, C. Effects of Wood Species on Durability and Chemical Changes of Fungal Decayed Wood Plastic Composites. *Compos. Part A Appl. Sci. Manuf.* **2011**, *42*, 501–510. [CrossRef]
17. Bouafif, H.; Koubaa, A.; Perré, P.; Cloutier, A. Effects of Fiber Characteristics on the Physical and Mechanical Properties of Wood Plastic Composites. *Compos. Part A Appl. Sci. Manuf.* **2009**, *40*, 1975–1981. [CrossRef]
18. Sandberg, D.; Kutnar, A.; Karlsson, O.; Jones, D. *Wood Modification Technologies: Principles, Sustainability, and the Need for Innovation*, 1st ed.; CRC Press: Boca Raton, FL, USA, 2021.
19. Ayrilmis, N.; Jarusombuti, S.; Fueangvivat, V.; Bauchongkol, P. Effect of Thermal-Treatment of Wood Fibres on Properties of Flat-Pressed Wood Plastic Composites. *Polym. Degrad. Stab.* **2011**, *96*, 818–822. [CrossRef]
20. Pelaez-Samaniego, M.R.; Yadama, V.; Lowell, E.; Espinoza-Herrera, R. A Review of Wood Thermal Pretreatments to Improve Wood Composite Properties. *Wood Sci. Technol.* **2013**, *47*, 1285–1319. [CrossRef]
21. Karinkanta, P.; Ämmälä, A.; Illikainen, M.; Niinimäki, J. Fine Grinding of Wood—Overview from Wood Breakage to Applications. *Biomass Bioenergy* **2018**, *113*, 31–44. [CrossRef]
22. Ribeiro, J.; Godina, R.; Matias, J.; Nunes, L. Future Perspectives of Biomass Torrefaction: Review of the Current State-Of-The-Art and Research Development. *Sustainability* **2018**, *10*, 2323. [CrossRef]
23. Chiou, B.-S.; Cao, T.; Valenzuela-Medina, D.; Bilbao-Sainz, C.; Avena-Bustillos, R.J.; Milczarek, R.R.; Du, W.-X.; Glenn, G.M.; Orts, W.J. Torrefaction Kinetics of Almond and Walnut Shells: Effects of Inorganic Species. *J. Therm. Anal. Calorim.* **2018**, *131*, 3065–3075. [CrossRef]
24. McCaffrey, Z.; Torres, L.; Flynn, S.; Cao, T.; Chiou, B.-S.; Klamczynski, A.; Glenn, G.; Orts, W. Recycled Polypropylene-Polyethylene Torrefied Almond Shell Biocomposites. *Ind. Crops Prod.* **2018**, *125*, 425–432. [CrossRef]
25. Volfson, S.I.; Fayzullin, I.Z.; Musin, I.N.; Fayzullin, A.Z.; Grachev, A.N.; Pushkin, S.A. The Physicomechanical and Rheological Characteristics of Wood–Polymer Composites Based on Thermally and Mechanically Modified Filler. *Int. Polym. Sci. Technol.* **2017**, *44*, 49–54. [CrossRef]
26. Vold, J.L.; Ulven, C.A.; Chisholm, B.J. Torrefied Biomass Filled Polyamide Biocomposites: Mechanical and Physical Property Analysis. *J. Mater. Sci.* **2015**, *50*, 725–732. [CrossRef]
27. Lu, W.; Yu, W.; Han, X.; Cai, H.; Gao, F. Torrefaction Pretreatment Facilitated Solvents-Resistant and Stable Wood-Plastic Composites. *Ind. Crops Prod.* **2022**, *177*, 114454. [CrossRef]
28. Berthet, M.-A.; Commandré, J.-M.; Rouau, X.; Gontard, N.; Angellier-Coussy, H. Torrefaction Treatment of Lignocellulosic Fibres for Improving Fibre/Matrix Adhesion in a Biocomposite. *Mater. Des.* **2016**, *92*, 223–232. [CrossRef]
29. Karger-Kocsis, J. (Ed.) *Polypropylene Structure, Blends and Composites*; Springer: Berlin/Heidelberg, Germany, 1995; Volume 3.
30. Yang, H.; Yan, R.; Chen, H.; Lee, D.H.; Zheng, C. Characteristics of Hemicellulose, Cellulose and Lignin Pyrolysis. *Fuel* **2007**, *86*, 1781–1788. [CrossRef]
31. Moustafa, H.; Guizani, C.; Dupont, C.; Martin, V.; Jeguirim, M.; Dufresne, A. Utilization of Torrefied Coffee Grounds as Reinforcing Agent To Produce High-Quality Biodegradable PBAT Composites for Food Packaging Applications. *ACS Sustain. Chem. Eng.* **2017**, *5*, 1906–1916. [CrossRef]
32. Kuka, E.; Cirule, D.; Kajaks, J.; Andersone, I.; Andersons, B. Wood Plastic Composites Made with Thermally Modified Birch Wood Residues. *Int. Wood Prod. J.* **2016**, *7*, 225–230. [CrossRef]
33. Esteves, B.; Velez Marques, A.; Domingos, I.; Pereira, H. Chemical Changes of Heat Treated Pine and Eucalypt Wood Monitored by FTIR. *Maderas Cienc. Tecnol.* **2013**, *15*, 245–258. [CrossRef]
34. Cheng, S.; Huang, A.; Wang, S.; Zhang, Q. Effect of Different Heat Treatment Temperatures on the Chemical Composition and Structure of Chinese Fir Wood. *BioResources* **2016**, *11*, 4006–4016. [CrossRef]

35. Zouari, M.; Marrot, L.; DeVallance, D.B. Effect of Demineralization and Ball Milling Treatments on the Properties of Arundo Donax and Olive Stone-Derived Biochar. In *International Journal of Environmental Science and Technology*; Springer: Berlin/Heidelberg, Germany, 2023. [CrossRef]
36. Chaudemanche, S.; Perrot, A.; Pimbert, S.; Lecompte, T.; Faure, F. Properties of an Industrial Extruded HDPE-WPC: The Effect of the Size Distribution of Wood Flour Particles. *Constr. Build. Mater.* **2018**, *162*, 543–552. [CrossRef]
37. Ramos-Carmona, S.; Pérez, J.F.; Pelaez-Samaniego, M.R.; Barrera, R.; Garcia-Perez, M. Effect of Torrefaction Temperature on Properties of Patula Pine. *Maderas Cienc. Tecnol.* **2017**, *19*, 39–50. [CrossRef]
38. Devallance, D.B.; Oporto, G.S.; Quigley, P. Investigation of Hardwood Biochar as a Replacement for Wood Flour in Wood–Polypropylene Composites. *J. Elastomers Plast.* **2016**, *48*, 510–522. [CrossRef]
39. Chen, W.-H.; Lin, B.-J.; Colin, B.; Pétrissans, A.; Pétrissans, M. A Study of Hygroscopic Property of Biomass Pretreated by Torrefaction. *Energy Procedia* **2019**, *158*, 32–36. [CrossRef]
40. Ortiz-Barajas, D.L.; Arévalo-Prada, J.A.; Fenollar, O.; Rueda-Ordóñez, Y.J.; Torres-Giner, S. Torrefaction of Coffee Husk Flour for the Development of Injection-Molded Green Composite Pieces of Polylactide with High Sustainability. *Appl. Sci.* **2020**, *10*, 6468. [CrossRef]
41. Chiou, B.-S.; Valenzuela-Medina, D.; Wechsler, M.; Bilbao-Sainz, C.; Klamczynski, A.K.; Williams, T.G.; Wood, D.F.; Glenn, G.M.; Orts, W.J. Torrefied Biomass-Polypropylene Composites. *J. Appl. Polym. Sci.* **2015**, *132*. [CrossRef]

polymers

Review

# Silver Bionanocomposites as Active Food Packaging: Recent Advances & Future Trends Tackling the Food Waste Crisis

**Federico Trotta [1,*], Sidonio Da Silva [1], Alessio Massironi [1], Seyedeh Fatemeh Mirpoor [2], Stella Lignou [2], Sameer Khalil Ghawi [2] and Dimitris Charalampopoulos [2]**

[1] Metalchemy Limited., 71-75 Shelton Street, London WC2H 9JQ, UK; sids@metalchemy.tech (S.D.S.); am@metalchemy.tech (A.M.)
[2] Department of Food and Nutritional Sciences, University of Reading, P.O. Box 226, Whiteknights, Reading RG6 6AP, UK; s.lignou@reading.ac.uk (S.L.); s.khalilghawi@reading.ac.uk (S.K.G.); d.charalampopoulos@reading.ac.uk (D.C.)
* Correspondence: ft@metalchemy.tech

**Abstract:** Food waste is a pressing global challenge leading to over $1 trillion lost annually and contributing up to 10% of global greenhouse gas emissions. Extensive study has been directed toward the use of active biodegradable packaging materials to improve food quality, minimize plastic use, and encourage sustainable packaging technology development. However, this has been achieved with limited success, which can mainly be attributed to poor material properties and high production costs. In the recent literature, the integration of silver nanoparticles (AgNPs) has shown to improve the properties of biopolymer, prompting the development of bionanocomposites. Furthermore, the antibacterial properties of AgNPs against foodborne pathogens leads towards food shelf-life improvement and provides a route towards reducing food waste. However, few reviews have analyzed AgNPs holistically throughout a portfolio of biopolymers from an industrial perspective. Hence, this review critically analyses the antibacterial, barrier, mechanical, thermal, and water resistance properties of AgNP-based bionanocomposites. These advanced materials are also discussed in terms of food packaging applications and assessed in terms of their performance in enhancing food shelf-life. Finally, the current barriers towards the commercialization of AgNP bionanocomposites are critically discussed to provide an industrial action plan towards the development of sustainable packaging materials to reduce food waste.

**Keywords:** bionanocomposites; food packaging; silver nanoparticles; bioplastics; colloidal silver; green chemistry; biopolymers; food waste; AgNPs

## 1. Introduction

Food packaging is a critical component of food technology that deals with the protection and preservation of diverse food products [1]. It has been reported that food packaging represented a global market size of £303 Billion in 2021 with a compound annual growth rate (CAGR) of 5.5% until 2030, and formed about 69% of the overall consumer packaging market [2]. Petrochemical plastics have achieved widespread success as packaging materials in the sector, owning 99% of the market share due to optimal properties such as oxygen barrier capabilities, high tensile, and tear strength. Other characteristics, such as a high Water Vapor Transmission Rate and biodegradability, is less prevalent in packaging, with biodegradability found in only 0.64% of all plastic materials. All these properties guard against external degradation agents and prevent the internal loss of nutrition in food products, assuring food quality at every level of the supply chain, from producers to end users (Figure 1).

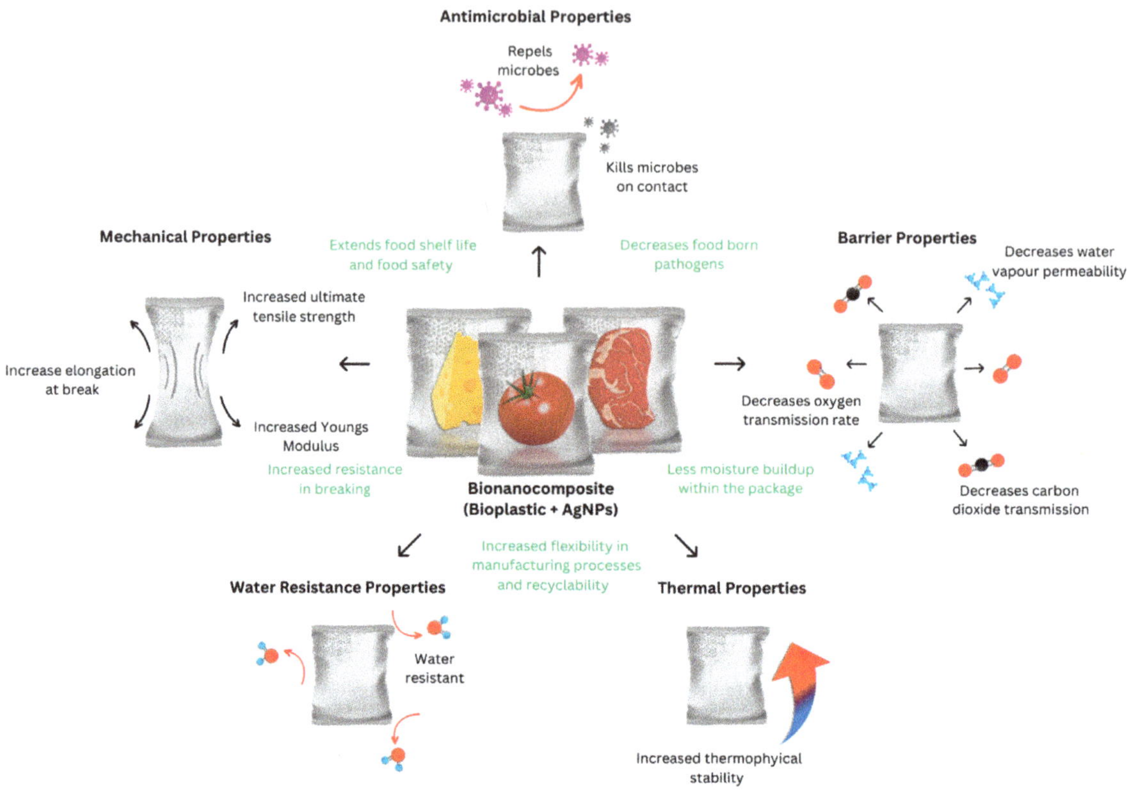


**Figure 1.** Visual abstract outlining the enhanced properties of bionanocomposites. Green text summarizes the key benefits of the bionanocomposites for food packaging applications.

As shown in Figure 2, synthetic plastic polymers such as polypropylene (21%), polyethylene (18%), polyvinyl chloride (17%), high-density polyethylene (15%), and polyethylene terephthalate dominate the food packaging market globally [1]. Between 1950 and 2015, an estimated 7.8 billion tons of plastics were created worldwide, with approximately 4.6 billion tons ending up in landfill or being wasted [3]. Polyethylene, the most manufactured and discarded synthetic polymer globally, is the major generator of two greenhouse gases—methane and ethylene [4]. Methane emissions contribute to climate change and can harm aquatic life by changing the oxygen levels and pH of water, whereas ethylene emissions can be hazardous to plants and animals and have an impact on crops and biodiversity, and there is some evidence that it may play a role in cancer development [4].

Plastics account for 10% of the global oil output, with single-use plastics accounting for more than one third of all plastics produced in 2017 [5]. Several environmentally hazardous disposal methods, such as incineration and landfill, are currently being employed to deal with the overflow of plastics [4]. As a result, sustainable, safe, and non-toxic food packaging options are highly desirable to ensure a transition to more environmentally friendly packaging materials in the food industry.

Bioplastics represent an innovative category of plastics derived from natural sources, such as chitosan, agar, alginate, and polylactic acid (PLA), among others. They are positioned as an environmentally friendly alternative to non-biodegradable synthetic plastics due to their reduced reliance on fossil fuels, faster biodegradability, and lower carbon footprint [6]. Bioplastics are made of biopolymers and biodegradable reinforcing agents [7]. The ability of bioplastics to return to the ecosystem, either through the natural breakdown of organic waste by microorganisms or composting, rather than accumulating in landfills, is an important differentiating factor compared to non-biodegradable synthetic plastics.

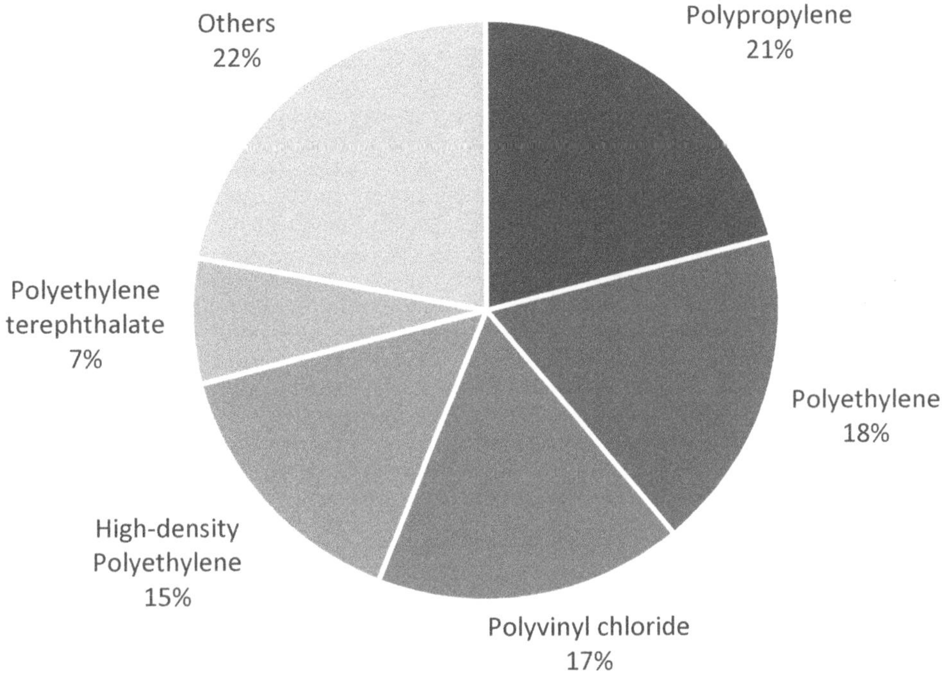


**Figure 2.** Global market distribution of synthetic plastics [1].

Even though bioplastics provide an alternative to synthetic packaging, they are regularly combined with petrochemical plastics. This is due to their weak mechanical qualities and moisture sensitivity, which are listed as contributing causes to their restricted utilization in food packaging [6]. To overcome the challenges around bioplastics, extensive research has been conducted to embed nanomaterials in food packaging materials, leading to the development of active packaging materials. Several metallic nanoparticles, most notably silver, aluminum, and zinc, have been shown to improve qualities such as tensile strength, Water Vapor Permeability, and biocidal activity [8]. Moreover, silver nanoparticles (AgNPs) have been found to have a strong antibacterial effect against foodborne pathogens such as bacteria, parasites, and viruses [9].

The antimicrobial activity of AgNPs in food packaging can help tackle two major global challenges:

- Food and beverage waste: Excess food production used to compensate for waste could be used to help feed the 811 million people worldwide experiencing chronic undernourishment [10];
- Foodborne infections: 550 million cases and 230,000 deaths worldwide each year could be avoided by providing sustainable and effective food packaging technology [9].

These challenges are aligned with the United Nations Sustainable Development Goal 12 to reduce food waste along production and supply chains in order to promote a more sustainable economic model [11]. Food waste is typically generated by food products which have a short shelf-life; studies have shown a wide spectrum of microorganisms being responsible for food deterioration [9], increasing the challenge to finding a one-stop solution to prevent highly nutritious foods from degrading as quickly, especially in warmer climates. As a result, $1.2 trillion is wasted globally each year from food and beverage detorioration [10]. It is believed that over 1.3 billion tons, or one third of all food produced for human use, is wasted annually [10]. This accounts for up to 10% of global greenhouse gas emissions [10].

While the main focus of current research has been to investigate the incorporation of AgNPs in bioplastics to improve their physical properties, only a limited amount of studies have looked into how these improvements affect food packaging in a holistic manner.

Due to the excellent performance of silver bionanocomposites in antibacterial activity and their ability to improve the physico-chemical properties of bioplastics, the purpose of this literature review is to critically examine key technological advances that are relevant to food packaging. These properties include the antimicrobial activity, barrier properties (Water Vapor Transmission Rate (WVTR, WVP), and Oxygen Transmission Rate (OTR)), mechanical properties (Ultimate Tensile Strength (UTS) and Elongation at Break (EaB)), thermal properties, and water resistance (WS, CA) of bionanocomposites containing AgNPs. Specifically, we will focus on the combination of AgNPs with agar and/or PLA and their applications as active materials in food packaging, since they represent the most exploited and promising materials for the development of advanced bioplastics. We also explore the latest developments in the antibacterial activities of silver bionanocomposites. In the last sections of this critical review, we discuss the challenges and environmental impact of silver bionanocomposites, aligning our discussion with current and evolving regulatory frameworks. The nature of this fast-paced and advanced field often leads to fragmented and conflicting literature; thus, this literature review serves to consolidate and clarify the current state of knowledge, contributing to the further understanding of silver bionanocomposites. Additionally, we critically analyze three barriers to the commercialization of these materials including a scalability, regulatory, and environmental analysis outlining the key obstacles that need to be overcome for these advanced materials to become widely available in the industry.

## 2. Silver Nanoparticles as an Active Additive

A common approach to enhancing food safety is to embed an active ingredient within a packaging material which not only inhibits microbial growth, but also enhances Water Vapor Permeability to extend the shelf-life [9]. This is achievable through nanomaterials such as AgNPs, which have a higher surface area-to-volume ratio with respect to their bulk counterpart, allowing them to easily interact with and bond to other materials.

Hence, when embedding silver nanoparticles in biopolymers, they interact with:

- Gases, such as oxygen and carbon dioxide, which increase barrier capabilities [12];
- The polymer matrix of the film, forming a network of strong bonds that improve mechanical and barrier properties through weak and covalent interaction, assuring their adhesion within the biopolymers [13];
- Bacteria and other microorganisms, which inhibit their growth [9]. The mechanism of action of AgNPs against bacteria is illustrated in Figure 3;
- UV radiation, reducing the UV penetration through the biopolymer by means of their strong scattering behaviour [14].

It is important to note that the intake of AgNPs into mammalian cells is size dependent, with the aggregation of smaller-sized nanoparticles (<10 nm) causing cytotoxicity to cells [15]. Therefore, packaging manufacturers can counteract the effects of AgNPs in human cells by tuning the size of AgNPs as well as embedding them into bioplastic polymer matrices to form bionanocomposite materials.

The antimicrobial activity of AgNPs, as presented in Figure 3, is influenced by key parameters, including the shape, size, and surface charge of the nanoparticle. Enhanced antimicrobial activity has been demonstrated in spherical and triangular-shaped AgNPs in comparison to cubic, platelet, decahedron, and other shapes, as increasing the surface area increases NPs' reactivity with the microorganism's cell membrane [16,17]. Moreover, previous research has demonstrated high antimicrobial properties for AgNPs sized between 1 and 30 nm. Additionally, the AgNPs' surface charge, conferred by their coating agent, influences their interaction with biological molecules. This includes their uptake by microorganism cells, a crucial aspect governing their antimicrobial mechanism [16].

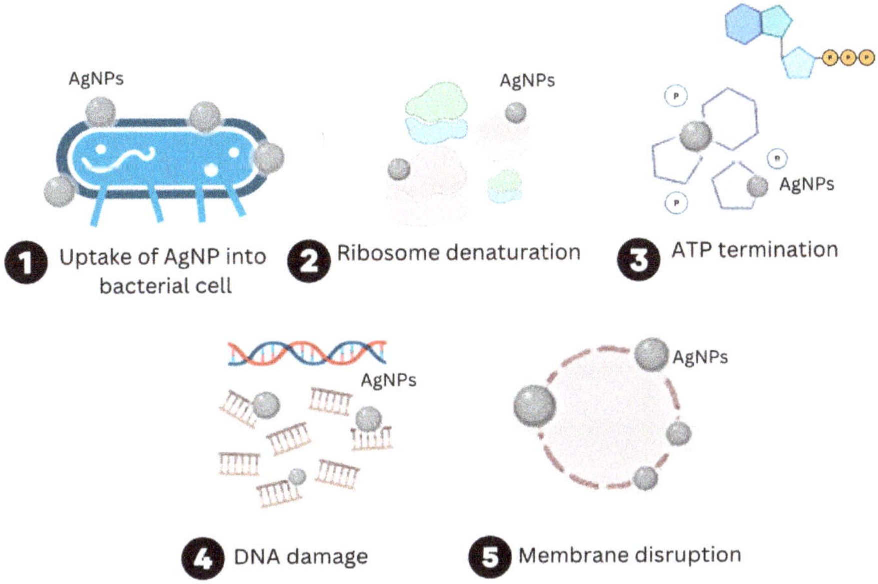


**Figure 3.** Various antimicrobial activity mechanisms of AgNPs. (1) Entry of the AgNPs into the cell membrane of the bacterial cells; (2) ribosome denaturation; (3) ATP termination; (4,5) membrane disruption; DNA damage; (5) rupture of the cell membrane [9].

The synthesis method of AgNPs will influence their final physical properties, therefore impacting their antimicrobial efficacy. AgNP production methods can be divided into three synthesis routes: physical, chemical, and biological, as summarized in Figure 4 [18]. To achieve control over AgNPs' size and morphology, a number of chemical and biological approaches constitute viable options. However, compared to chemical methodologies, biological methods are rapidly becoming the preferred synthesis route due to the absence of hazardous chemicals in their production [18]. Furthermore, biological methods do not require the addition of stabilizing agents during AgNPs' synthesis, since many of the commonly investigated natural compounds can act both as reducing and capping agents. This enables the development of simpler and more cost-effective reaction strategies compared to chemical methods.

Hence, the selection of the AgNPs' synthesis method, their characteristics, and their concentration within the bionanocomposite are critical for understanding how to extend food shelf-life, as AgNPs exert different antimicrobial activities and physical properties depending on the bionanocomposite formulation.

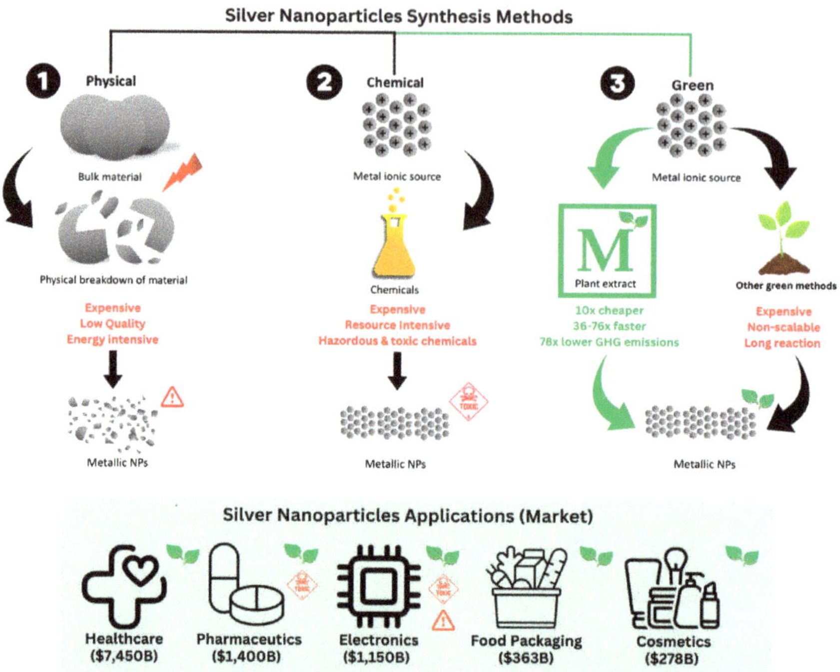


**Figure 4.** Methods for AgNPs synthesis. (1) Physical: In this method, the bulk material, such as silver foil, is broken down utilizing high energy and material resources; however, the produced AgNPs lack uniform size and shape, the most used methods comprise evaporation-condensation and laser ablation; (2) Chemical: This process commonly involves the metal ionic sources being reduced by a reducing agent and stabilized by a capping agent to produce AgNPs of defined size and shape. However, these use toxic chemicals such as sodium borohydride; (3) Biological: These methods either use microbes or plant extracts to carry out bio-reduction of ionic solutions as well as use biological molecules to stabilize the final AgNPs to produce defined shape/size NPs, without the use of hazardous chemicals [18].

## 3. Formulation and Bionanocomposites Manufacturing

Bionanocomposites are composite materials comprising two fundamental components: biopolymers that constitute the bioplastic matrix and embedded nanostructures capable of imparting unique properties which re-enforce the polymeric material. These nanostructures can be organic, encompassing polysaccharides, proteins, or synthetic colloids, or inorganic, including substances such as silica, noble metal oxides, or ceramics. Various bioplastic formulations have been produced in the literature. In particular, polysaccharide-based bionanocomposites are the most investigated due to their unique chemical–physical properties, and relative low cost provided by the abundance of their main sources. Indeed, polysaccharides are naturally abundant, are generally non-toxic and circular, and meet the criteria for packaging production [19]. These biopolymers, unlike lipids which are commonly subjected to peroxidation reactions leading to the loss of their main structure, exhibit a higher thermal stability. However, they are highly susceptible to moisture and have limited mechanical resilience [19]. These include agar [20], chitosan [13], and hydroxypropyl methylcellulose (HPMC) [21] in combination with multiple components such as gum tragacanth/HMPC/beeswaxes (GT/HMPC/BW) [22] and agar/banana powder [23]. Understanding their formulation is critical in order to develop optimal bionanocomposite materials, as their formulation will impact the final material properties as well as our understanding of how AgNPs interact with the polymers themselves.

Physical properties that affect food safety and shelf life must be considered when designing food packaging.

To address these issues, three solutions have been suggested:

1. The addition of different reinforcing chemicals into polysaccharide matrices such as cellulose, lignocellulose, or micro/nanocrystals [24];
2. The combination of different polymers which produce blends or multilayer films, as presented in Table 1 [25];
3. The inclusion of inorganic additives, such as AgNPs [19,25].

Furthermore, plasticizers are commonly used in bioplastics to overcome brittleness and prevent cracking and chipping of the biopolymer during handling and storage. Plasticizers are substances with a low molecular weight and high volatility, such as glycerol or sorbitol. These can reduce intermolecular interactions and boost polymeric chain mobility, resulting in a drop in the material glass transition temperature via protein structural change [26]. It has been demonstrated that adding oleic acid to glycerol improves the mechanical and barrier properties of edible films and coatings [27]. This mixture creates polyglycerol-esters through the esterification of pendant hydroxyl groups with fatty acids. Polyglycerol-esters are commonly employed as additives, modifiers, and emulsifiers in products that contain immiscible food ingredients [28]. The main bioplastic formulations are presented in the table presented in Table 1.

The selection of components is critical to the food packaging quality of the bionanocomposite, where the base formulation of the bioplastic will not only define the starting point of the physical, chemical, and biological properties of the film, but is also critical in understanding the interaction of the AgNPs with the bioplastic formulation itself, which also depend on how the AgNPs are integrated into the biopolymer matrix.

AgNPs can be embedded into bionanocomposites using a variety of techniques, including in situ synthesis [29], solution blending (melt blending and solution casting) [30], and electrospinning [31]. The first method involves the synthesis of AgNPs within the biopolymer itself by adding a silver salt to a solution containing a reducing agent before its addition into a bioplastic material. Melt blending mixes AgNPs with a bioplastic in its molten state. On the other hand, in the solution casting method, a bioplastic is dissolved in a solvent, allowing the nanoparticles to be dispersed in the solution and mixed with the polymer before being cast into sheet materials. Finally, AgNPs can be electrospun to produce nanofibers and then integrated into a biopolymer solution. The benefits and drawbacks of each method are summarized in Table 1.

**Table 1.** Summary of bionanocomposites manufacturing processes, outlining main benefits and drawbacks.

| Bionanocomposite Manufacturing Method | Benefits | Drawbacks |
|---|---|---|
| In situ intercalative polymerization [32] | Improved mechanical, thermal and barrier properties | Only works for low-viscosity polymers. |
| | Easy to automate | Expensive equipment required. |
| | Cost-effective materials required | Difficulties in controlling the polymerization, leading to variations in the properties of the bionanocomposite. |
| Solution intercalation [32] | Homogeneous dispersion. | Environmental concerns around solvent use. |
| | Increased interlayer spacing of the nanofiller, allowing for greater polymer penetration and adhesion to the filler surface. | Time-consuming processing time. |
| | Efficient processing. | Nanofiller compatibility issues, could lead to poor dispersion and suboptimal properties of the bionanocomposite. |

**Table 1.** *Cont.*

| Bionanocomposite Manufacturing Method | Benefits | Drawbacks |
|---|---|---|
| Template synthesis [32] | Precise control over size and shape. | Template removal is challenging and can damage or alter the properties of the final product. |
| | Tailored properties. | Sensitivity to reaction conditions lead to low optimised reactions and limited the range of materials. |
| | High yield. | High cost starting materials. |
| Melt intercalation [32] | Improved thermal stability of bionanocomposite. | Limited control over intercalation, can result in variations of final product properties. |
| | Improved barrier properties. | Reduced mechanical strength of end-product. |
| | Reduced cost compared to other methods. | |

Overall, the choice of embedding technique will depend on the specific packaging material applications and the desired properties of the final product as well as the economic viability of such a process. For example, thermal extrusion methods will be the preferred choice for thermoplastic polymers such as Poly (Lactic Acid) (PLA) bionanocomposite production but not for agar-cellulose materials for which in-situ synthesis and mixing are preferred due to the high water affinity of AgNPs once stabilised by hydrophilic capping agents.

A fundamental aspect to assure an increase in bionanocomposites activities is represented by the AgNPs' uniform distribution within the biocomposites, in order to maintain an homogeneous action over the bionanocomposites' surface and avoid AgNP aggregations, which can cause a lack of activities. To achieve homogeneity and uniform dispersion, manufacturers employ different techniques during the processing of the material:

- The use of dispersant or surfactant [33];
- Mechanical mixing during the extrusion process for thermoplastic polymers [34];
- Sonication to break down agglomerates and ensure proper dispersion [35];
- The fine tuning of processing conditions, such as temperature, pressure, and mixing time to optimize AgNP dispersion.

Finally, the AgNPs' high stability at different conditions, in particular with regard to pH and temperature, is a required feature when they are embedded within bionanocomposites used in food packaging materials to ensure food safety, regulatory compliance, consumer acceptance, and the long-term performance of the packaging. AgNP aggregates can lead to changes in material performance, compromising the integrity and functionality of the packaging [36].

## 4. Characterization of Bionanocomposites

### *4.1. Antimicrobial Activity of Ag Bionanocomposite Materials*

AgNPs have been proven to be broadly efficient against common food bacteria such as *Escherichia coli*, *Staphylococcus aureus*, and *Pseusomonas aeruginosa*. The average minimum inhibitory concentration (MIC) of AgNPs for *E. coli* was found to be 10.85 μg/mL [30]. Similarly, AgNPs' MIC of S. aureus was found to be 14.39 μg/mL [30]. Finally, the average AgNP MIC of *P. aerugionsa* was reported to be 6.41 μg/mL [37]. This body of evidence implies that S. aureus is the most vulnerable of these widespread bacterial species to AgNPs, and have the potential to cause serious food poisoning, especially in milk and cheeses [9]. AgNPs have also shown antimicrobial efficacy against a variety of fungi and viruses often found in food [38], including *Cladosporium*, *Aspergillus*, and *Norovirus*, which are prevalent in carbohydrate-based foods, chilled meat, leafy greens, and shellfish, respectively.

A thorough review of bionanocomposite using AgNPs and their antimicrobial activity is presented in Table 2. AgNPs have a strong bactericidal rate on common food pathogens such as *S. aureus*, *E. coli*, *Listeria monocytogenes*, and *Salmonella typhimurium*. To further optimize the use of AgNPs, the stronger the antibacterial effect of the AgNPs, the lower the concentration of AgNPs that is required to achieve the same result [30].

**Table 2.** Use of silver nanoparticles in food packaging biomaterials. It has been demonstrated that bionanocomposite materials can help prevent the development of common foodborne pathogens such as *S. aureus*, *E. coli*, *S. typhimurium*, and *L. monocytogenes* at low concentrations.

| Polymer Used | Size of AgNPs (nm) | Concentration | Food Tested | Strain Tested | Bactericidal Rate (%) | Ref. |
|---|---|---|---|---|---|---|
| Chitosan | 8.05 | 1% w/v | Data not reported or not investigated | *S. aureus* | 20.5% | [5] |
| | | | | *E. coli* | 23.7% | |
| Polylactic acid | 4.5 | 1% wt/wt | Strawberries | *S. aureus* | 97.04% | [32] |
| | | | | *E. coli* | 95.85% | |
| | 2.7 | 1% wt/wt AgNPs | Fresh apple and apple juice | *L. monocytogenes* | 100% | [33] |
| | | | | *S. typhymurium* | 100% | |
| Alginate | 20 | 0.8% | Strawberries | *S. aureus* | 91.26% | [34] |
| | | | | *E. coli* | 92.01% | |

The literature has generally outlined that smaller particle sizes increase the surface area-to-volume ratio, therefore having a higher antimicrobial activity [16]. This is shown by comparing the results of Argudín et al. [39] and Elgorban et al. [40], for which the same AgNP concentration of 1% (wt/wt) was integrated into PLA [41,42]. The results revealed that a smaller particle size of 2.7 nm had a 100% bactericidal rate on *L. monocytogenes* and *S. typhymurium* [39]. For particle sizes of 4.5 nm, the bactericidal rates were reported to be 97.04% and 95.85% for *S. aureus* and *E. coli* [39]. However, it must be noted that the difference reported is still small and the tested bacterial strains were different.

Moreover, a study by Scialabba [10] applied a concentration of 1% w/v of AgNPs to chitosan, employing a particle size of 8.05 nm. The bactericidal rate on *S.aureus* and *E.coli* was 20.5% and 23.7%, respectively [10]. Argudín et al. [39] used a particle size of 4.5 nm in PLA which demonstrated a 76.54% increase in antibacterial effectiveness compared to a particle size of 8.05 nm in chitosan [10] for *S. aureus*, as shown in Table 2. However, the difference in bionanocomposites and processing methods differed and these may play a significant role in AgNPs' antibacterial activity.

The antimicrobial efficiency of AgNP bionanocomposites and in general of bionanocomposites is commonly investigated through different methodologies such as:

- Zone of Inhibition (Agar Diffusion Method): A bioplastic film or disc is placed on an agar seeded with microorganisms and the antimicrobial activity is indicated by the zone of inhibition [37];
- Direct contact test (ASTM E2149 [43]): This test aims to evaluate the antimicrobial activity of the bionanocomposites in direct contact with a suspension of microorganisms. After a selected contact time, the microorganisms are recovered and counted to determine the reduction in viability [44];
- Suspension test (ASTM E2180): Specifically used for hydrophobic bionanocomposites, this method is designed to evaluate the antimicrobial activity of incorporating an active agent in the bionanocomposite, such as ASTM E2149. The bionanocomposite is soaked in a microorganism suspension and the activity is calculated as the percentage of reduced vitality [45].

Microorganisms are the primary culprits behind food spoilage, as they accelerate degradation reactions, alter pH levels, and produce toxins that lead to changes in taste, texture, and appearance, rendering the food unpalatable. Hence, reducing their presence in food products brings several benefits to producers, such as a higher flexibility and time management of food products, as well as to consumers, with better food quality overall and less food spoilage, leading to reduced waste and improving the products' shelf-life and possible storage periods [46,47].

### 4.2. Enhanced Physical Properties of Bionanocomposite Materials

AgNPs can also improve the physical and functional properties of bioplastic materials, resulting in improved food protection and preservation [41]. The addition of AgNPs into the bioplastic matrix showed improved barrier permeability, mechanical properties, thermophysical stability, and water resistance [41,42]. This is critical since the primary function of food packaging is to extend the food product's shelf-life by avoiding unfavorable changes caused by microbial breakdown, chemical pollutants, temperature change, air, moisture, light, and external factors [21]. The impact of AgNPs on the physical properties of bioplastic material is critically examined in the following sections.

#### 4.2.1. Membrane Barrier Properties

The barrier properties of packaging contribute to food shelf-life. For example, oxygen scavenging and $CO_2$ emission from the pack are moisture-dependent, which is governed in part by the packaging barrier capability. As previously established in the literature, the efficiency of these barrier properties is predominantly contingent upon the primary polymer constituents of the bioplastic materials. The choice of polymers dictates the permeability of the packaging to gases and moisture, directly impacting food's quality and safety during storage. However, recent advancements in food packaging technology have introduced a transformative factor: the incorporation of AgNPs. AgNPs have garnered significant attention due to their unique properties, which include antimicrobial and barrier-enhancing effects. When AgNPs are integrated into the bioplastic matrix, they interact with the polymer chains, reinforcing the barrier capabilities of the packaging material [47]. AgNPs have shown promising activity in reducing the impact of external factors, particularly the adverse effects of moisture and oxygen on packed food items. This enhancement translates to a tangible reduction in food waste, as the shelf life of products is extended. Additionally, it mitigates potential health risks for consumers, as it minimises the chances of microbial growth and oxidation within the packaging.

There are several methods to characterise the barrier properties of packaging films, and the most common are: Water Vapour Transmission Rate (WVTR), Water Vapor Permeability (WVP), and Oxygen Transmission Rate (OTR).

The Water Vapor Transmission Rate (WVTR) is a measurement that quantifies the rate at which water vapour or moisture permeate through a material or barrier over a specific period of time. The WVTR depends highly on the packaging film, thickness, resin composition, and the polymers mixed as well as their physio-chemical properties such as the hydrophilic/hydrophobic degree. This parameter can be improved with the integration of AgNPs by reducing the porosity and increasing the hydrophobicity of the film [48]. Food texture, nutritional, and flavour profiles are all affected by water vapour levels, which have an impact on food quality and safety. When the moisture content of food changes, the rates of lipid oxidation, microbial development, and browning fluctuate [47].

Bahrami noticed a slight but considerable reduction in moisture content with an 8% AgNP concentration, decreasing from 0.2859% to 0.2757% during the study of TG-HPMC-BW films [29]. The increased crosslinking in the biopolymer network caused by the electrostatic contact between the nanoparticles and the OH-groups of glucosyl was responsible for the diffusion of water molecules in the film samples [29]. The moisture content presence was similarly reduced in the agar/banana powder film when AgNPs were integrated, with both blending compositions experiencing a 2% moisture reduction [23].

However, in particular situations, AgNPs may inhibit the intermolecular hydrogen link between polymers. This was observed in the chitosan–starch blend, where the nanoparticles promoted water vapour adsorption at the hydrophilic regions of polysaccharide molecules, and eventually moisture penetration, increasing the WVTR by 24.4% at a 3.79% AgNP concentration [47].

AgNPs can influence the WVTR by means of different physical, mechanical, and chemical interactions with both water molecules and the biocomposites. Indeed, besides the described phenomenon, AgNPs can form tortuous paths for water vapour molecules, making it more difficult for them to pass through the biocomposite and thus reducing the WVTR. Moreover, the inherent hydrophobicity of silver core reduces the biocomposite polar behaviour, thus reducing its interaction with water molecules.

The WVRT of bionanocomposites is commonly quantified through the gravimetric method (ASTM F1249) [49]).

Water Vapor Permeability (WVP) determines how effectively a packaging material can control the moisture transfer between packaged food and its surrounding environment, and its regulation is fundamental to preserve food quality. Its value is measured by means of standard procedures [50].

The incorporation of AgNPs positively affects the WVP, resulting in a decrease in the WVP in different compositions. For example, the WVP of the TG-HPMC-BW film fell significantly to approximately half of its original value [51]. This was explained by the presence of AgNPs in the matrix, which prevents the biopolymer chains from moving, resulting in a reduced WVP [12]. Similar behaviour was observed in J. Rhim et al. [20], in which agar, starch, and pectin-based films obtained a 25% reduction with 2% AgNPs. Moreover, Ortega et al. [44] reported a 45% WVP reduction at 28 ppm, and Shankar et al. obtained a 9% reduction at 100 ppm [42]. Although similar results were expected in the agar/banana film, the WVP was increased by 43% through the addition of AgNPs to the sample containing the highest banana powder ratio. This could be due to the reduced compatibility of banana powder and AgNPs, as well as the larger size of the AgNPs detected during SEM imaging [23]. Moura et al. [21] investigated and confirmed this trend while testing different AgNP particle sizes, with 100 nm AgNPs having a 15.8% higher WVP than 41 nm AgNPs in the same cellulose-based composite. Therefore, this shows how smaller-sized AgNPs with larger surface areas enhance the WVP, independently of the concentration of AgNPs employed.

As a result, the addition of AgNPs can improve the composite material's WVP, with the highest improvement of 53% being observed in the TG-HPMC-BW film, delivering superior results for food preservation through moisture retention [51]. However, the best performing WVP composition of $1.6 \times 10^{-10}$g m$^{-1}$ s$^{-1}$ Pa$^{-1}$ was a starch-based bionanocomposite at 28.6 ppm AgNPs [52].

The Oxygen Transmission Rate (OTR) refers to the measurement of the amount of oxygen that passes through a specific material over a given period. It is a crucial parameter in packaging industries, especially for products that are sensitive to oxygen exposure, such as food. The OTR is typically expressed in cc/m$^2$/day and indicates the material's barrier properties against oxygen. Measuring the OTR of a film involves oxygen permeation analyzers, which can create a controlled environment. One side of the material is exposed to a high concentration of oxygen, while the other side is exposed to a vacuum or a controlled low oxygen concentration. By monitoring the oxygen permeating through the material, the analyzer calculates the OTR. This measurement is crucial in selecting appropriate packaging materials to ensure the freshness and quality of food products.

Some food shelf-life metrics, such as off-odour, aerobic plate count, and colour parameters, can be affected by the OTR in packaging [53]. It was determined that moderate OTR values (2000–7000 mL $O_2$ m$^{-2}$ per 24 h) resulted in the greatest packaging performance. This range of values is commonly seen in synthetic plastics, but bioplastics exhibit a range in the hundreds [54]. The OTR is commonly measured at room temperature at a controlled relative humidity by means of standard procedure [55].

Similarly, to the WVP, the OTR of AgNPs containing bionanocomposite demonstrated decreased transmission rates when compared to the original film. As the concentration of

AgNPs increased from 0% to 20% in a chitosan–starch-based film, the OTR decreased from 2.39 mL $O_2$ $m^{-2}$ per 24 h to 1.48 mL $O_2$ $m^{-2}$ per 24 h [19]. Dairi et al. [56] identified this behaviour on a cellulose acetate bionanocomposite, for which the inclusion of 5% AgNP organoclay resulted in a 13.6% drop in the OTR [56]. This behaviour is based on an increase in the diffusion route length, which causes gas molecules to flow more slowly through the matrix [55,56].

It can be concluded that the integration of AgNPs into packaging materials offers a versatile means to enhance barrier properties by reducing moisture and oxygen permeation [56]. However, the effects of AgNPs can vary depending on the polymer matrix, characterizing the importance of careful consideration when implementing AgNP-based strategies for food packaging.

#### 4.2.2. Mechanical Properties

The mechanical properties of bionanocomposites, such as tensile strength (TS) and elongation at break (EaB), are generally improved, and are substantially dependent on the AgNP concentration [19,53–55]. The improved mechanical capabilities of bionanocomposites can be attributed to their high stiffness and aspect ratio, as well as their strong affinity via contact between the polymer matrix and disseminated nanoparticles. The mechanical properties of nanoparticles in biopolymer food packaging are critical to control, as they allow the food packaging material to resist breakdown and mechanical disruption during manufacturing, transit, and storage [57]. Mechanical characterizations of bionanocomposites offer essential information regarding the composite's suitability for its intended applications. The characterization procedures adhere to standardized protocols as outlined by ASTM guidelines [58].

Ultimate Tensile Strength (UTS) measures the maximum stress a material can withstand without breaking. In the food packaging industry, UTS is a crucial technique, as it determines the strength and durability of packaging materials, ensuring they can withstand various handling and transportation conditions, thus maintaining the integrity of packaged products.

Figure 5 summarises two opposite UTS behaviours of the bionanocomposites of PLA and agar polymer matrices, which are dependent on the AgNP concentration. It is important to understand that these two bionanocomposites are used for different packaging applications, as PLA is a hard plastic and the agar-based matrix is flexible plastic. As a hard plastic, PLA will present, by default, a very high UTS, to which the AgNPs' integration presents a decreasing effect of its structural strength, getting almost 15% lower strength at a 1% AgNP concentration [29]. This is not an isolated case, as it has also been reported that the addition of AgNPs to TG-HPMC-BW reduced the original material's UTS to nearly half its value [51].

The UTS of the bionanocomposite can be improved through empirical experimentation to determine the optimal AgNP concentration, as observed in agar-based bionanocomposite materials. Although agar-based biocomposites' initial strength is lower compared to hard plastic materials, the addition of AgNPs (1% concentration) increases the UTS by 8% [20]. This improvement is also seen for chitosan–starch-based films, where the UTS was increased to 69.6 Mpa at a 5.2% AgNP concentration [19]. The UTS of the bionanocomposite material is not only dependent on the base biopolymer and the concentration of AgNPs, but also on the size of the nanoparticles. This was demonstrated by Moura et al., [21] who found that 41 nm-sized nanoparticles increased the UTS of HPMC by 13% while 100 nm particles improved the UTS by only 9.8%.

Therefore, when developing a bionanocomposite, careful consideration of the material design is required when selecting the base material, concentration, and size of the nanoparticles, as this will significantly impact their strength.

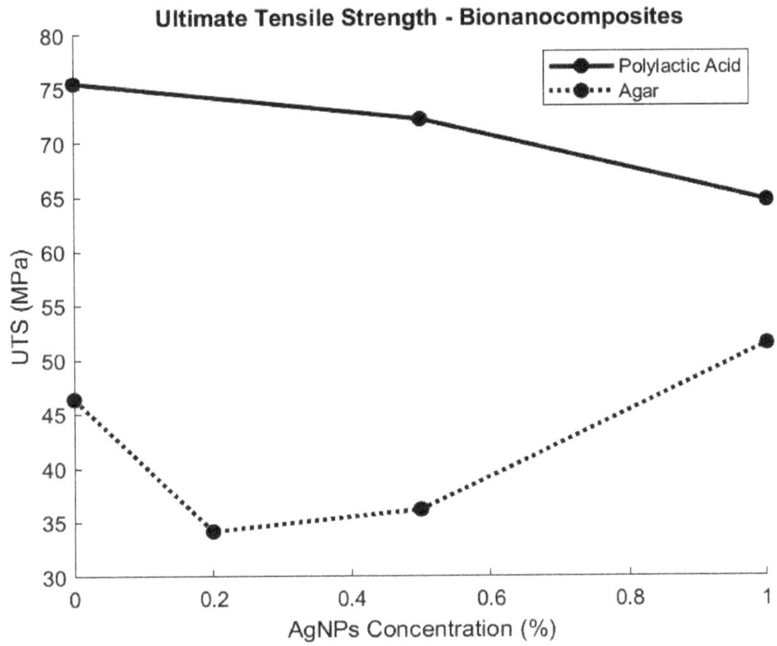


**Figure 5.** Ultimate Tensile Strength (UTS) of PLA and agar bionanocomposites containing AgNPs [29,52].

Another key packaging property is the elongation at break (EaB), the material's ability to stretch before breaking. In food packaging, a high elongation at break is vital as it indicates the flexibility of the packaging material. This property allows packaging to withstand deformations and movements during storage and transportation, preventing tears or ruptures and ensuring the safety and quality of packaged food products.

The EaB property of the bionanocomposite, similar to UTS, is dependent on the base material, concentration, and size of the nanoparticles. For instance, for materials such as PLA and chitosan/gelatin/polyethylene glycol (CH/GE/PEG), when AgNPs are embedded, the EaB can be reduced by 53%, a significant decrease [30,59]. This enhances the formulation of PLA as a hard plastic by increasing its resistance to external forces, as shown in Figure 6. However, flexible plastic films have the opposite outcome, and the EaB can be increased up to 79% with AgNPs in the case of starch biopolymer films [60,61].

In many cases, the addition of AgNPs results in a trade-off between the UTS and EaB [59]. This is visible in HPMC films with different nanoparticle sizes, as both 100 nm and 41 nm nanoparticles decrease the EaB but increase the UTS of the film [51].

Therefore, the specific mechanical requirements will have to be considered when developing the bionanocomposite, and priorities will have to be defined to account for UTS and EaB trade-offs, as illustrated in Table 2.

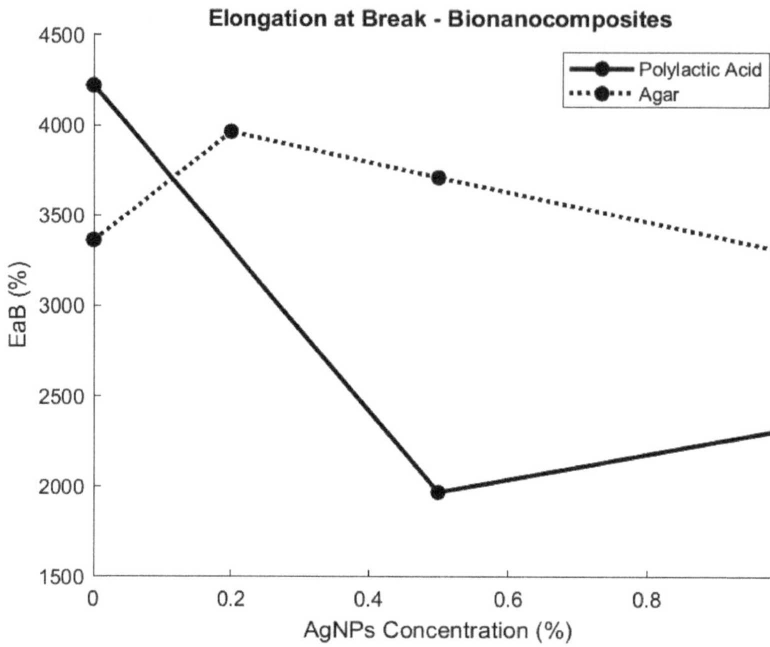


**Figure 6.** Elongation at Break (EaB) variation in bionanocomposites with AgNP concentration. EaB is a measure of a material's ability to stretch or deform before breaking, for which hard plastics, such as PLA, typically have a low elongation at break, indicating that they are relatively inflexible and brittle. Flexible plastics such as agar-based composites typically have a higher elongation at break, indicating higher elasticity before breaking [60,61].

#### 4.2.3. Thermal Properties

The thermal properties of bionanocomposites are critical, for example in the industrial thermoforming process for the production of packaging materials. Mixing biopolymers with AgNPs can enhance the thermal and dimensional stability of biopolymers. The enhanced stiffness and reduced thermal expansion that are characteristic of AgNPs have been identified as factors that enhance the dimensional stability of bionanocomposites containing both polymers and nanofillers. Because they have a greater modulus and a lower thermal expansion coefficient than the polymer matrix, bionanocomposites have been reported to have increased dimensional stability [62].

Youssef et al. [13] investigated chitosan embedded with silver and zinc oxide nanoparticles, with silver nanocomposite demonstrating greater thermal stability and reaching a 10% weight loss difference compared to the control composite. Corn starch-based and agar/banana powder bionanocomposites have also shown improved thermal stability, with a 5% and 7% decrease in weight loss, respectively [23,63]. Previous research has shown improvements in PLA thermal stability, respectively, by 8% at 1% wt/wt AgNPs and 6% at 0.3 wt% AgNPs. This improvement was due to the formation of a strong interfacial interaction between the AgNPs and the PLA matrix [64].

As a result, incorporating AgNPs into bioplastic composites has the potential to improve the material's thermal stability. This is critical for food packaging materials if recycling is a viable option after disposal; high-temperature-resistant materials are more likely to be recycled and reused for packaging, a preferred route to incineration and landfill [65].

#### 4.2.4. Water Resistance Properties

Water resistance is an important characteristic of biodegradable films used in food packaging because, in some situations, the packaging may be exposed to humidity and water during food storage, and the key functions of the packaging may be compromised as a result of the high water activity. Understanding and optimising properties, such as solubility and hydrophobicity, are paramount for ensuring the effectiveness and reliability of biodegradable films in real-world food packaging scenarios.

Food packaging can interact with water in a variety of ways during transit to storage systems such as freezers and fridges. As a result, packaging must be water resistant to a certain extent. Films formed entirely of bioplastics (e.g., chitosan, starch, and sodium alginate) have high solubility rates in water at room temperature, with chitosan and starch having solubility rates of 76% and 21%, respectively [25].

However, in the case of chitosan/gelatin (CH/GE) Ag bionocomposites, the solubility increased by 10% at 0.025% AgNPs concentration, which was explained by the presence of water-soluble phytochemicals such as carbohydrates, alkaloids, and tannins capping AgNPs prepared with Mussaenda Frondosa leaf extract [66]. Alginate films with melanin from watermelon seeds were integrated with zinc oxide and AgNPs to reduce the solubility of the original film. Alginate films had over an 80% solubility, which was decreased to 74% by adding 0.25% melanin. Although zinc oxide nanoparticle integration improved it by 2%, AgNPs had a significant influence, resulting in a nearly 10% drop in water solubility [67]. However, Ortega et al. [44] found that AgNPs significantly augmented the solubility of starch-based films at both 25 °C and 100 °C. Although the bionanocomposite exhibited an 8% lower water solubility at 25 °C, the control film, originally 100% soluble, saw an increase to an average of 45% solubility with just 23 ppm AgNPs at 100 °C [52].

Clear conclusions cannot be drawn due to a lack of consistency in the research and the small volume of research in this area. However, since the increase in solubility with the AgNP concentration was explained by the presence of phytochemicals, the WS might be influenced by the synthesis method and other compounds present in the formulation. Furthermore, the incorporation of AgNPs may result in a more ordered structure in the polymer matrix, preserving the integrity and enhancing the solubility resistance [52].

The Contact Angle (CA) is directly related to the hydrophobicity of the film. A higher CA indicates greater hydrophobicity. When a packaging material is hydrophobic, it repels water-based liquids, preventing moisture from penetrating the package. This is crucial in food packaging, as moisture intrusion can lead to spoilage, mould growth, and reduced food product quality [68]. As a result, a CA target of >90° is set in order to achieve hydrophobic packaging.

Silver nanoparticles have been shown to improve the hydrophobicity of bioplastics in some cases. For example, the CA of agar films was increased by 100% at only a 2% AgNP concentration [23]. This increase in hydrophobicity was also observed in pure chitosan films, which improved the films' CA by 12% [27]. As a result, the film's water resistance was improved.

However, in contrast, blended films of chitosan–starch and agar/banana powder have also shown a lower CA with the addition of AgNPs. This suggests that the biopolymer components within the matrix interacted to a lesser extent, indicating altered surface properties in these composite films. The nanoparticles lowered the CA in the chitosan starch blend film by 15 o at a 20% AgNP concentration [25], and only by 30 in the agar/banana powder mix film [23]. The increased surface roughness due to the presence of AgNPs at the surface could also explain a higher water CA in Ag bionanocomposites. The increased surface roughness creates more microscopic pockets and uneven terrain on the film's surface, which disrupts water expansion through the film's surface.

Considering the above, AgNPs in blended films are likely to increase the hydrophobicity by minimizing the interactions between the biopolymers and by increasing the surface roughness. Plus, the inclusion of an emulsifier component can further enhance the hydrophobicity of individual films. In summary, the water solubility of AgNP bionanocomposites is influenced by various factors, leading to complex outcomes. AgNPs have shown potential in enhancing hydrophobicity, but the effects are influenced by nanoparticle concentration, film composition, and surface roughness changes. These findings emphasise the need for standardised approaches in studying AgNPs' impact on film properties. Further research is essential to unlock the full potential of AgNPs in efficient and sustainable food packaging materials.

## 5. Bionanocomposites Application as Food Packaging Materials

A variety of foods, including fruits, vegetables, meats, and cheeses, have been tested with bionanocomposites incorporating AgNPs [69]. Silver-based bionanocomposites have been shown to be effective in improving the shelf life of various types of food products, as shown in Table 3.

**Table 3.** Shelf-life extension of food products packaged in bionanocomposites.

| Polymer | Size of AgNPs (nm) | Concentration | Food Tested | Shelf-Life Increase vs. Control | Ref. |
|---|---|---|---|---|---|
| Chitosan | 80 +/− 11 nm | | Meat | 1 Week | [69] |
| | | 5% w/wcs | Pork | 6 days | [70] |
| | | | Litchis fruit | 2 days | [71] |
| Polylactic acid | 2.5–6.5 nm | 1% wt/wt | Strawberries | 4 days | [72] |
| Alginate | 5–40 nm | 0.25, 0.50, and 1.00 μg/mL | Fior di Latte cheese | 5 days | [52] |
| | | 50, 60, 70, and 80 μg/mL | Carrot and pear | Up to 10 days | [73] |

Highly wasted fruits and vegetables, such as strawberries and carrots, have shown that their shelf life was extended by up to 4 days and 10 days, respectively, compared to untreated samples [72,73]. Another study on the effects of AgNP bionanocomposites on meat, a high-cost and environmentally impactful food product, showed that the storage time could be extended by up to 7 days compared to control [74]. These examples demonstrate the potential of AgNP bionanocomposites to improve the shelf-life of perishable food products and reduce food waste, as premature expiry throughout the food supply chain is a leading cause of waste. Moreover, this application could reduce or eliminate the number of preservatives commonly added directly into the bulk of food, reducing the cost of food and increasing its nutritional quality [75].

## 6. Barriers to Commercial Rollout

### *6.1. Limited Scalability of Bionanocomposites*

The scalability of bionanocomposites depends on various factors such as the manufacturing method, the properties of the virgin material, and the intended application. However, scalability is often a significant challenge for most bionanocomposite production techniques, as summarized in Table 1.

One scalability barrier is based on the complexity of some of these processes. For example, solution intercalation and in situ polymerization methods often require precise control over the reaction conditions and may involve multiple steps. Changes to the conditions during polymerization can lead to variations in the properties of the bionanocomposite [31,76].

Scalability also depends on the availability and cost of the raw materials, such as the AgNPs and the biopolymer feedstock. For example, producing large quantities of high-quality AgNPs is expensive and time-consuming. Plus, biopolymers are often more expensive than synthetic polymers, since they are typically derived from natural sources and require specific processing and purification steps, which can add to the production costs [77].

Additionally, these processes have challenging optimisation processes. For instance, template synthesis can be sensitive to reaction conditions such as temperature, pH, and solvent choice. The reaction temperature significantly affects the rate of the chemical reactions involved. Higher or lower temperatures can speed up or slow down the process, impacting the final structure's size, shape, and uniformity. Different reactions have optimal pH ranges where they occur most efficiently. Deviating from this range can lead to incomplete or unwanted reactions, affecting the template synthesis process. Some reactions occur only in specific solvents. The solvent also influences the solubility of reactants and products, affecting the reaction equilibrium and, consequently, the final structure.

In summary, the success of template synthesis relies on maintaining several precise conditions. Any deviation from the optimal conditions might result in undesired structures or incomplete reactions. This can make it difficult to optimise reactions and may limit the range of materials that can be synthesised.

In summary, while some bionanocomposite production methods may be more scalable than others, scalability remains a challenge for most techniques and requires the careful consideration of various factors such as the synthesis method, starting materials, and intended end-use.

### 6.2. Safety & Regulation of Bionanocomposites

As research into the application of nanotechnology in the food sector advances, so does the potential of nanotechnology in the food industry, and hence human exposure to these materials [62]. While multiple studies have indicated that consumers are more ready to accept the use of nanotechnology in food packaging than in food products [62], there is still concern about AgNP migration into food, which poses a risk to consumers' health above threshold levels [78].

Silver is present in traces of everyday foods, and adults are estimated to eat between 20–80 μg per day [79]. Hence, the migration of silver from the packaging to the food could be of concern as it could further increase this dietary exposure, meaning that the exact silver migration must be quantified [79]. Hence, regulations are in place by governing bodies such as the European Food Safety Authority (EFSA) and the United States food and drug association (USFDA) to regulate the use of AgNPs in active packaging materials [6].

AgNPs should not exceed 0.05 mg/L in water and 0.05 mg/kg in food, according to the EFSA [6]. This means that analysing the migration profile of silver is crucial. It helps ensure the effectiveness of its antibacterial properties and ensures compliance with current regulations [6]. In 2011, the EFSA published a report requiring producers to undergo in vitro genotoxicity, absorption, distribution, metabolism, and excretion testing [6]. Similarly, the USFDA issued a paper with advice for manufacturers of food additives and food contact chemicals [6]. According to the USFDA, companies should conduct research and give a hazard profile for each packaging material containing nanomaterials. Moreover, Canada has no limitations on nanoparticles as additives, with many other countries having little to no food-contact material regulations [7].

The absence of strict regulations on nanoparticles in food-contact materials fosters an environment that is conducive to innovation. This freedom from stringent limitations allows researchers and industries to explore novel applications of nanotechnology in food packaging and other sectors. Innovations in materials science, such as advanced coatings and barrier technologies enabled by nanoparticles, hold the potential to revolutionise food packaging, enhancing the shelf-life, freshness, and overall product quality.

However, this innovation comes with the responsibility of ensuring consumer safety. While the flexibility in regulations promotes creativity and experimentation, it also emphasises the need for robust research on the safety and environmental impact of these emerging technologies. Collaborative efforts between industries, regulatory bodies, and scientific communities can strike a balance between encouraging innovation and safeguarding public health, paving the way for groundbreaking advancements in the field of nanotechnology while ensuring consumer well-being.

Concerning the use of silver nanoparticles in food-contact materials, the EFSA completed a risk assessment in 2021 confirming that the use of AgNPs in polymeric matrices is safe at concentrations of up to 0.025% w/w, corresponding to a total migration of Ag ions that is less than 50 μg/silver per kg of food [61]. As demonstrated by the EFSA study, when 0.025% AgNPs were integrated into polymers employed as food-contact materials, the Ag migration was recorded as 6 μg/kg of food, which is far below the threshold of the 50 μg/kg food-specific migration limit [61].

A study conducted by Echegoyen and Nerín concluded that the migration of Ag from food packaging (testing low-density PE and polypropylene with AgNPs) was increased by 1.43-fold and 50-fold when protecting acidic foods and if the packaging was microwaved, respectively [80]. This indicates that the migration rate is influenced by the type of food packaged and the heating conditions of the packaging. However, this particular study's results revealed that the migration levels detected were below the threshold set by the EFSA [81]. Moreover, Cushen et al. [82] concluded that the percentage of nanofiller incorporated in the polymer film (testing polyethylene with AgNPs on chicken breast) accelerated Ag migration in comparison to other parameters such as the nanoparticle size, storage time, or temperature conditions.

Most studies to date have been carried out on synthetic nanocomposites which are by definition highly hydrophobic, enabling low migration levels of AgNPs. However, bioplastics tend to have a higher water absorption capacity, which can make them more prone to AgNP migration [83,84]. Hence, further research on the leaching of AgNPs from bionanocomposites is therefore required.

It is evident that many factors are in play in the migration of AgNP towards food products. Hence, each material containing AgNPs must be evaluated independently with a specific food product stimulant to fully understand the migration risks associated with these materials in order to meet regulatory requirements.

### 6.3. Environmental Considerations of Bionanocomposites

Around 40% of food packaging materials are made from plastic, with nearly 99% of these plastics being synthetic. Of the remaining 1%, 64% are non-biodegradable while the rest are biodegradable. The use of biodegradable plastics is often seen as a viable alternative to synthetic plastics, but many of these options only break down at high temperatures or when treated in specific industrial composting conditions. While their decomposition is faster than synthetic plastics, their end-of-life environmental impact is still significant due to the production of methane gas from composting, which is 25 times more potent than $CO_2$, according to the US Environmental Protection Agency [84].

To evaluate the environmental impact of bioplastics and synthetic plastics, life-cycle assessment (LCA) is a tool used to determine the overall material impact at each stage of their life cycle [85]. This process considers factors such as global warming, human toxicity, abiotic depletion, eutrophication, and acidification, as well as Land-Use Change (LUC)-related emissions, which are important factors to consider when land is converted for composting or biofuel feedstock production [85].

Studies have shown that the use of bioplastics can significantly reduce carbon dioxide emissions compared to petroleum-derived plastics, in the case of PLA by 50–70% [86]. However, the disposal methods currently available, such as incineration or landfilling, are not ideal and the bioplastic's emissions are significantly dependent on the manufacturing process, for which some are resource-intensive [87]. For example, a study completed by Qinqin Xia indicated that polyhydroxyalkanoates (PHAs) bioplastics could have a reducing effect on global warming potential ($-8 \times 10^{-5}$ kg $CO_2$e per $cm^3$/MPa) by almost 2× that of PET LCA, depending on the study's current manufacturing process [88]. Moreover, in some studies, other categories of the LCA are significantly higher when comparing bioplastics with petroleum plastics. For example, the ozone depletion potential of PLA can be as high as $30 \times 10^{-12}$ kg CFCe$^{-11}$ per $cm^3$/MPa, while polypropilene (PP) is almost half this value [89]. This difference depends mainly on the selected biopolymer. For example, polybutylene adipate terephthalate (PBAT)'s ozone depletion value is similar to PP, but polybutylene succinate (PBS) has a three-times higher impact.

Moreover, the integration of nanoparticles in bioplastics adds to their environmental impact, with the upstream production of bulk silver being the dominant factor in nearly every environmental impact category. When AgNPs are applied to bioplastics, the overall environmental burdens are highly sensitive to the synthesis route of the AgNPs. For instance, when applied in wound dressing, the AgNPs could contribute to 66–88% of the global warming impact category depending on the synthesis process [90]. Although the bio-based chemical reduction route was found to have improvements in ozone depletion potential and ecotoxicity, chemical reduction methods are still mainstream and further work into developing green synthesis routes of AgNPs are still required at the commercial scale. A schematic visualization of the life cycle of Ag bionanocomposites is depicted in Figure 7.

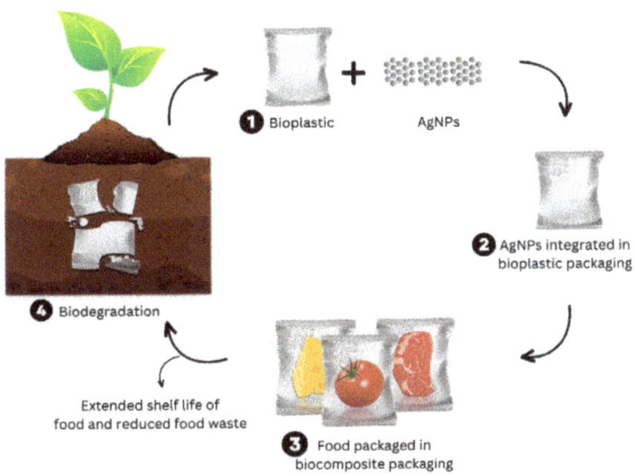


**Figure 7.** Life cycle of Ag Bionanocomposites. (1–2) Bionanocomposites are materials made by combining natural biopolymers and AgNPs. (3) Bionanocomposites are used as food packaging to extend food shelf-life and improve food safety. (4) Bionanocomposites are then broken down through microbial action in soil, reducing packaging waste compared to synthetic alternatives as well as serving as feedstock for further biomass growth leading towards biomaterial circularity.

Therefore, the replacement of synthetic plastics with bioplastics is challenging, and trade-offs must be considered. The combination of AgNPs and bioplastics with current commercial and large-scale manufacturing presents a significant environmental barrier to a successful and meaningful replacement of synthetic options [91], prompting further research to better understand the full life-cycle impact of these materials.

#### 6.3.1. Bioplastic Fate in the Environment

The accumulation of petrochemical plastic waste in landfills and marine environments poses significant challenges for both living organisms and ecosystems. Additionally, microplastics resulting from the degradation of plastic products have been discovered in unexpected locations, including soil, oceans, seas, drinking water, and even in arctic and mountain regions [92]. This alarming trend, coupled with the escalating levels of $CO_2$ emissions in the atmosphere, has compelled the scientific community to explore novel alternatives to traditional plastics, such as biodegradable bioplastics and advanced bionanocomposites.

Bioplastics' biodegradation can occur under specific conditions depending on their environment and the physical-chemical nature of the bioplastic. In this section, it is important to highlight the difference between the degradation and biodegradation of (bio)plastic materials [93]. Degradation refers to the fragmentation of the polymer chain due to factors like heat, moisture, or enzymes, leading to a loss of polymer structure and the formation of subspecies that may differ from the original monomers [94]. On the other hand, biodegradation involves the complete breakdown of bioplastics into compounds, primarily $CO_2$, water, nitrogen ($N_2$), or hydrogen ($H_2$), carried out by living species such as microorganisms, algae, or more complex organisms [95]. Unlike the residues of conventional plastics, the products of bioplastic degradation should not be toxic and should serve as a source of nutrients for other living organisms [95]. Besides the specific environments where the biodegradation takes place, the bioplastic polymer structure plays a key role during this process. Factors such as surface charge, hydrophilicity, molecular weight, crystallinity, and mechanical and thermal properties significantly influence degradation, particularly when microorganisms are involved [96].

For instance, polylactic acid (PLA), one of the most extensively studied bioplastics, biodegrades under specific conditions of humidity (>60% moisture) and temperature (>50 °C), and in the presence of specific bacteria [97]. The degradation process involves the ester bonds of lactic acid units being hydrolysed. The presence of a semi-crystalline structure in PLA increases its melting temperature, affecting its biodegradability, which predominantly occurs in its amorphous regions [98]. Researchers have explored ways to manage these parameters by combining the PLA with other polymers, such as polysaccharides (e.g., starch, agar, cellulose, and chitosan) or other biopolymers like PHA and collagen [64].

Polymer biodegradation can be tuned and improved by the incorporation of active molecules. In a study conducted by Ramos and colleagues, PLA active bionanocomposite films incorporating a flavor (vanillin) and citrate-stabilised AgNPs (at a concentration of 0.01 mg $kg^{-1}$ of total dry weight) obtained through chemical methods as additives were investigated [99]. A thermal analysis using differential scanning calorimetry (DSC) and thermogravimetric (TG) tests revealed that the introduction of flavour and AgNPs resulted in a decrease in both the glass transition temperature (Tg) and the melting temperature (Tm) of PLA. This decrease in thermal stability demonstrates the potential for improving the degradation process of PLA composite films by altering the polymer's physical–chemical properties [99]. The inclusion of AgNPs into biocomposites besides the antimicrobial and thermal improvement of polymer matrix has shown promise in successfully accelerating the degradation rate of PLA-based active nanocomposite films, demonstrating promising future uses for the improvement of biocomposites' biodegradation rate.

#### 6.3.2. AgNPs Fate in the Environment

The degradation of silver bionanocomposites involves the individual degradation processes of each component. After discussing polymer matrix degradation, it is important to focus on AgNPs embedded in the composites. AgNPs do not undergo biodegradation processes, and their fate and environmental transport are influenced by several factors. Their mobility in different environments is linked to water chemistry, including the pH of the suspension and the AgNPs' physical–chemical features such as size, shape, and the presence of capping agents [100,101]. They may exist as isolated particles in suspension, or they may aggregate, especially in environments with a high ionic strength [102], undergoing partial oxidation, leading to the release of Ag+ ions [103]. Furthermore, AgNPs may react with natural substances present in soils and waters such as sulphide, chloride, and others, altering their original properties [104]. AgNPs can also adsorb charged species in the environment through electrostatic interactions which could lead to their aggregation and the loss of colloidal status [105]. The behaviour of AgNPs is thus significantly influenced by their surface properties and the surrounding environment, including capping agents, methods of synthesis, electrolyte composition, solution ionic strength, pH, and the presence of natural organic matter (NOM) [106].

The importance of synthesis methods in AgNPs' environmental impact is well-supported by life cycle assessment (LCA) investigations, which have demonstrated that physical methods tend to impose higher environmental impacts when compared to chemical and biological synthesis routes, while also reducing costs and energy requirements [107,108].

However, the complexity of this system makes it challenging to definitively anticipate the fate of AgNPs, given the multitude of influencing factors, and in particular, considering the wide number of methods of synthesis and capping agents. Furthermore, the existing literature on this topic does not always provide a consistent picture regarding their environmental impact, fate, and potential toxicity. Researchers continue to seek a definitive answer regarding whether AgNP toxicity arises from the nanoparticles themselves, Ag+ ions, or chemicals involved during their synthesis. In comparison to the well-established field of bioplastic degradation, the study of AgNPs' fate in the environment is relatively new and still requires further data to better understand the key factors influencing their behaviour.

#### 6.3.3. AgNPs Environmental Considerations

Notwithstanding the complexity of predicting their fate in the environment, the presence of AgNPs, in particular in soil, appears to offer advantages for seed growth and germination in both laboratory- and field-condition experiments, as they function as growth enhancers, exerting positive effects on various plants [102]. For instance, the use of 20 nm citrate-capped silver nanoparticles obtained through biological methods has been reported to significantly improve bean growth and yield, especially in adverse climatic conditions. Prazak et al. demonstrated a strong beneficial germination effect at various tested concentrations (0.25 and 1.25 mg/mL), coupled with efficient antimicrobial activity [109]. Additional studies have demonstrated the beneficial effects of biosynthesized AgNPs on *Triticum aestivum*, commonly known as wheat. These studies have shown improvements in shoot length, fresh and dry shoot weight, chlorophyll levels, carbohydrate content, and protein content [110]. Similar outcomes have been obtained on different plants such as *Pisum sativum* (common pea) and *Phaseolus vulgaris* (common bean) with AgNPs synthetised through biological methods and stabilised by natural capping agents [111,112].

Conversely, other studies have demonstrated the toxicity of AgNPs on various plant species and organisms [102]. In a work by Yin, it has been demonstrated that AgNPs stabilised by Arabic gum synthetized by means of NaBH4 reduction displayed a negative effect when tested at 40 mg/L against Lolium multiflorum. The available data on the toxicity of AgNPs in soil are relatively few and mainly obtained for the AgNP contents of 102 –105 μg/kg soil, which are particularly higher if compared to silver bionanocomposite formulations [113,114]. Numerous studies in the literature consistently highlight a common observation: AgNPs' synthesis using NaBH4 and/or stabilised by non-natural capping agents consistently exhibits inhibitory effects on various plant species due to the toxicity of the synthetic chemical method [115]. This phenomenon persists across diverse colloidal characteristics, including shape, size, and concentration, resulting in a notable reduction in plant growth [105,116–118].

The method of synthesis does not only influence the AgNPs' fate during their degradation but also their possible toxicity against organisms. While chemical and physical methods required toxic reducing and stabilising agents, biological "green" procedures only involve the use of natural and eco-friendly stabilising and reducing agents [77], which confer higher stability and unique biocompatibility key aspects during the bionanocomposites' biodegradation process [108]. Their enhanced stability avoids particle aggregation and the loss of colloidal behaviour responsible for their possible toxicity in the environment, contributing to AgNPs' positive effects on plants and lower toxicity [119,120]. Moreover, non-biodegradable stabilizing agents commonly exploited through chemical and physical methods may enhance the AgNPs' environmental toxicity, supporting the necessity of the development of green methods for metallic nanoparticles [121].

#### 6.3.4. Silver Bionanocomposites Degradation

To comprehend the biodegradability of silver bionanocomposites, it is imperative to investigate the individual behaviours of both components, as well as their combined behaviour within the composite material. The biodegradation of silver bionanocomposites is commonly assessed using soil burial tests, similar to those employed for plastic and bioplastic materials, as defined by ISO 20200:2015 [122]. Briefly, the bioplastic is buried in the soil to simulate its deterioration, with the soil's moisture level maintained by periodic watering. To assess (bio)degradation, samples are periodically taken out (every 15 days) over a 90-day period, and the final dry weight is recorded to determine the weight loss.

The presence of AgNPs, as discussed in this review, can potentially modify the physical–chemical properties of the polymer and influence its degradability due to interactions within the polymeric matrix. When incorporated into alginate films, the introduction of AgNPs leads to a slight decrease in bioplastic biodegradation behaviour, resulting in a 10% reduction in weight loss after 90 days compared to native alginate films [123]. Similar outcomes have been observed with other bioplastics, where the interaction of AgNPs with polysaccharides enhances material properties while slightly reducing the biodegradation rate. In a study conducted by Ediyilyam and co-workers, the reduced biodegradation rate was attributed to the release of AgNPs following biopolymer degradation, which led to the natural reduction in growth of bacterial strains present in the soil [124]. Specifically, their research on chitosan-based bioplastics reinforced with environmentally synthesised 20 nm AgNPs observed a 5% reduction in the biodegradation rate compared to pure chitosan films after a 7-day period. This phenomenon has been observed not only with polysaccharides but also with other biocompatible polymers such as PVA [125], PVA combined with banana peel [126], and various polyesters [127].

Despite the inherent challenges associated with this emerging field, the initial findings concerning their biodegradation show significant promise and opportunity. The integration of AgNPs results in only a slight reduction in biodegradability while conferring unique properties that can only be achieved through the use of the noble metal colloids.

## 7. Silver Bionanocomposites Discussion of Open Challenges & Opportunities

When silver nanoparticles are integrated into bioplastics for food packaging applications, these properties, together with a substantial antibacterial action against common foodborne pathogens such as *E. coli*, *S. Aureus*, and *P. aeruginosa*, result in an enhanced food shelf-life. Furthermore, the tunability of bioplastic films by changing the composition by inserting different additives is an appealing concept for food-packaging producers, who may tune the material to guarantee multiple end-uses. Additionally, there is an opportunity to explore novel additives and compositions that can further enhance the performance of these bioplastic films, potentially extending their applications to novel areas beyond food preservations such as cosmetics, medical devices, and pharmaceuticals.

It should be noted that the physicochemical properties and dosage of nanomaterials in foods and food packaging dictate their ultimate fate and safety. The safe use of nanotechnology in food packaging involves rigorous characterization through standard testing, as previously discussed, to quantify the nanoparticle migration and guarantee that it does not exceed the regulatory limits. Moreover, several challenges and considerations accompany the utilisation of silver nanoparticles in food packaging. These challenges encompass scalability issues, environmental impact concerns, cost-effectiveness, and low yields in the production process. Addressing these challenges is essential to ensure the scalability, affordability, and sustainability of incorporating silver bionanocomposites into the food packaging industry. There is an opportunity for research and development efforts to focus on overcoming these challenges and enhancing the viability of silver nanoparticles in food packaging. This could involve innovations in production methods, energy efficiency, and novel cost-effective manufacturing techniques. These multifaceted challenges require dedicated funding for research and development to facilitate the broader and faster adoption of these advanced materials to tackle global challenges such as food waste. Environmental issues are also an important consideration. This includes potential impacts on ecosystems and concerns about the long-term environmental fate of these materials, as reported in Section 6 of this review. Safeguarding the environment in the use of nanotechnology in food packaging involves rigorous characterization through standard testing to assess nanoparticle migration and guarantee that it does not exceed regulatory limits. As the industry advances, there is a growing focus on mitigating the environmental footprint and embracing sustainable processes and practices. One promising avenue is the commercialisation of biological synthesis methods for silver nanoparticles. These approaches offer the potential to reduce the environmental impact associated with traditional production methods. Biological synthesis not only provides a greener and more eco-friendly alternative but also aligns with the broader trend of sustainable and responsible nanotechnology. Further silver technologies which demonstrated good antimicrobial activities and low toxicities are Silver-Based Coordination Polymers (Ag-MOF) (136). In Ag-MOFs, organic ligands encapsulate the metal centre, enabling a controlled, gradual release of metallic species as natural cations, allowing a controlled release of Ag+ ions. However, despite their promising results in terms of stability and antibacterial actions, their integration within bioplastics for food packaging has still not been investigated and could represent a novel alternative to obtaining advanced materials (136).

By actively researching and implementing environmentally friendly approaches through green synthesis and novel technologies, the industry can work towards minimising its ecological footprint and fostering sustainable practices in the development and widespread commercialisation of nanotechnology in food packaging. This not only addresses existing environmental concerns but also positions the industry as a leader in eco-friendly and novel food preservation technologies, which can lead to significant market advantages and consumer trust in products that prioritise environmental responsibility as well as a positive economic impact. Collaborations with environmental organizations can help to verify and communicate the industry's commitment to eco-friendly practices, fostering a positive public image and promoting industry-wide sustainability initiatives.

## 8. Conclusions and Future Prospects

In this review, it has been reported that the use of silver nanoparticles can improve the antibacterial, mechanical, physical, water-barrier, and thermal properties of bioplastics. As a result, they are becoming increasingly attractive additives to bioplastic formulations in order to generate an ideal food packaging material which has similar properties to synthetic plastics with the added benefits of being biodegradable and non-toxic.

The incorporation of silver nanoparticles increased the ultimate tensile strength by 10–15% across a variety of bioplastics [29–32], improved the thermal stability through higher melting temperatures [62–64], and reduced the Water Vapor Permeability [41,42]. These qualities combine to give the material greater strength and resistance to breaking, and less moisture build-up. The practical application of silver bionanocomposites as food packaging materials has revealed that the shelf-life of meat can be increased by up to 7 days [74]. The shelf-life of food has been demonstrated to increase with the concentration of silver nanoparticles and to differ amongst bioplastic kinds and food products tested. As a result, the ideal concentration of silver nanoparticles can be tuned, providing the necessary increase in food shelf-life while also adhering to safety and regulatory legislation. Moreover, the development of active bioplastics, made from natural resources and grafted with silver nanoparticles, will contribute to reducing the problem of plastic pollution. On the other hand, the addition of silver nanoparticles to the matrix of the film improves multiple bioplastics properties, and, in particular, their antibacterial activity, which makes these active bioplastics a suitable candidate for active food packaging that contributes to less food waste. These bioplastics are a sustainable replacement for highly polluting petroleum-based plastics.

In the future, research should place a strong emphasis on further characterising silver bionanocomposites and gaining a deeper understanding of how these nanoparticles affect key material properties. This research should aim to establish standardised reporting procedures, creating a robust foundation for the development of an entirely new industry centred around these advanced materials.

This forward-looking approach offers food and packaging manufacturers, and, more broadly, fast-moving consumer goods companies (FMCG), a unique opportunity to refine their material properties, with the ultimate objective of achieving optimal product shelf-lives, ultimately fighting the food waste pandemic.

**Author Contributions:** F.T.: Conceptualization, Funding Acquisition, Methodology, Project administration, Resources, Supervision, Writing—review and editing; S.D.S.: Analysis, Data Curation, Visualisation, Writing—original draft preparation, review and editing; A.M.: Writing—formal analysis, Methodology, original draft preparation, review and editing; S.F.M.: Writing—review and editing; S.L.: Writing—review and editing; S.K.G.: Writing—review and editing; D.C.: Conceptualization, Funding Acquisition, Methodology, Project Administration, Resources, Supervision, Writing—review and editing. All authors have read and agreed to the published version of the manuscript.

**Funding:** This research was funded by Innovate UK, Grant Project No 10075083.

**Institutional Review Board Statement:** Not applicable.

**Data Availability Statement:** Data will be made available on request.

**Acknowledgments:** The authors express their gratitude to Precious Jolugbo, Alna Dony, and Vajiha Naveed for their valuable contributions in writing and editing this paper. Their expertise and dedication significantly enhanced the quality of this research. Authors are thankful for funding this work by a grant of Better Food for All, funded by Innovate UK (10075083).

**Conflicts of Interest:** The funders had no role in the design of the study; in the collection, analyses, or interpretation of data; in the writing of the manuscript; or in the decision to publish the results. Federico Trotta, Alessio Massironi and Sidonio Da Silva are employees of Metalchemy Limited.

## Nomenclature

| | |
|---|---|
| AgNP(s) | Silver nanoparticle(s) |
| CA | Contact angle |
| CH/GE | Chitosan/Gelatin |
| CH/GE/PEG | Chitosan/Gelatin/Polyethylene glycol |
| DSC | Differential scanning calorimetry |
| EaB | Elongation at break |
| EFSA | European Food Safety Authority |
| GT/HMPC/BW | Gum tragacanth/HMPC/Beeswaxes |
| HPMC | Hydroxypropyl Methylcellulose |
| LCA | Life cycle assessment |
| LDPE | Low-density polyethylene |
| LUC | Land Use Change |
| MIC | Minimum Inhibitory Concentration |
| NOM | Natural organic matter |
| OTR | Oxygen transmission rate |
| PBAT | Polybutylene adipate terephthalate |
| PBS | Polybutylene succinate |
| PHAs | Polyhydroxyalkanoates |
| PLA | Poly-lactic acid |
| PP | Polypropilene |
| TG-HPMC-BW | Tragacanth/Hydroxypropyl methylcellulose/Beeswax |
| TG | Thermogravimetric |
| Tg | Glass transition temperature |
| Tm | melting temperature |
| TS | Tensile Strength |
| USFDA | United states food and drug association |
| UTS | Ultimate tensile strength |
| WS | Water solubility |
| WVP | Water vapour permeability |
| WVTR | Water transmission rate |

## Appendix A

**Table A1.** Main bioplastic formulations, with short description, advantages and disadvantages of each component.

| Bioplastic | Components | Description | Advantages | Disadvantages | Ref. |
|---|---|---|---|---|---|
| TG/HPMC/BW | Gum tragacanth (TG) | Used as emulsifier, thickener, stabiliser and texturant additive. | High water binding ability, efficient suspending action and effective surface-activity | Poor film preparation | [22] |
| | Hydroxypropyl Methylcellulose (HPMC) | Improve film-forming property | Abundant availability, biodegradability, thermal stability, process ability and excellent film-forming capability | Poor moisture barrier | |
| | Beeswaxes (BW) | Improve moisture barrier and flexibility | Hydrophobicity and firmly packed crystalline structure | | |
| | Oleic Acid | Emulsifier | | | |
| | Glycerol | Plasticizer | Increases flexibility (Elongation at break), low molecular weight non-volatile substance, hydrophilic | Lower moisture retention, decreases UTS, Elasticity modulus, lower oxygen water vapour permeability | |
| Agar/banana powder | Agar | | Good film-forming, abundance, renewability and biocompatability | Low mechanical, water resistance properties | [23] |
| | Banana Powder | To improve water barrier, UV Screening effect, antioxidant and antimicrobial activity. | Potential to increase hydrophobicity, functional properties to secure food safety and extend its life. Good film-forming. Can help reduce metal ions to nanoparticles. | | |
| | Glycerol | Gelling, stabilisers, emulsifiers, and thickening agents | Temperature stability, firmness directly proportional to agar concentration, good clarity, low adhesiveness and inert. Abundant, cheap. Low hydroscopic property | Low mechanical properties | |
| Agar | Agar | Gelling, stabilisers, emulsifiers, and thickening agents | Temperature stability, firmness directly proportional to agar concentration, good clarity, low adhesiveness and inert. Abundant, cheap. Low hydroscopic property | Low mechanical properties | [20] |
| | Glycerol | Plasticizer | | | |

**Table 1.** *Cont.*

| Bioplastic | Components | Description | Advantages | Disadvantages | Ref. |
|---|---|---|---|---|---|
| HPMC | HPMC | | Good film-forming, good for film or coating. Odorless, water-soluble, and tasteless. Moderate moisture and oxygen permeability. | Low mechanical and barrier properties | [21] |
| Chitosan–Starch | Starch (Rice and Waxy Rice) | Abundant, non-toxic, renewable source, suitable for film-formation. Cheap, easily biodegradable | Sensitive to Moisture, poor mechanical properties | | [25] |
| | Chitosan | To improve mechanical, water vapor barrier, antimicrobial attributes, reduce cost and biodegradability of starch composite | Abundant, non-toxic, renewable source, suitable for film-formation. Antimicrobial activity | | |
| | Acetic Acid | To dissolve Chitosan flakes | | | |
| Chitosan/Gelatin | Chitosan | To improve mechanical properties of Gelatin | Biocompatible, biodegradable, antimicrobial. Better mechanical and gas barrier properties | | [82] |
| | Gelatin | Emulsifier | Gel formation, texturizing, thickening. Good film forming. Absorption of UV light | Low thermal strength, elasticity, mechanical properties | |
| | Acetic Acid | To dissolve Chitosan flakes | | | |
| Chitosan | Chitosan | | Antimicrobial activity, edible, excellent oxygen and carbon dioxide barrier | Only acid soluble | [13] |
| | Formic or Acetic acid | To dissolve Chitosan | | | |
| | Poly Ethylene Glycol (PEG) | | | | |
| | Hydroxyl Ethyl Cellulose | | Water soluble, non-ionic, high compatibility | | |

**Table 2.** Mechanical properties of bionanocomposites.

| Bioplastic | Concentration of AgNPs (wt%) | Ultimate Tensile Strength (MPa) | Elongation at Break (%) | Ref. |
|---|---|---|---|---|
| TG/HPMC/BW | 0% | 75.5 | 42.2 | [30] |
| | 0.5% | 72.2 | 19.7 | |
| | 1% | 64.7 | 23.2 | |
| CH/GE/PEG | 0% | 24.47 | 4.48 | [82] |
| | 0.0075% | 25.80 | 4.34 | |
| | 0.0125% | 26.30 | 4.29 | |
| | 0.0250% | 26.40 | 4.12 | |
| Agar | 0% | 46.38 | 33.64 | [52] |
| | 0.2% | 34.17 | 39.66 | |
| | 0.5% | 36.17 | 37.13 | |
| | 1% | 51.49 | 33.02 | |
| | 2% | 53.44 | 36.46 | |
| Sugar palm starch | 0% | 0.010 | 86 | [85] |
| | 1% | 0.015 | 165 | |
| | 2% | 0.020 | 197 | |
| | 3% | 0.410 | 415 | |
| | 4% | 0.300 | 376 | |

## References


1. Jabeen, N.; Majid, I.; Nayik, G.A. Bioplastics and Food Packaging: A Review. *Cogent Food Agric.* **2015**, *1*, 1117749. [CrossRef]
2. Geueke, B.; Groh, K.; Muncke, J. Food packaging in the circular economy: Overview of chemical safety aspects for commonly used materials. *J. Clean. Prod.* **2018**, *193*, 491–505. [CrossRef]
3. Hannah, R.; Max, R. Plastic Pollution. 2018. Available online: https://ourworldindata.org/plastic-pollution (accessed on 1 September 2023).
4. Sarah-Jeanne, R.; Sara, F.; Samuel, T.W.; David, M.K. Production of Methane and Ethylene from Plastic in the Environment. *PLoS ONE* **2017**, *13*, e0200574. [CrossRef]
5. Jefferson, M. Whither Plastics?—Petrochemicals, Plastics and Sustainability in a Garbage-Riddled World. *Energy Res. Soc. Sci.* **2019**, *56*, 101229. [CrossRef]
6. Youssef, A.M.; El-Sayed, S.M. Bionanocomposites Materials for Food Packaging Applications: Concepts and Future Outlook. *Carbohydr. Polym.* **2018**, *193*, 19–27. [CrossRef]
7. Kumar, S.; Thakur, K.S. Bioplastics-Classification, Production and Their Potential Food Applications. *J. Hill Agric.* **2017**, *8*, 118–129. [CrossRef]
8. Nagalakshmaiah, M.; Afrin, S.; Malladi, R.P.; Elkoun, S.; Robert, M.; Ansari, M.A. Chapter 9—Biocomposites: Present trends and challenges for the future. In *Green Composites for Automotive Applications*; Koronis, G., Silva, A., Eds.; Woodhead Publishing: Sawston, UK, 2019; pp. 197–215.
9. Anvar, A.A.; Ahari, H.; Ataee, M. Antimicrobial Properties of Food Nanopackaging: A New Focus on Foodborne Pathogens. *Front. Microbiol.* **2021**, *12*, 690706. [CrossRef]
10. FAO. *Food Wastage Footprint & Climate Change*; FAO: Rome, Italy, 2015; pp. 1–4.
11. Ensure Sustainable Consumption and Production Patterns. Available online: https://www.unep.org/explore-topics/sustainable-development-goals/why-do-sustainable-development-goals-matter/goal-12 (accessed on 1 September 2023).
12. He, Y.; Li, H.; Fei, X.; Peng, L. Carboxymethyl Cellulose/Cellulose Nanocrystals Immobilized Silver Nanoparticles as an Effective Coating to Improve Barrier and Antibacterial Properties of Paper for Food Packaging Applications. *Carbohydr. Polym.* **2021**, *252*, 117156. [CrossRef]
13. Youssef, A.M.; Abou-Yousef, H.; El-Sayed, S.M.; Kamel, S. Mechanical and antibacterial properties of novel high performance chitosan/nanocomposite films. *Int. J. Biol. Macromol.* **2015**, *76*, 25–32. [CrossRef]
14. Wang, X.; Wang, S.; Liu, W.; Wang, S.; Zhang, L.; Sang, R.; Hou, Q.; Li, J. Facile Fabrication of Cellulose Composite Films with Excellent UV Resistance and Antibacterial Activity. *Carbohydr. Polym.* **2019**, *225*, 115213. [CrossRef]
15. Zhang, T.; Wang, L.; Chen, Q.; Chen, C. Cytotoxic potential of silver nanoparticles. *Yonsei Med. J.* **2014**, *55*, 283–291. [CrossRef]

16. Zorraquín-Peña, I.; Cueva, C.; Bartolomé, B.; Moreno-Arribas, M.V. Silver Nanoparticles against Foodborne Bacteria. Effects at Intestinal Level and Health Limitations. *Microorganisms* **2020**, *8*, 132. [CrossRef] [PubMed]
17. Dong, Y.; Zhu, H.; Shen, Y.; Zhang, W.; Zhang, L. Antibacterial Activity of Silver Nanoparticles of Different Particle Size against *Vibrio natriegens*. *PLoS ONE* **2019**, *14*, e0222322. [CrossRef] [PubMed]
18. Massironi, A.; Morelli, A.; Grassi, L.; Puppi, D.; Braccini, S.; Maisetta, G.; Esin, S.; Batoni, G.; Della Pina, C.; Chiellini, F. Ulvan as Novel Reducing and Stabilizing Agent from Renewable Algal Biomass: Application to Green Synthesis of Silver Nanoparticles. *Carbohydr. Polym.* **2019**, *203*, 310–321. [CrossRef] [PubMed]
19. Aleksanyan, K.V. Polysaccharides for Biodegradable Packaging Materials: Past, Present, and Future (Brief Review). *Polymers* **2023**, *15*, 451. [CrossRef]
20. Rhim, J.W.; Wang, L.F.; Hong, S.I. Preparation and Characterization of Agar/Silver Nanoparticles Composite Films with Antimicrobial Activity. *Food Hydrocoll.* **2013**, *33*, 327–335. [CrossRef]
21. De Moura, M.R.; Mattoso, L.H.C.; Zucolotto, V. Development of cellulose-based bactericidal nanocomposites containing silver nanoparticles and their use as active food packaging. *J. Food Eng.* **2012**, *109*, 520–524. [CrossRef]
22. Bahrami, A.; Rezaei Mokarram, R.; Sowti Khiabani, M.; Ghanbarzadeh, B.; Salehi, R. Physico-mechanical and antimicrobial properties of tragacanth/hydroxypropyl methylcellulose/beeswax edible films reinforced with silver nanoparticles. *Int. J. Biol. Macromol.* **2018**, *129*, 1103–1112. [CrossRef]
23. Orsuwan, A.; Shankar, S.; Wang, L.F.; Sothornvit, R.; Rhim, J.W. Preparation of Antimicrobial Agar/Banana Powder Blend Films Reinforced with Silver Nanoparticles. *Food Hydrocoll.* **2016**, *60*, 476–485. [CrossRef]
24. Pratama, J.H.; Amalia, A.; Rohmah, R.L.; Saraswati, T.E. The Extraction of Cellulose Powder of Water Hyacinth (*Eichhornia crassipes*) as Reinforcing Agents in Bioplastic. *AIP Conf. Proc.* **2020**, *22319*, 10000.
25. Yoksan, R.; Chirachanchai, S. Silver Nanoparticle-Loaded Chitosan-Starch Based Films: Fabrication and Evaluation of Tensile, Barrier and Antimicrobial Properties. *Mater. Sci. Eng. C* **2010**, *30*, 891–897. [CrossRef]
26. Tarique, J.; Sapuan, S.M.; Khalina, A. Effect of Glycerol Plasticizer Loading on the Physical, Mechanical, Thermal, and Barrier Properties of Arrowroot (*Maranta Arundinacea*) Starch Biopolymers. *Sci. Rep.* **2021**, *11*, 13900. [CrossRef] [PubMed]
27. Aguirre-Loredo, R.Y.; Rodríguez-Hernández, A.I.; Chavarría-Hernández, N. Physical properties of emulsified films based on chitosan and oleic acid. *CyTA J. Food* **2014**, *12*, 305–312. [CrossRef]
28. Díaz-Lasprilla, A.M.; Mercado, R.A.; Ramírez-Caballero, G.E. Glycerol polymerization degree effect on the emulsifying properties of polyglycerol esters. *J. Appl. Polym. Sci.* **2021**, *138*, 50566. [CrossRef]
29. Bonardd, S.; Saldías, C.; Ramírez, O.; Radić, D.; Recio, F.J.; Urzúa, M.; Leiva, A. A novel environmentally friendly method in solid phase for in situ synthesis of chitosan-gold bionanocomposites with catalytic applications. *Carbohydr. Polym.* **2019**, *207*, 533–541. [CrossRef]
30. Kraśniewska, K.; Galus, S.; Gniewosz, M. Biopolymers-Based Materials Containing Silver Nanoparticles as Active Packaging for Food Applications–A Review. *Int. J. Mol. Sci.* **2020**, *21*, 698. [CrossRef]
31. Cui, J.; Yu, X.; Shen, Y.; Sun, B.; Guo, W.; Liu, M.; Chen, Y.; Wang, L.; Zhou, X.; Shafiq, M.; et al. Electrospinning Inorganic Nanomaterials to Fabricate Bionanocomposites for Soft and Hard Tissue Repair. *Nanomaterials* **2023**, *13*, 204. [CrossRef] [PubMed]
32. Cuce, P.M.; Cuce, E.; Guclu, T. Methods of fabrication of bionanocomposites. In *Bionanocomposites for Food Packaging Applications*; Woodhead Publishing: Sawston, UK, 2022; pp. 7–29. [CrossRef]
33. Díez-Pascual, A.M.; Díez-Vicente, A.L. Poly(3-hydroxybutyrate)/ZnO Bionanocomposites with Improved Mechanical, Barrier and Antibacterial Properties. *Int. J. Mol. Sci.* **2014**, *15*, 10950–10973. [CrossRef] [PubMed]
34. Parra, D.F.; Marchini, L.G.; Komatsu, L.G.H.; de Oliveira, C.B.; Oliani, W.L.; Rangari, V.K. AgNPs@ZnO Hybride Nanoparticles Infused Thermoplastic Polyester Elastomer and Their Biocide Effect. *SN Appl. Sci.* **2021**, *3*, 392. [CrossRef]
35. Zhu, L.; Zhu, W.; Hu, X.; Lin, Y.; Machmudah, S.; Wahyudiono; Kanda, H.; Goto, M. PVP/Highly Dispersed AgNPs Nanofibers Using Ultrasonic-Assisted Electrospinning. *Polymers* **2022**, *14*, 599. [CrossRef]
36. Singh, R.; Dutt, S.; Sharma, P.; Sundramoorthy, A.K.; Dubey, A.; Singh, A.; Arya, S. Future of Nanotechnology in Food Industry: Challenges in Processing, Packaging, and Food Safety. *Glob. Challenges* **2023**, *7*, 2200209. [CrossRef] [PubMed]
37. Thapliyal, A.; Chandra, A. Antibacterial and anticancer potential of silver nanoparticles synthesized using gallic acid in bentonite/starch bio-nanocomposites. *Int. J. Appl. Pharm.* **2018**, *10*, 178–189. [CrossRef]
38. Bhardwaj, S.K.; Bhardwaj, N.; Kumar, V.; Bhatt, D.; Azzouz, A.; Bhaumik, J.; Kim, K.-H.; Deep, A. Recent progress in nanomaterial-based sensing of airborne viral and bacterial pathogens. *Environ. Int.* **2021**, *146*, 106183. [CrossRef]
39. Argudín, M.Á.; Mendoza, M.C.; Rodicio, M.R. Food Poisoning and *Staphylococcus Aureus* Enterotoxins. *Toxins* **2010**, *2*, 1751–1773. [CrossRef] [PubMed]
40. Elgorban, A.M.; El-Samawaty, A.E.-R.M.; Abd-Elkader, O.H.; Yassin, M.A.; Sayed, S.R.; Khan, M.; Adil, S.F. Bioengineered silver nanoparticles using *Curvularia pallescens* and its fungicidal activity against *Cladosporium fulvum*. *Saudi J. Biol. Sci.* **2017**, *24*, 1522–1528. [CrossRef]
41. Sen, M. Nanocomposite Materials. In *Nanotechnology and the Environment*; IntechOpen: London, UK, 2020.
42. Shankar, S.; Rhim, J.W. Polymer Nanocomposites for Food Packaging Applications. In *Functional and Physical Properties of Polymer Nanocomposites*; Wiley: Hoboken, NJ, USA, 2016; pp. 29–55.
43. *ASTM E2149*; Standard Test Method for Determining the Antimicrobial Activity of Antimicrobial Agents Under Dynamic Contact Conditions. ASTM International: West Conshohocken, PA, USA, 2020.

44. Ortega, F.; Giannuzzi, L.; Arce, V.B.; García, M.A. Active composite starch films containing green synthetized silver nanoparticles. *Food Hydrocoll.* **2017**, *70*, 152–162. [CrossRef]
45. *ASTM E2180*; Standard Test Method for Determining the Activity of Incorporated Antimicrobial Agent(s) in Polymeric or Hydrophobic Materials. ASTM International: West Conshohocken, PA, USA, 2018.
46. Emamifar, A.; Kadivar, M.; Shahedi, M.; Soleimanian-Zad, S. Evaluation of Nanocomposite Packaging Containing Ag and ZnO on Shelf Life of Fresh Orange Juice. *Innov. Food Sci. Emerg. Technol.* **2010**, *11*, 742–748. [CrossRef]
47. Mousavi, F.P.; Pour, H.H.; Nasab, A.H.; Rajabalipour, A.A.; Barouni, M. Investigation into Shelf Life of Fresh Dates and Pistachios in a Package Modified with Nano-Silver. *Glob. J. Health Sci.* **2015**, *8*, 134–144. [CrossRef]
48. Water Vapor Transmission Rate Flair Flexible Packaging. Available online: https://www.flairpackaging.com/pages/home/resources/packaging_101/WaterVaporTransmissionRate(WVTR)/1 (accessed on 1 September 2023).
49. *ASTM F1249-13*; Standard Test Method for Water Vapor Transmission Rate Through Plastic Film and Sheeting Using a Modulated Infrared Sensor. ASTM International: West Conshohocken, PA, USA, 2013.
50. *ASTM E96-1*; Standard Test Methods for Water Vapor Transmission of Materials. ASTM International: West Conshohocken, PA, USA, 2014.
51. Erem, A.D.; Ozcan, G.; Erem, H.; Skrifvars, M. Antimicrobial activity of poly(l-lactide acid)/silver nanocomposite fibers. *Text. Res. J.* **2013**, *83*, 2111–2117. [CrossRef]
52. Gammariello, D.; Conte, A.; Buonocore, G.G.; Del Nobile, M.A. Bio-Based Nanocomposite Coating to Preserve Quality of Fior Di Latte Cheese. *J. Dairy Sci.* **2011**, *94*, 5298–5304. [CrossRef]
53. Istiqola, A.; Achmad, S. A Review of Silver Nanoparticles in Food Packaging Technologies: Regulation, Methods, Properties, Migration, and Future Challenges. *J. Chin. Chem. Soc.* **2020**, *67*, 1942–1956. [CrossRef]
54. Olmos, D.; Pontes-Quero, G.M.; Corral, A.; González-Gaitano, G.; González-Benito, J. Preparation and Characterization of Antimicrobial Films Based on LDPE/Ag Nanoparticles with Potential Uses in Food and Health Industries. *Nanomaterials* **2018**, *8*, 60. [CrossRef] [PubMed]
55. *ASTM D3985-17*; Standard Test Method for Oxygen Gas Transmission Rate Through Plastic Film and Sheeting Using a Coulometric Sensor. ASTM International: West Conshohocken, PA, USA, 2017.
56. Dairi, N.; Ferfera-Harrar, H.; Ramos, M.; Garrigós, M.C. Cellulose Acetate/AgNPs-Organoclay and/or Thymol Nano-Biocomposite Films with Combined Antimicrobial/Antioxidant Properties for Active Food Packaging Use. *Int. J. Biol. Macromol.* **2019**, *121*, 508–523. [CrossRef] [PubMed]
57. Motelica, L.; Ficai, D.; Ficai, A.; Oprea, O.C.; Kaya, D.A.; Andronescu, E. Biodegradable Antimicrobial Food Packaging: Trends and Perspectives. *Foods* **2020**, *9*, 1438. [CrossRef] [PubMed]
58. Nourbakhsh, A.; Ashori, A. Wood Plastic Composites from Agro-Waste Materials: Analysis of Mechanical Properties. *Bioresour. Technol.* **2010**, *101*, 2525–2528. [CrossRef]
59. Braga, L.R.; Pérez, L.M.; Soazo, M.d.V.; Machado, F. Evaluation of the Antimicrobial, Antioxidant and Physicochemical Properties of Poly(Vinyl Chloride) Films Containing Quercetin and Silver Nanoparticles. *LWT* **2019**, *101*, 491–498. [CrossRef]
60. Abreu, A.S.; Oliveira, M.; De Sá, A.; Rodrigues, R.M.; Cerqueira, M.A.; Vicente, A.A.; Machado, A.V. Antimicrobial Nanostructured Starch Based Films for Packaging. *Carbohydr. Polym.* **2015**, *129*, 127–134. [CrossRef]
61. European Food Safety Authority. Safety Assessment of the Substance Silver Nanoparticles for Use in Food Contact Materials. Available online: https://www.efsa.europa.eu/en/efsajournal/pub/6790 (accessed on 1 September 2023).
62. Singh, T.; Shukla, S.; Kumar, P.; Wahla, V.; Bajpai, V.K. Application of Nanotechnology in Food Science: Perception and Overview. *Front. Microbiol.* **2017**, *8*, 1501. [CrossRef] [PubMed]
63. Ji, N.; Liu, C.; Zhang, S.; Xiong, L.; Sun, Q. Elaboration and Characterization of Corn Starch Films Incorporating Silver Nanoparticles Obtained Using Short Glucan Chains. *LWT* **2016**, *74*, 311–318. [CrossRef]
64. Lv, S.; Zhang, Y.; Gu, J.; Tan, H. Biodegradation Behavior and Modelling of Soil Burial Effect on Degradation Rate of PLA Blended with Starch and Wood Flour. *Colloids Surf. B Biointerfaces* **2017**, *159*, 800–808. [CrossRef]
65. Chiang, T.-C.; Liu, H.-L.; Tsai, L.-C.; Jiang, T.; Ma, N.; Tsai, F.-C. Improvement of the mechanical property and thermal stability of polypropylene/recycled rubber composite by chemical modification and physical blending. *Sci. Rep.* **2020**, *10*, 2432. [CrossRef]
66. Sreelekha, E.; George, B.; Shyam, A.; Sajina, N.; Mathew, B. A Comparative Study on the Synthesis, Characterization, and Antioxidant Activity of Green and Chemically Synthesized Silver Nanoparticles. *BioNanoScience* **2021**, *11*, 489–496. [CrossRef]
67. Łopusiewicz, Ł.; Macieja, S.; Śliwiński, M.; Bartkowiak, A.; Roy, S.; Sobolewski, P. Alginate Biofunctional Films Modified with Melanin from Watermelon Seeds and Zinc Oxide/Silver Nanoparticles. *Materials* **2022**, *15*, 2381. [CrossRef] [PubMed]
68. Franzolin, M.R.; Lopes, I.S.; Courrol, D.d.S.; Barreto, S.d.S.; Courrol, L.C. Synthesis, Characterization, Antimicrobial Activity, and Toxicity Evaluation of Aminolevulinic Acid-Silver and Silver-Iron Nanoparticles for Potential Applications in Agriculture. *RSC Adv.* **2022**, *12*, 30094–30103. [CrossRef] [PubMed]
69. Ediyilyam, S.; George, B. Chitosan/Gelatin/Silver Nanoparticles Composites Films for Biodegradable Food Packaging Applications. *Polymers* **2021**, *13*, 18. [CrossRef]
70. Pandey, V.K.; Upadhyay, S.N.; Niranjan, K.; Mishra, P.K. Antimicrobial biodegradable chitosan-based composite Nano-layers for food packaging. *Int. J. Biol. Macromol.* **2020**, *157*, 212–219. [CrossRef]

71. Santos, U.P.; Campos, J.F.; Torquato, H.F.V.; Paredes-Gamero, E.J.; Carollo, C.A.; Estevinho, L.M.; De Picoli Souza, K.; Dos Santos, E.L. Antioxidant, Antimicrobial and Cytotoxic Properties as Well as the Phenolic Content of the Extract from *Hancornia Speciosa* Gomes. *PLoS ONE* **2016**, *11*, e0167531. [CrossRef] [PubMed]
72. Wu, Z.; Zhou, W.; Pang, C.; Deng, W.; Xu, C.; Wang, X. Multifunctional chitosan-based coating with liposomes containing laurel essential oils and nanosilver for pork preservation. *Food Chem.* **2019**, *295*, 16–25. [CrossRef] [PubMed]
73. Wu, Z.; Huang, X.; Li, Y.-C.; Xiao, H.; Wang, X. Novel chitosan films with laponite immobilized Ag nanoparticles for active food packaging. *Carbohydr. Polym.* **2018**, *199*, 210–218. [CrossRef]
74. Johnson, D.L.; Wang, Y.; Stealey, S.T.; Alexander, A.K.; Kaltchev, M.G.; Chen, J.; Zhang, W. Biosynthesis of silver nanoparticles using upland cress: Purification, characterisation, and antimicrobial activity. *Micro Nano Lett.* **2020**, *15*, 110–113. [CrossRef]
75. Turan, D.; Gunes, G.; Kilic, A. Perspectives of Bio-Nanocomposites for Food Packaging Applications. In *Bionanocomposites for Packaging Applications*; Springer: Berlin/Heidelberg, Germany, 2017.
76. Pires, J.R.; Rodrigues, C.; Coelhoso, I.; Fernando, A.L.; Souza, V.G. Current Applications of Bionanocomposites in Food Processing and Packaging. *Polymers* **2023**, *15*, 2336. [CrossRef]
77. Massironi, A.; Morelli, A.; Puppi, D.; Chiellini, F. Renewable Polysaccharides Micro/Nanostructures for Food and Cosmetic Applications. *Molecules* **2020**, *25*, 4886. [CrossRef]
78. Jain, A.; Ranjan, S.; Dasgupta, N.; Ramalingam, C. Nanomaterials in food and agriculture: An overview on their safety concerns and regulatory issues. *Crit. Rev. Food Sci. Nutr.* **2018**, *58*, 297–317. [CrossRef]
79. Mohammed Fayaz, A.; Balaji, K.; Girilal, M.; Kalaichelvan, P.T.; Venkatesan, R. Mycobased Synthesis of Silver Nanoparticles and Their Incorporation into Sodium Alginate Films for Vegetable and Fruit Preservation. *J. Agric. Food Chem.* **2009**, *57*, 6246–6252. [CrossRef]
80. Echegoyen, Y.; Nerín, C. Nanoparticle release from nano-silver antimicrobial food containers. *Food Chem. Toxicol.* **2013**, *62*, 16–22. [CrossRef]
81. Carbone, M.; Donia, D.T.; Sabbatella, G.; Antiochia, R. Silver nanoparticles in polymeric matrices for fresh food packaging. *J. King Saud Univ. Sci.* **2016**, *28*, 273–279. [CrossRef]
82. Cushen, M.; Kerry, J.; Morris, M.; Cruz-Romero, M.; Cummins, E. Evaluation and Simulation of Silver and Copper Nanoparticle Migration from Polyethylene Nanocomposites to Food and an Associated Exposure Assessment. *J. Agric. Food Chem.* **2014**, *62*, 1403–1411. [CrossRef]
83. Song, R.; Murphy, M.; Li, C.; Ting, K.; Soo, C.; Zheng, Z. Current development of biodegradable polymeric materials for biomedical applications. *Drug Des. Dev. Ther.* **2018**, *24*, 3117–3145. [CrossRef]
84. Abdul Kareem, T.; Anu Kaliani, A. Synthesis and thermal study of octahedral silver nano-plates in polyvinyl alcohol (PVA). *Arab. J. Chem.* **2011**, *4*, 325–331. [CrossRef]
85. Samanta, P.; Dey, S.; Ghosh, A.R.; Senapati, T. Life Cycle Assessment of Bioplastics. In *Encyclopedia of Green Materials*; Springer Nature: Singapore, 2022; pp. 1–6.
86. Atiwesh, G.; Mikhael, A.; Parrish, C.C.; Banoub, J.; Le, T.-A.T. Environmental impact of bioplastic use: A review. *Heliyon* **2021**, *7*, e07918. [CrossRef]
87. Chidambarampadmavathy, K.; Karthikeyan, O.P.; Heimann, K. Sustainable bio-plastic production through landfill methane recycling. *Renew. Sustain. Energy Rev.* **2017**, *71*, 555–562. [CrossRef]
88. Asunis, F.; De Gioannis, G.; Francini, G.; Lombardi, L.; Muntoni, A.; Polettini, A.; Pomi, R.; Rossi, A.; Spiga, D. Environmental life cycle assessment of polyhydroxyalkanoates production from cheese whey. *Waste Manag.* **2021**, *132*, 31–43. [CrossRef]
89. Vasi, S.; Ceccio, G.; Cannavò, A.; Pleskunov, P.; Vacík, J. Study of Wettability of Polyethylene Membranes for Food Packaging. *Sustainability* **2022**, *14*, 5863. [CrossRef]
90. Kanmani, P.; Rhim, J.-W. Physicochemical properties of gelatin/silver nanoparticle antimicrobial composite films. *Food Chem.* **2014**, *148*, 162–169. [CrossRef]
91. Kumar, S.; Basumatary, I.B.; Mukherjee, A.; Dutta, J. An overview of natural biopolymers in food packaging. *Biopolym.-Based Food Packag. Innov. Technol. Appl.* **2022**, *8*, 1–28.
92. Kibria, M.G.; Masuk, N.I.; Safayet, R.; Nguyen, H.Q.; Mourshed, M. *Plastic Waste: Challenges and Opportunities to Mitigate Pollution and Effective Management*; Springer International Publishing: Berlin/Heidelberg, Germany, 2023; Volume 17, ISBN 0123456789.
93. Polman, E.M.N.; Gruter, G.J.M.; Parsons, J.R.; Tietema, A. Comparison of the Aerobic Biodegradation of Biopolymers and the Corresponding Bioplastics: A Review. *Sci. Total Environ.* **2021**, *753*, 141953. [CrossRef]
94. Emadian, S.M.; Onay, T.T.; Demirel, B. Biodegradation of Bioplastics in Natural Environments. *Waste Manag.* **2017**, *59*, 526–536. [CrossRef]
95. Bioplastic, E. Biodegradable or Not Biodegradable, That Is the Question. Available online: https://www.european-bioplastics.org/biodegradable-or-not-biodegradable-that-is-the-question/ (accessed on 1 September 2023).
96. Cucina, M.; Carlet, L.; De Nisi, P.; Somensi, C.A.; Giordano, A.; Adani, F. Degradation of Biodegradable Bioplastics under Thermophilic Anaerobic Digestion: A Full-Scale Approach. *J. Clean. Prod.* **2022**, *368*, 133232. [CrossRef]
97. Yu, M.; Zheng, Y.; Tian, J. Study on the Biodegradability of Modified Starch/Polylactic Acid (PLA) Composite Materials. *RSC Adv.* **2020**, *10*, 26298–26307. [CrossRef]
98. Taib, N.A.A.B.; Rahman, M.R.; Huda, D.; Kuok, K.K.; Hamdan, S.; Bakri, M.K.B.; Julaihi, M.R.M.B.; Khan, A. *A Review on Poly Lactic Acid (PLA) as a Biodegradable Polymer*; Springer: Berlin/Heidelberg, Germany, 2023; Volume 80, ISBN 0123456789.

99. Ramos, M.; Beltran, A.; Fortunati, E.; Peltzer, M.A.; Cristofaro, F.; Visai, L.; Valente, A.J.M.; Jiménez, A.; Kenny, J.M.; Garrigós, M.C. Controlled Release of Thymol from Poly(Lactic Acid)-Based Silver Nanocomposite Films with Antibacterial and Antioxidant Activity. *Antioxidants* **2020**, *9*, 395. [CrossRef]
100. Sagee, O.; Dror, I.; Berkowitz, B. Transport of Silver Nanoparticles (AgNPs) in Soil. *Chemosphere* **2012**, *88*, 670–675. [CrossRef]
101. Mahdi, K.N.M.; Peters, R.J.B.; Klumpp, E.; Bohme, S.; van der Ploeg, M.; Ritsema, C.; Geissen, V. Silver Nanoparticles in Soil: Aqueous Extraction Combined with Single-Particle ICP-MS for Detection and Characterization. *Environ. Nanotechnol. Monit. Manag.* **2017**, *7*, 24–33. [CrossRef]
102. Mehmood, A. Brief Overview of the Application of Silver Nanoparticles to Improve Growth of Crop Plants. *IET Nanobiotechnol.* **2018**, *12*, 701–705. [CrossRef] [PubMed]
103. Khan, M.; Khan, M.S.A.; Borah, K.K.; Goswami, Y.; Hakeem, K.R.; Chakrabartty, I. The Potential Exposure and Hazards of Metal-Based Nanoparticles on Plants and Environment, with Special Emphasis on ZnO NPs, $TiO_2$ NPs, and AgNPs: A Review. *Environ. Adv.* **2021**, *6*, 100128. [CrossRef]
104. Wei, L.; Lu, J.; Xu, H.; Patel, A.; Chen, Z.S.; Chen, G. Silver Nanoparticles: Synthesis, Properties, and Therapeutic Applications. *Drug Discov. Today* **2015**, *20*, 595–601. [CrossRef] [PubMed]
105. Musante, C.; White, J.C. Toxicity of Silver and Copper to Cucurbita Pepo: Differential Effects of Nano and Bulk-Size Particles. *Environ. Toxicol.* **2012**, *27*, 510–517. [CrossRef]
106. Bélteky, P.; Rónavári, A.; Igaz, N.; Szerencsés, B.; Tóth, I.Y.; Pfeiffer, I.; Kiricsi, M.; Kónya, Z. Silver Nanoparticles: Aggregation Behavior in Biorelevant Conditions and Its Impact on Biological Activity. *Int. J. Nanomed.* **2019**, *14*, 667–687. [CrossRef]
107. Pourzahedi, L.; Eckelman, M.J. Comparative Life Cycle Assessment of Silver Nanoparticle Synthesis Routes. *Environ. Sci. Nano* **2015**, *2*, 361–369. [CrossRef]
108. Khan, M.R.; Urmi, M.A.; Kamaraj, C.; Malafaia, G.; Ragavendran, C.; Rahman, M.M. Green Synthesis of Silver Nanoparticles with Its Bioactivity, Toxicity and Environmental Applications: A Comprehensive Literature Review. *Environ. Nanotechnol. Monit. Manag.* **2023**, *20*, 100872. [CrossRef]
109. Prażak, R.; Święciło, A.; Krzepiłko, A.; Michałek, S.; Arczewska, M. Impact of Ag Nanoparticles on Seed Germination and Seedling Growth of Green Beans in Normal and Chill Temperatures. *Agriculture* **2020**, *10*, 312. [CrossRef]
110. Hanan, L.; Mohamed Ghareib, M.T. Phytosynthesis of Silver Nanoparticles Using Leaf Extracts from *Ocimum Basilicum* and *Mangifira Indica* and Their Effect on Some Biochemical Attributes of *Triticum aestivum*. *Gesunde Pflanz.* **2017**, *69*, 39–46.
111. El-Batal, A.I.; Gharib, F.A.E.L.; Ghazi, S.M.; Hegazi, A.Z.; El Hafz, A.G.M.A. Physiological Responses of Two Varieties of Common Bean (*Phaseolus Vulgaris* L.) to Foliar Application of Silver Nanoparticles. *Nanomater. Nanotechnol.* **2016**, *6*, 13. [CrossRef]
112. Mehmood, A.; Murtaza, G. Impact of Biosynthesized Silver Nanoparticles on Protein and Carbohydrate Contents in Seeds of *Pisum Sativum* L. *Crop Breed. Appl. Biotechnol.* **2017**, *17*, 334–340. [CrossRef]
113. Zhang, X.; Dang, D.; Zheng, L.; Wu, L.; Wu, Y.; Li, H.; Yu, Y. Effect of Ag Nanoparticles on Denitrification and Microbial Community in a Paddy Soil. *Front. Microbiol.* **2021**, *12*, 785439. [CrossRef] [PubMed]
114. Ihtisham, M.; Noori, A.; Yadav, S.; Sarraf, M.; Kumari, P.; Brestic, M.; Imran, M.; Jiang, F.; Yan, X.; Rastogi, A. Silver Nanoparticle's Toxicological Effects and Phytoremediation. *Nanomaterials* **2021**, *11*, 2164. [CrossRef] [PubMed]
115. Mazumdar, H.; Ahmed, G.U. Phytotoxicity Effect of Silver Nanoparticles on *Oryza sativa*. *Int. J. ChemTech Res.* **2011**, *3*, 1494–1500.
116. Dimkpa, C.O.; McLean, J.E.; Martineau, N.; Britt, D.W.; Haverkamp, R.; Anderson, A.J. Silver Nanoparticles Disrupt Wheat (*Triticum aestivum* L.) Growth in a Sand Matrix. *Environ. Sci. Technol.* **2013**, *47*, 1082–1090. [CrossRef] [PubMed]
117. Wang, J.; Koo, Y.; Alexander, A.; Yang, Y.; Westerhof, S.; Zhang, Q.; Schnoor, J.L.; Colvin, V.L.; Braam, J.; Alvarez, P.J.J. Phytostimulation of Poplars and *Arabidopsis* Exposed to Silver Nanoparticles and Ag+ at Sublethal Concentrations. *Environ. Sci. Technol.* **2013**, *47*, 5442–5449. [CrossRef]
118. Yin, L.; Cheng, Y.; Espinasse, B.; Colman, B.P.; Auffan, M.; Wiesner, M.; Rose, J.; Liu, J.; Bernhardt, E.S. More than the Ions: The Effects of Silver Nanoparticles on *Lolium multiflorum*. *Environ. Sci. Technol.* **2011**, *45*, 2360–2367. [CrossRef] [PubMed]
119. Anandalakshmi, K.; Venugobal, J.; Ramasamy, V. Characterization of Silver Nanoparticles by Green Synthesis Method Using Pedalium Murex Leaf Extract and Their Antibacterial Activity. *Appl. Nanosci.* **2016**, *6*, 399–408. [CrossRef]
120. Mensah, R.A.; Trotta, F.; Briggs, E.; Sharifulden, N.S.; Silva, L.V.B.; Keskin-erdogan, Z.; Diop, S.; Kureshi, A.K.; Chau, D.Y.S. A Sustainable, Green-Processed, Ag-Nanoparticle-Incorporated Eggshell-Derived Biomaterial for Wound-Healing Applications. *J. Funct. Biomater.* **2023**, *14*, 450. [CrossRef] [PubMed]
121. Trotta, F. Composition and Method. Patent GB2598715B, 23 November 2022.
122. Plastics—Determination of the Degree of Disintegration of Plastic Materials under Simulated Composting Conditions in a Laboratory-Scale Test. Available online: https://www.iso.org/standard/63367.html (accessed on 1 September 2023).
123. Yadav, M.; Liu, Y.K.; Chiu, F.C. Fabrication of Cellulose Nanocrystal/Silver/Alginate Bionanocomposite Films with Enhanced Mechanical and Barrier Properties for Food Packaging Application. *Nanomaterials* **2019**, *9*, 1523. [CrossRef] [PubMed]
124. Bai, Y.; Liu, H.; Ma, X.; Tai, X.; Wang, W.; Du, Z.; Wang, G. Synthesis, Characterization and Physicochemical Properties of Glycosyl-Modified Polysiloxane. *J. Mol. Liq.* **2018**, *266*, 90–98. [CrossRef]
125. Mathew, S.; Jayakumar, A.; Kumar, V.P.; Mathew, J.; Radhakrishnan, E.K. One-Step Synthesis of Eco-Friendly Boiled Rice Starch Blended Polyvinyl Alcohol Bionanocomposite Films Decorated with in Situ Generated Silver Nanoparticles for Food Packaging Purpose. *Int. J. Biol. Macromol.* **2019**, *139*, 475–485. [CrossRef] [PubMed]

126. Khan, H.; Raghuvanshi, S.; Saroha, V.; Singh, S.; Baba, W.N.; Mudgil, P.; Dutt, D. Biotransformation of Banana Peel Waste into Bacterial Nanocellulose and Its Modification for Active Antimicrobial Packaging Using Polyvinyl Alcohol with In-Situ Generated Silver Nanoparticles. *Food Packag. Shelf Life* **2023**, *38*, 101115. [CrossRef]
127. Muthukumar, C.; Thiagamani, S.; Krishnasamy, S. *Polymer Based Bio-Nanocomposites: Properties, Durability and Applications*; Springer: Berlin/Heidelberg, Germany, 2022; ISBN 9789811685774.

polymers

MDPI

Article

# The Evolution of the Intrinsic Flexural Strength of Jute Strands after a Progressive Delignification Process and Their Contribution to the Flexural Strength of PLA-Based Biocomposites

**Francisco J. Alonso-Montemayor *, Francesc X. Espinach, Quim Tarrés, Manel Alcalà, Marc Delgado-Aguilar and Pere Mutjé ***

LEPAMAP-PRODIS Research Group, University of Girona, C/Maria Aurèlia Capmany 61, 17003 Girona, Spain; francisco.espinach@udg.edu (F.X.E.); joaquimagusti.tarres@udg.edu (Q.T.); manuel.alcala@udg.edu (M.A.); m.delgado@udg.edu (M.D.-A.)

* Correspondence: francisco.alonso@udg.edu (F.J.A.-M.); pere.mutje@udg.edu (P.M.)

**Abstract:** Biocomposites from poly-(lactic acid) (PLA) and jute strands were prepared, and their flexural strength was analyzed. Jute strands were submitted to a progressive delignification process and the resulting morphology, composition, and crystallinity index were evaluated. Then, PLA biocomposites comprising 30 wt% of jute strands were produced and characterized under flexural conditions. The delignification processes decreased the lignin content and progressively increased the cellulose content. All this resulted in an enhancement of the composite flexural strength. A modified rule of mixtures, and the relation between tensile and flexural properties were used to determine the intrinsic flexural strength (of the jute strands) and their correlation with their physic-chemical characteristics. Equations correlating the intrinsic flexural strength with the crystallinity index, the cellulose content, and the microfibril angle were proposed. These equations show the impact of these properties over the intrinsic properties of the fibers and can help researchers to select appropriate fibers to obtain accurate properties for the composites. Jute strands show their value as reinforcement by increasing the flexural strength of the matrix by 70% and being less expensive and more environmentally friendly than mineral reinforcements. Together with the profitability and the environmental advantages, the mechanical results suggest that these PLA biocomposites are suitable for specific products of different market sectors.

**Keywords:** jute strands; delignification polylactic acid; biocomposites; flexural strength behavior; intrinsic flexural strength

## 1. Introduction

The growing environmental awareness of society is shifting industries towards higher utilization of bio-based and biodegradable raw materials to manufacture their products [1]. In this context, natural fibers are playing an important role for composite manufacturers, either to be used as a substitute for mineral fibers or to lessen the final product's cost. Anyhow, the decision will be subjected to the composite or biocomposite performance. It is known that the mechanical properties of composites mainly function as a result of the intrinsic properties of their components, the fiber-matrix interface, the fiber content, and the surface morphological characteristics of the fibers [2]. At a critical fiber length, the stress is transferred from the matrix to the fibers, resulting in a strengthening of the matrix [3]. In principle, plant fiber-reinforced composites should have similar applications as short glass fiber (GF)-reinforced composites (e.g., structural, automotive, or furniture sectors) [4].

PLA is one of the polymeric matrices that has called researcher interest in the last few years. PLA is a transparent, highly crystalline, and industrially compostable polymer. This polymer can be obtained from petroleum sources or bioresources. Thus, the production of biobased PLA involves fewer fossil resources as compared to polymers exclusively obtained from petroleum sources [5]. In order to ensure a robust environmental impact of the matrix, the researchers must be aware of its origin and the chemical process used to obtain it [1,5]. In addition, PLA exhibits high mechanical strength, thermoplastic behavior, biocompatibility, and good processability [6]. PLA has several industrial applications such as in packaging, textiles, biomedicine, and structures [1].

Cellulose fibers can be classified as non-wood fibers and wood fibers [7]. From an origin point of view, the lignocellulosic fibers can be obtained from strands of annual plants (e.g., abaca, flax, hemp, jute, sisal, and cotton), wood fibers (e.g., softwood and hardwood), fibers obtained from agricultural residues (e.g., bagasse, corn, and colza), forest residues (e.g., young wood obtained mainly from pruning fruit trees), and secondary fibers from recycled paper and board. These lignocellulosic fibers are characterized by a specific chemical composition (cellulose, hemicellulose, lignin, and extractives) with a determined length-diameter ($l/d$) ratio. In turn, the lignocellulosic fibers can be single strands or bundles of strands, depending on the ability of the extraction method to remove the glue components (e.g., pectin and lignin) binding the single strands. Differentiating single strands from the bundles is important since the strands exhibit higher intrinsic strength and stiffness than the bundles. Subsequent exposition of the strands to sodium hypochlorite solutions led to significant variations in their chemical composition since practically all lignin and a considerable fraction of hemicelluloses can be removed after a critical delignification stage using a non-selective solvent such as sodium hypochlorite [8], instead of sodium chlorite, which preserves the hemicelluloses. The removal of the soft and amorphous phases from the fibers not only strengthens these, but also allows the strong and highly crystalline cellulose microfibrils to interact with the matrix, enhancing the composite mechanical performance [1].

Among the strands, jute is one of the most promising and highly commercially available fibers to reinforce PLA. Jute strand-reinforced polymer composites have already been used to manufacture automobile interior decoration and architectural furnishing [9]. Jute belongs to the *Malvaceae* family, *Corchorus* spp., which comprises around 100 species, and is currently one of the cellulose fibers with the highest production rate and has an inherently low strain at break of about 1.7% that may provide high mechanical strength to the reinforced composites [6]. Together with hemp, jute has better physical and mechanical properties when compared to other natural fibers. Besides, using cheap reinforcement can deliver economic competitiveness to the composite [10].

Short fiber-reinforced composites show anisotropic behavior. Usually, the tensile properties are lower than the flexural properties. As Figure 1 depicts, during a flexural test the composite is subjected to compressive and tensile forces above and below the neutral axis, respectively [11,12]. Depending on the application, cellulose fiber-reinforced composites must be designed to offer the best response to loads following a preferable direction. This is of great importance when a novel material is introduced in the market since it needs to meet the industrial requirements of specific applications. Specifically, flexural properties are significantly important and relevant for engineers when predicting the potential of the material to be applied in structural, semi-structural, construction, and other similar commercial areas. This is explained by the fact that flexural or bending conditions are far more common than situations with tensile loads only, making designers especially interested in predicting the behavior of the materials at flexural loads [13].

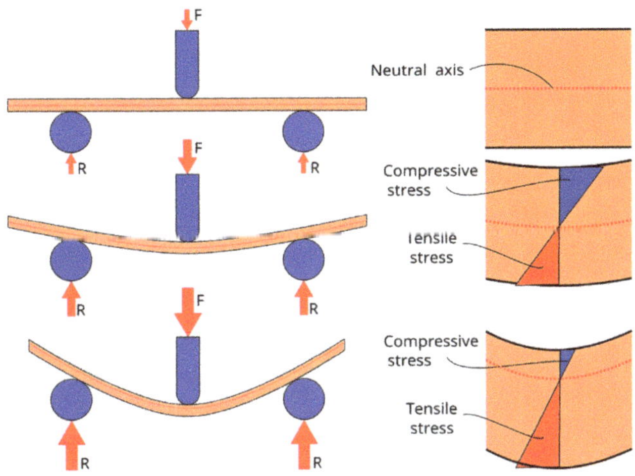


**Figure 1.** Evolution of the position of the neutral axis against specimen strain, showing the areas of the section of the specimen under compressive and tensile stresses. (R refers to the reaction at the supports and F refers to the applied loads".

By the analysis of the composite flexural strength ($\sigma_f^C$) it is possible to model the mechanical behavior to obtain the intrinsic fiber flexural strength ($\sigma_f^F$) and the flexural coupling factor ($f_C$) as main outcomes. The modified rule of mixtures is depicted according to Equation (1) [14]:

$$\sigma_f^C = f_C \sigma_f^F V^F + \left(1 - V^F\right) \sigma_f^{m*} \quad (1)$$

where, $\sigma_f^C$ is the composite flexural strength, $\sigma_f^F$ the intrinsic fiber flexural strength, $\sigma_f^{m*}$ is the contribution of the matrix to the $\sigma_f^C$ at the composite break, $V^F$ the fiber volume fraction, and $f_C$ the flexural coupling factor. From this equation, the flexural strength of a composite can be predicted using its dependency on the matrix characteristics, fiber content, fiber distribution, and morphology of the fiber. It is also worth mentioning that the intrinsic properties of the strands will depend on their origin, soil characteristics, and climatic conditions; these conditions will affect their chemical composition and morphology, including parameters such as cellulose content (Cel%), crystallinity index (CI), degree of polymerization, density ($\rho^F$), diameter ($D$), and microfibrillar angle (MFA) of cellulose at the cell wall of the strands [14].

In this work, jute strands were subjected to different delignification stages using a sodium hypochlorite solution. The strands were chemically characterized, and their crystallinity index was determined. Later, PLA composites of PLA comprising 30 wt% of these jute strands were prepared and tested under flexural conditions. A modified rule of mixtures (Equation (1)) was used to obtain the intrinsic flexural strength of the jute strands. The correlation between the intrinsic flexural strength of the strands and their physic-chemical characteristics (CI, Cel%, and MFA) is explored and different equations are proposed. To the best of our knowledge, this correlation has been seldom reported in the literature and never for jute strands.

## 2. Materials and Methods

### 2.1. Materials

The composites were prepared utilizing polylactic acid (PLA) (Ingeo Biopolymer 3251D), with a melt flow index (MFI) of 35 g/10 min at 190 °C and a load of 2.16 kg, which was kindly provided by Natureworks LLC (Blair, NE, USA). Jute strands were kindly supplied by CELESA S.A. (Tarragona, Spain). The raw jute strands have a diameter and mean weighted length of 22.9 μm and 353.0 μm, respectively [15]. All the chemical reagents used for strand characterization, extraction, and bleaching were supplied by Scharlau, S.L. (Sentmenat, Spain).

## 2.2. Delignification Treatments

The raw jute strands (J.0) were milled using a knife mill and passed through a 1 mm mesh. The jute strands were subsequently delignified with a solution with 8% sodium hypochlorite at 70 °C for 30 min, per stage. This is to say, that raw jute strands (named J.0) were subjected to four subsequent delignification stages (named J.1 to J.4, respectively). The number following "J" indicates the number of delignification stages. After each delignification stage, the strands were washed with distilled water to remove the residual sodium hypochlorite and dried at 105 °C. Figure 2 shows the flowchart of the experimental procedure. The mixing process, the preparation of the biocomposites, and the characterization procedures are addressed in further sections.

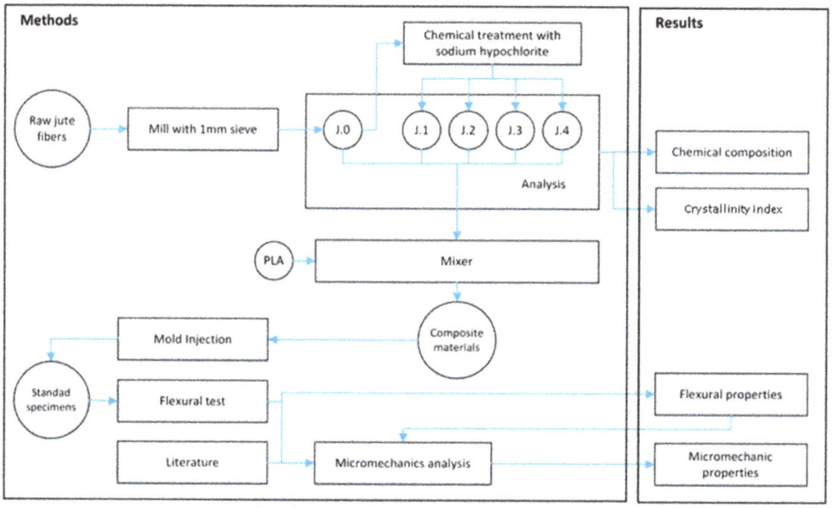


**Figure 2.** Flowchart of the experimental procedure.

## 2.3. Kappa Number and Chemical Composition

The Kappa number (KN) of the jute strands was determined following the ISO 302:2004 to evaluate the lignin presence. The TAPPI standards T204 cm-97, T413 om-93, and T-222 [16–18] were used to determine the content of lignin, extractives, and ashes. The amount of holocellulose was estimated from the difference between the total weight and the sum of lignin, extractives, and ashes. The Cel% was evaluated according to the TAPPI-T-212-om-12 standard [19].

## 2.4. The Preparation of the Fully Biodegradable Composites

The jute strands were homogeneously incorporated at 30 wt% to the PLA matrix using a Gelimat multi-kinetic mixer. Both strands and matrix were incorporated into the mixing chamber at low speed (300 rpm), which was gradually increased up to 2500 rpm. At this stage, the polymer matrix was molten and blended with jute strands. The process lasted from 3 to 4 min. Each material was discharged and milled using a knife mill. Before injection molding, the pellets were dried in an oven to remove moisture. Standard specimens for tensile testing were obtained in an Allrounder-220M injection machine fabricated by ARBURG (Eschweiler, Germany) according to UNE-EN ISO 178:2001 standard [20]. Before flexural testing, specimens were placed in a climatic chamber at 23 °C and 50% relative humidity for 48 h.

### 2.5. Flexural Characterization

The characterization of the bending properties of the PLA composite specimens was carried out following the UNE-EN ISO 178:2001 standard dimensions [21]. Ten samples of each PLA composite, reinforced with J.0, J.1, J.2, J.3, and J.4 were tested using an Instron 1122 universal testing machine equipped with a 5 kN load cell. A load element with a radius of 5 mm and supports located at 50 mm between them was used. ANOVA analyses of the results were made with R® and RCommander at a 95% confidence interval.

### 2.6. Densities Determination

According to ISO 1183-1 [21], a pycnometer was used to obtain the empirical values needed to determine the density of the composites ($\rho^C$). Distilled water was used as a reference liquid. The composite density was calculated by Equation (2) [22].

$$\rho^C = \frac{w^C}{V - w^{H_2O}\rho^{H_2O}} \tag{2}$$

where $V$ is the total volume of the pycnometer and $\rho^{H_2O}$ is the density of the water. The weight of the composite and water were represented by $w^C$ and $w^{H_2O}$, respectively. The $\rho^F$ identity described by Equation (3) was deduced from the Equation (2) [22].

$$\rho^F = \frac{w^F \rho^C \rho^M}{w^C \rho^M - w^M \rho^C} \tag{3}$$

where, $\rho^M$ is the density of the matrix. The weight of the fiber and the matrix were signed as $w^F$ and $w^M$, respectively. The $\rho^F$ calculation was used to estimate the $V^F$, which is a parameter that can be calculated from Equation (4) [22].

$$V^F = \left(1 + \frac{\rho^F(1 - \text{Fiber content(wt\%)})}{\rho^M x^F}\right)^{-1} \tag{4}$$

### 2.7. Fiber Recovering from Composites

The obtained biocomposites were submitted to Soxhlet extraction in the presence of dichloromethane, dissolving PLA with the purpose of recovering the fibers inside the matrix. This extraction was performed with grinded biocomposites for 24 h. Finally, the fibers were washed with distilled water and stored at 4 °C for morphological analysis [14].

### 2.8. The Morphological Analysis of Fibers

Morphological analysis was carried out using MorFi analyzer (Techpap SAS, Grenoble, France), equipped with CCD video camera. About 30,000 fibers were analyzed by the software MorFi v9.2. Among other parameters, this software was able to calculate the mean fiber length, the mean diameter and the fines percentage (fibers shorter than 76 µm). All characterizations were performed in triplicate [14].

### 2.9. The Evaluation of the Intrinsic Flexural Strength of the Reinforcements and Their Contribution to the Flexural Strength of the Composites

In addition, the well-known limitations associated with the measurement of the intrinsic properties of strand-reinforced composites would imply [22] that the use of mathematical models to estimate such intrinsic properties was chosen since the fiber deformation inside a composite is considerably lower than the minimum strain considered by any empirical approach (e.g., the single fiber tensile test has a minimal strain of 0.6 mm, while the fiber strain inside the polymer matrix is around of 0.35 mm). The literature presents a wide variety of models that can be used for that purpose. Nonetheless, one of the more elegant, due mainly to its simplicity, is a modified rule of mixtures. Nevertheless, the $\sigma_f^F$ and $f^C$ are unknown values, leaving Equation (5) unsolved. However, a fiber flexural strength factor (FFSF) can be estimated throughout Equation (5) to know the neat contribution of the fibers to the $\sigma_f^C$. As can be seen, the FFSF can be calculated from known parameters [23].

$$\mathrm{FFSF} = \frac{\sigma_f^C - (1 - V^F)\sigma_f^{m*}}{V^F} = f_C \sigma_f^F \tag{5}$$

Anyway, the $\sigma_f^F$ of the fibers remains unknown and it cannot be experimentally obtained for non-wood fibers and single strands. The first way used to estimate the $\sigma_f^F$ is $\sigma_f^F = \sigma_t^F \frac{\mathrm{FFSF}}{\mathrm{FTSF}}$ which is based on the ratio of FFSF and the fiber tensile strength factor (FTSF), where $\sigma_t^F$ is the fiber tensile strength [24]. These factors represent the contribution of the fibers to the respective composite strength values. Meanwhile, the second way $\sigma_f^{F\bullet} = \sigma_t^F \frac{\sigma_f^C}{\sigma_t^C}$, is based on the ratio of the $\sigma_f^C$ and the composite tensile strength ($\sigma_t^C$) [25]. Finally, once $\sigma_f^F$ or $\sigma_f^{F\bullet}$ are known, the respective flexural coupling factors ($f_C$ and $f_C^\bullet$) can be calculated from Equation (1). It is known that the morphology of the fibers, mainly their length distribution and diameter, impacts the mechanical properties of a composite. In this case, the authors link the tensile and flexural properties of the composites and the fibers to obtain the intrinsic flexural strength of the fibers. The intrinsic tensile strength of such fibers was obtained by solving Kelly and Tyson's equation that takes into account the mean diameter and the length distribution of the reinforcements. Then, while such morphologic parameters are not implicitly used in the model used to obtain the intrinsic flexural strength of the reinforcements, they are explicitly used because the value of the intrinsic tensile strength was obtained by taking into account such morphologic properties [25].

## 3. Results and Discussion

Understanding the microstructure and chemical composition of the cellulose fibers is needed to design and develop reinforced polymer composites. The delignification of virgin jute generates four types of strands that have been evaluated in terms of their bending resistance. As Table 1 shows, the progressive delignification not only produced changes in the surface of the strand, affecting the strength of its interface with PLA but also in the chemical composition of the strands. It is important to highlight that the lignin content follows a lineal correlation (0.996 $R^2$) with the KN described by 0.4743(KN) + 0.0591, which is useful for predictions.

**Table 1.** Structural properties and chemical composition of the raw and delignified jute strands.

| Stages | $\rho^F$ (g cm$^{-3}$) | $D$ (μm) [1] | KN [1] | Lignin (%) | Hemicellulose (%) | Cellulose (%) | CI (%) |
|---|---|---|---|---|---|---|---|
| J.0 | 1.48 | 22.90 | 27.3 ± 2.5 | 13.0 | 18.3 | 66.6 | 65.0 |
| J.1 | 1.50 | 22.80 | 20.0 ± 1.7 | 9.2 | 16.1 | 69.8 | 68.1 |
| J.2 | 1.51 | 22.60 | 14.8 ± 1.5 | 7.5 | 15.8 | 77.3 | 70.0 |
| J.3 | 1.53 | 22.55 | 7.7 ± 0.8 | 3.9 | 13.6 | 82.7 | 79.0 |
| J.4 | 1.54 | 22.40 | 2.1 ± 0.6 | 0.8 | 13.1 | 86.9 | 80.0 |

[1] Recovered from [15].

Broadly speaking, the cellulose content increases gradually, while a fraction of the hemicelluloses is carried along with the lignin. Therefore, each value of composite flexural strength must be considered relative to a specific composite. That is to say, the generated strands have a different chemical composition and $\rho^F$ values, working with the same reinforcement weight load, that increase as the lignin and hemicelluloses contents decrease, and while the cellulose content and the CI increase, as indicated in Table 1. The evolution of the $D$ of the strands indicates that these were slightly compacted. This compaction can be associated with the structural rearrangement implied by the chemical changes caused by the delignification stages.

### 3.1. The Flexural Properties of PLA Matrix and Jute Strand-Reinforced Composites

Table 2 shows the flexural properties of PLA and its composites, reinforced with 30 wt% of jute strands and delignified strands, which were assessed in terms of $\sigma_f^C$ and the matrix flexural strength ($\sigma_f^M$), elongation at break ($\varepsilon_f^M$) and the composite ($\varepsilon_f^C$) under flexural load, and the $\sigma_f^{M*}$. Additionally, for further analysis, the $V^F$ of the reinforcing jute strands was calculated for each delignification stage.

**Table 2.** Flexural mechanical properties of the PLA composites reinforced with jute strands (J.0) and delignified strands (J.1–J.4) and comparison with their tensile mechanical properties and of reported GF-reinforced PLA composites.

| Stages | wt% | $V^F$ | $\sigma_f^M \vee \sigma_f^C$ (MPa) [4] | $\sigma_t^M \vee \sigma_t^C$ (MPa) [1,4] | $\varepsilon_f^M \vee \varepsilon_f^C$ (MPa) [4] | $\sigma_f^{M*}$ (MPa) |
|---|---|---|---|---|---|---|
| NA | 0 | 0 | 68.35 ± 0.90 [a] | 49.8 ± 1.54 [a] | 3.27 ± 0.40 [a] | NA |
| J.0 | 30 | 0.264 | 82.15 ± 3.20 [b] | 54.7 ± 2.75 [b] | 2.42 ± 0.20 [b] | 61.52 |
| J.1 | 30 | 0.262 | 100.05 ± 2.40 [c] | 68.6 ± 0.54 [c] | 2.62 ± 0.04 [b] | 65.46 |
| J.2 | 30 | 0.260 | 99.35 ± 2.90 [c] | 70.0 ± 0.84 [d] | 2.48 ± 0.04 [b] | 62.75 |
| J.3 | 30 | 0.258 | 104.55 ± 1.70 [d] | 72.9 ± 1.08 [e] | 2.54 ± 0.04 [b] | 63.94 |
| J.4 | 30 | 0.257 | 100.55 ± 2.70 [c] | 71.7 ± 1.53 [e] | 2.55 ± 0.05 [b] | 64.13 |
| GF [2] | 30 | 0.173 | 140 | 92 | 0.90 | NA |
| GF [3] | 30 | 0.173 | 145 | 114 | NA | NA |

[1] Recovered from [15]. [2] and [3] are values, respectively, obtained from the data reported by [26] ($\sigma_f^M$ = 82 MPa and $\sigma_t^M$ = 43 MPa) and [27] ($\sigma_f^M$ = 108 MPa and $\sigma_t^M$ = 62 MPa). GF exhibited ($\rho^F$ = 2.55 g cm$^{-3}$). [4] Different letters a, b, c, d, and e represent the statistical difference (ANOVA, $p < 0.05$) between the properties of the materials.

The ANOVA analysis at a 95% confidence rate shows that the first stage of delignification had a statistically significant impact on the flexural strength of the composites. A second delignification treatment show no significant impact on the same property. A third delignification treatment returned flexural strength for the composites with significant differences to the other composites. A fourth delignification treatment, showed a similar impact to that of a one or two stage treatment. All of the composites show significative differences in the flexural strength of the matrix. In the case of tensile strength, the delignification treatment showed an impact on the tensile strength of the composites up to three stages. The results obtained with the fourth stage treatment is statistically similar to a two or three stage treatment. All the delignification stages showed a similar impact on the strain at the break of the composites.

The enhancement of the $\sigma_f^C$ of the PLA composites with the incorporation of delignified jute strands can be attributed to a stronger interphase between the matrix and the jute strands achieved by the removal of lignin and hemicelluloses after the exposition to the sodium hypochlorite solution [28]. The fiber-matrix interface is produced by the capacity of the PLA functional groups to interact with those of the jute strands fusing hydrogen bonds and van der Waals interactions, as Figure 3 shows [29]. It has been explained that the higher aspect ratio and cellulose content in the delignified strands provide enhanced composite flexural properties since increased active surface and more cellulose content mean higher hydroxyl groups available at the fiber surface for bonding with the matrix [3].

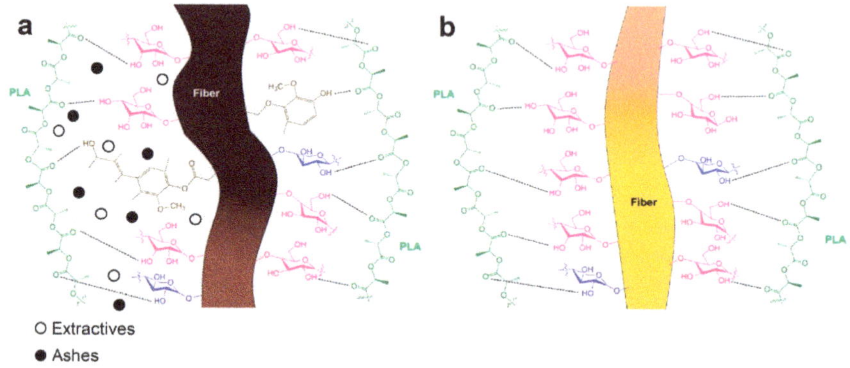


**Figure 3.** Schematic of possible chemical and electrostatic interactions between the PLA matrix and a raw jute strand with 14% lignin (**a**) and a fully delignified jute strand (**b**).

Before any delignification process, lignin, extractives, and holocellulose are usually found on the most superficial layer of the strands [30]. According to the Börås–Gantenholm model, which represents the chemical distribution on the surface of a typical chemi-thermo mechanical pulp fiber, the largest area corresponds to lignin (28%) and extractives (32%), while only 40% of the remaining surface was covered with holocellulose [31]. For fibers with a high number of extractives, the waxy substances and pectin should cap the functional groups of raw cellulose fibers [1,2]. From this, it can be deduced that the raw jute strands did not present enough extractives that hinder the interactions between their functional groups and those of the PLA since the flexural strength was enhanced with the incorporation of the jute strands prior to being subjected to sodium hypochlorite treatment.

The enhancement of the $\sigma_f^C$ of the PLA composites reinforced with jute strands is in line with the reported $\sigma_t^C$ trend [15], presenting a good coefficient of correlation of 0.989 $R^2$ and higher strength values. Similar enhancement trends have been reported for the $\sigma_f^C$ of PLA composites reinforced with 30 wt% of raw and alkali-treated hemp fibers [32].

For subsequent delignification stages, the incorporation led to a stabilization or "plateau" of the $\sigma_f^C$ around 100 MPa. Considering that sodium hypochlorite solutions remove surface lignin and hemicellulose from the cellulose fibers [1], it can be speculated that subsequently exposing the fibers to the sodium hypochlorite solution may remove a fraction of cellulose, reducing the fiber capacity to reinforce and leading to a plateau effect on the $\sigma_f^C$. However, the number of ruptures, voids, and dislocations within the reinforcing jute strands, and the changes in the strand's polymerization degree after being exposed to alkali treatment (from 2500 for virgin to 1020 for alkali-treated) [14,33–35] are factors that also determine flexural strength and also need to be considered. Anyway, it seems to be that the subsequent delignification of the jute strands is not required to enhance the $\sigma_f^C$.

On the other hand, it is important to highlight that the jute strand-reinforced PLA composites achieved around 70% of the $\sigma_f^C$ values of a commercially available and another reported PLA composite reinforced with 30 wt% GF, whose values of $\sigma_f^M$ were 82 MPa [26] and 62 MPa [27], respectively, and considerably higher than PP composites reinforced with 30 wt% of chopped GF ($\sigma_f^C = 79.70 \pm 0.80$ MPa, respectively) [36]. These results indicate that the delignified jute strand-reinforced PLA composites could compete with GF-reinforced PP composites and be a complementary option to GF-reinforced PLA composites, especially when other advantageous properties of the jute strand reinforcement are required (e.g., low density and high specific strength).

*3.2. The Evaluation and Analysis of the Intrinsic Flexural Properties*

Figure 4 shows the contribution of the PLA and jute strands to the $\sigma_f^C$. It was found that the delignified jute strands contributed more than 50% of the $\sigma_f^C$, while the contribution of the virgin jute strands was lesser. Moreover, it was found that the jute strands exhibited strength contributions comparable to the abovementioned reported GF-reinforced composites. This difference in contribution depends not only on the type of strand and $\sigma_f^F$ magnitude, but also on the reinforcing capability of the strands, which is expressed by the $f_C$. However, in the case of GF-reinforced PLA composites, it is necessary to consider that PLA could suffer hydrolysis during the processing of the composites [27] which could affect the estimation of $\sigma_f^F$ and $f_C$.

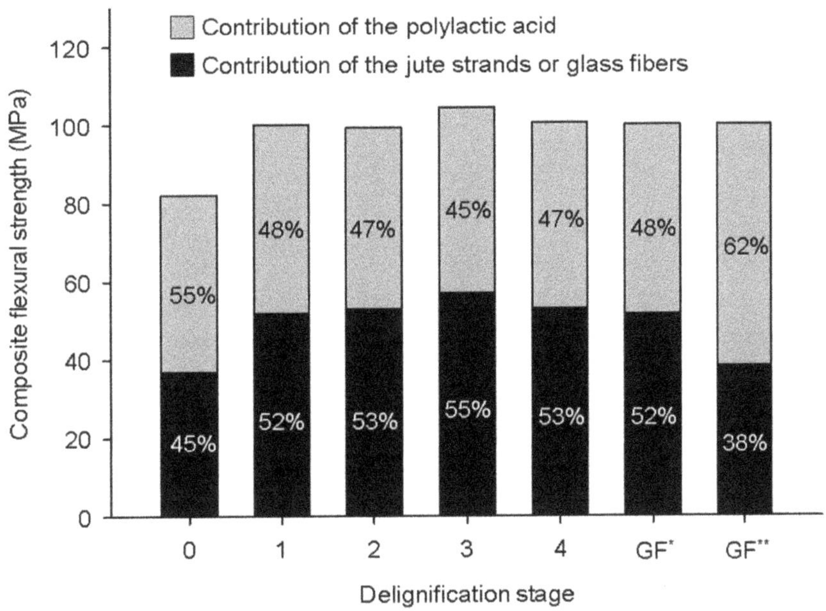


**Figure 4.** Percentage contributions of the PLA and delignified jute strands or GF to the $\sigma_f^C$. (GF* and GF** are values, respectively, obtained from the data reported by [26,27]).

After corroborating that the $\sigma_f^C$ and $\sigma_t^C$ present a good correlation, there are at least two methods to estimate $\sigma_f^F$ values apart from a modified rule of mixtures approach. These methods assume that similar $f_C$ values can be obtained from the tensile and flexural mechanical properties. Table 3 compiles the $\sigma_f^F$ and $\sigma_f^{F\bullet}$ obtained from Equations (6) and (7), respectively.

It was found that $\sigma_f^F$ and $f_C$, respectively, exhibit high correlations and similar percentages above 95% with $\sigma_f^{F\bullet}$ (0.992 $R^2$) and $f_C^\bullet$ (0.988 $R^2$). These $f_C$ values were considered high for cellulose fiber-reinforced polymers, implying that the fiber-matrix interface was strong in general. Meanwhile, for the GF-reinforced PLA composites, the Equation (7) approach cannot be replaced by Equation (6) since the $f_C$ was too different from $f_C^\bullet$, indicating that there could not be a good correlation between the tensile and flexural behavior. These results indicate that Equation (7), recently proposed by our research group, led to a more similar $\sigma_f^F$ estimation than Equation (6), which has been proved with other cellulose fiber-reinforced composites [37], when applied to delignified jute strands-reinforced PLA composites. Moreover, the results show that the delignification stages not only provoked chemical and morphological changes in the strands but also strengthened them. In the

LEPAMAP-PRODIS research group it was considered that fiber very well, well, normally, poorly, and badly bonded to the polymer matrix corresponding to the $f_C$ value ranges of 0.19–0.21, 0.17–0.18, 0.15–0.16, 0.13–0.14, and 0.12 or less, respectively. These ranges have been proposed considering the criterium of von Mises (IFSS $= \sigma_T^M/\sqrt{3}$) and Tresca (IFSS $= \sigma_T^M/2$) criteria that correspond to the very well-bonded and well-bonded expected $f_C$ values, where IFSS is the interfacial shear strength.

**Table 3.** Values of $\sigma_f^F$ of the jute strands reinforcing the PLA estimated through $\sigma_f^F = \sigma_t^F \frac{\text{FFSF}}{\text{FTSF}}$ or $\sigma_f^{F\bullet} = \sigma_t^F \frac{\sigma_f^C}{\sigma_t^C}$. The $f_C$ and $f_C^{\bullet}$ were estimated from $\sigma_f^F$ and $\sigma_f^{F\bullet\prime}$, respectively.

| Stages | wt% | $V^F$ | FFSF (MPa) | FTSF (MPa) [1] | $\sigma_t^F$ (MPa) [1] | $\sigma_f^F$ (MPa) | $f_C$ | $\sigma_f^{F\bullet}$ (MPa) | $f_C^{\bullet}$ |
|---|---|---|---|---|---|---|---|---|---|
| J.0 | 30 | 0.264 | 139.66 | 96.20 | 597 | 867 | 0.161 | 897 | 0.156 |
| J.1 | 30 | 0.262 | 197.48 | 128.60 | 726 | 1115 | 0.177 | 1059 | 0.187 |
| J.2 | 30 | 0.260 | 203.52 | 137.80 | 806 | 1190 | 0.171 | 1144 | 0.178 |
| J.3 | 30 | 0.258 | 221.34 | 147.50 | 846 | 1270 | 0.174 | 1213 | 0.182 |
| J.4 | 30 | 0.257 | 205.84 | 142.90 | 822 | 1184 | 0.174 | 1153 | 0.179 |
| GF [2] | 30 | 0.173 | 417.26 | 326.24 | 2400 | 3069 | 0.136 | 3652 | 0.114 |
| GF [3] | 30 | 0.173 | 321.87 | 362.58 | 2400 | 2131 | 0.151 | 3053 | 0.105 |

[1] Recovered from [15]. [2] and [3] are values, respectively, obtained from the data reported by [26,27]. GF exhibited. The $\sigma_t^F$ value of GF corresponds to sized fibers.

### 3.3. The Analysis of the Chemical Composition and Microfibril Angle of Jute Strands

The content of lignin and hemicellulose, the main amorphous phases each one with a density of 1.40 g cm$^{-3}$, decreased as the KN increased, as has been stated. Then, the amount of amorphous cellulose (density of 1.42 g cm$^{-3}$) decreases, while the crystalline cellulose (density of 1.60 g cm$^{-3}$) content increases after each delignification stage, as Table 4 shows. The densities of the lignin, hemicelluloses, and celluloses were recovered from [38]. However, other authors have indeed reported different densities for these components [39]. It was found that the CI of the jute strands increased with the subsequent delignification stages. Such behavior implies an increment of the crystalline regions of the fiber at the expense of the amorphous after each delignification stage. It is estimated that individual fibers composed of single cells with high cellulose content are obtained from the fiber bundles in higher amounts after each successive delignification stage [40]. The removal of lignin and hemicellulose was attributed to the presence of sodium hypochlorite during delignification [15]. Comparatively, the chemical composition of the raw jute strands is in line with that reported for different genotypes of *Corchorus capsularis* L. and *Corchorus olitorius* L. [41]. Jute is a strand (or phloem fiber) that contains a proportionally high amount of lignin in its fiber cell walls.

**Table 4.** Crystalline and amorphous cellulose content and MFA of the delignified jute strands.

| Stages | CI (%) | Cellulose (%) | | | MFA (°) | |
|---|---|---|---|---|---|---|
| | | Total | Crystalline | Amorphous | Constant | Variable |
| J.0 | 65.0 | 66.6 | 43.3 | 23.3 | 8 | 8 |
| J.1 | 68.1 | 69.8 | 47.5 | 22.3 | 8 | 9 |
| J.2 | 70.0 | 77.3 | 54.4 | 22.9 | 8 | 10 |
| J.3 | 79.0 | 82.7 | 65.3 | 17.4 | 8 | 11 |
| J.4 | 80.0 | 86.9 | 69.5 | 17.4 | 8 | 12 |

The cellulose fibers provide mechanical support to the plant. As Figure 5 depicts, the plant fiber cells have a central lumen surrounded by the cell wall and are connected by the middle lamellae. In turn, the cell wall consists of a primary section followed by several secondary layers that form the thickest section (S1, S2, and S3) [42].

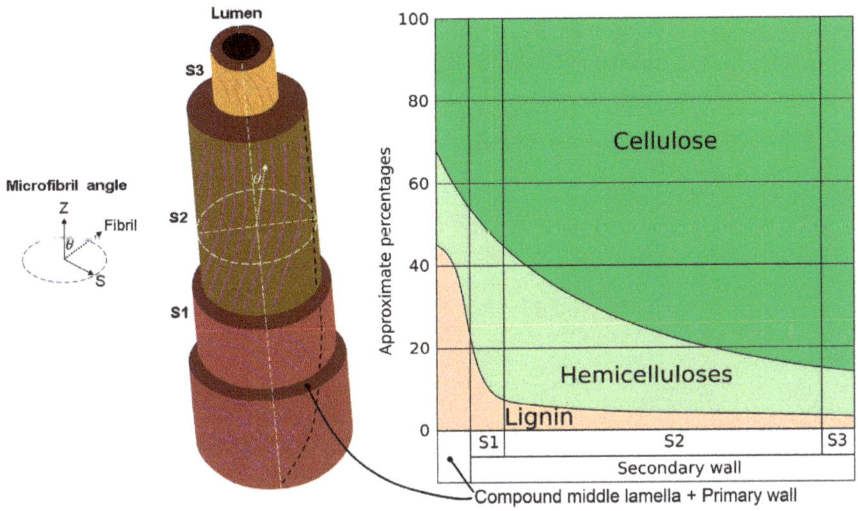


**Figure 5.** Schematic diagram of the three-dimensional structure and distribution of the main chemical components of a delignified fiber across the sections of the fiber cell wall, emphasizing the microfibril angle ($\theta$) corresponding to the cellulose microfibrils located in the S2 layer.

The cell wall layers contain lignin and hemicellulose regions intermixed with cellulose microfibrils disposed in a right-hand spiral. The angle between the fiber axis and the microfibrils in the S2 layer is denoted as the MFA, which is a key driver for the fiber and composite mechanical properties [39]. Thus, the cellulose microfibrils within the S2 are the main structural elements that govern the mechanical strength of the plant fibers, since it is estimated that the thickness of the S2 layer is more than a half of the cell wall [43]. The cell wall thickness scattering can be attributed to the different shapes and sizes of the fiber cells, as the reported micrographs reveal [41] If random cells are selected from the micrographs, a wide range of average wall thicknesses can be estimated for each cell, since their thickness varies considerably with the side measured (e.g., for three random cells and measuring the four sides of each one, the respective cell wall thickness was 16.67 ± 5.40 µm, 19.36 ± 4.22 µm, and 11.16 ± 3.26 µm). In response to this concern, another report estimated that the middle lamellae together with the lumen and the extreme cell wall layers (S1 and S3) of lignin-free cellulose fibers containing around 65 wt%, as is the case in the delignified jute strands, have relative thicknesses of 8%, while the S2 layer has 76% [43].

The individual cellulose fiber cells are composed of microfibrils whose orientation, specifically those from the S2 layer of the cell wall, have a significant influence on their fiber mechanical properties [44]. Indeed, cellulose microfibrils with small MFA facilitate the stress transfer from the matrix to the reinforcing fibers when the polymer composites are deformed [45]. Strands have the lowest MFA $< 10°$, concerning fiber from leaves (10°–25°) and seeds (30–50°), leading to their corresponding high, moderate, and low level of mechanical strengths [42]. It has been reported that the MFA of lignocellulosic fibers does not significantly or consistently change after being treated with alkaline solutions or subjected to other treatments. Specifically, the MFA of cabuya (strands of leaves), fragrant screw pine, and ichu (grass) fibers respectively went from 6.6° to 7.1°, 7.2° to 8.1°, and 7.5° to 5.4° after being subjected to alkali treatment [44,45]). However, there is controversy on this matter since some authors have reported a change in the MFA or improvement of the intrinsic mechanical properties associated with the change of MFA during alkali treatment [46–50]. Thus, further investigation is needed.

For these cases, it is explained that progressive removal of hemicelluloses by the alkali treatment might involve a rearrangement of the network of cellulose microfibrils since these could be more susceptible to swelling and shrinkage that disoriented them, impacting the fiber-matrix stress transfer [14,51]. According to a reported graphical model, the cellulose microfibrils are interlinked by the electrostatic interactions between the functional groups of the pectin located between the S2 cellulose microfibrils and of the hemicellulose coating these microfibrils. It is speculated that the S2 layer hemicellulose content is negatively correlated to the MFA (higher hemicellulose content, lower MFA) since a more important hemicellulose matrix implies a larger interfibrillar space, which can absorb load and allows more sliding between cellulose microfibrils [52].

The sensitiveness of the fiber strength concerning the intrinsic physicochemical properties has been discussed in terms of the cellulose fraction, the volumetric cellulose crystallinity, the MFA, and the lumen porosity. According to a broad reported analysis of the sensitivity of the strength of jute strands, whose cell wall thickness goes from 8.304–11.341 μm [41] and cellulose microfibril width is 28 ± 3 Å as estimated from reported micrographs [53], and several other lignocellulosic fibers to MFA and other microstructural parameters, it has been found that, in general, the variations in MFA have little effect on the fiber strength variability. However, it is necessary to consider that the reported model predictions depend on the literature recovered MFA, which for jute strands varies from 7° to 9°, but the most representative is 8°, according to a broad literature review [39]. In turn, it has also been reported that the Cel(%) has a positive and strong correlation with the specific strengths, while negatively correlated with the MFA. On the contrary, the hemicellulose content has a negative correlation with the specific strengths and MFA. Meanwhile, the lignin content is directly proportional to the MFA, but inversely proportional to the specific strengths [54,55]. Thus, an increasing MFA decreases the strength and stiffness of the cell wall but increases the strain at the break. This phenomenon allows plants to adjust the mechanical behavior of their tissues by shifting the MFA [56].

### 3.4. The Analysis of the Impact of Chemical Composition on the Intrinsic Flexural Strength

In 1986, Mukherjee and Satyanarayana [51] established a series of linear and quadratic equations that relate the $\sigma_f^F$ to Cel(%). Since the literature shows a proportional relationship between $\sigma_t^F$ and $\sigma_f^F$, the authors explored further equations correlating the intrinsic flexural strength of jute strands ($\sigma_f^F$ and $\sigma_f^{F\bullet}$) to MFA, Cel(%), and crystalline cellulose contents (CI · Cel(%)). The authors found that the highest coefficients of correlations were obtained for quadratic equations. The intrinsic flexural strength was positively correlated with the cellulose contents obtaining the following nonlinear regression quadratic Equations (6) and (7):

$$\sigma_f^F = -1.8184(\text{Cel\%})^2 + 292.94(\text{Cel\%}) - 10540, \text{ with } 0.91\ R^2 \tag{6}$$

$$\sigma_f^{F\bullet} = -1.286(\text{Cel\%})^2 + 209.3(\text{Cel\%}) - 7319.6, \text{ with } 0.95\ R^2 \tag{7}$$

These equations show the positive correlation between cellulose content and the $\sigma_f^F$ of the strands. Moreover, the authors explored the correlation with the crystalline cellulose content ($\psi$ = CI·Cel%), obtaining Equations (8) and (9):

$$\sigma_f^F = -1.2752(\psi)^2 + 155.02(\psi) - 3426.9, \text{ with } 0.94\ R^2 \tag{8}$$

$$\sigma_f^{F\bullet} = -0.9167(\psi)^2 + 112.89(\psi) - 2260.6, \text{ with } 0.98\ R^2 \tag{9}$$

Thus, the intrinsic flexural strength of jute strands was more strongly correlated to crystalline cellulose contents than to whole cellulose content. Figure 6 shows the regression curves of Equations (8) and (9).

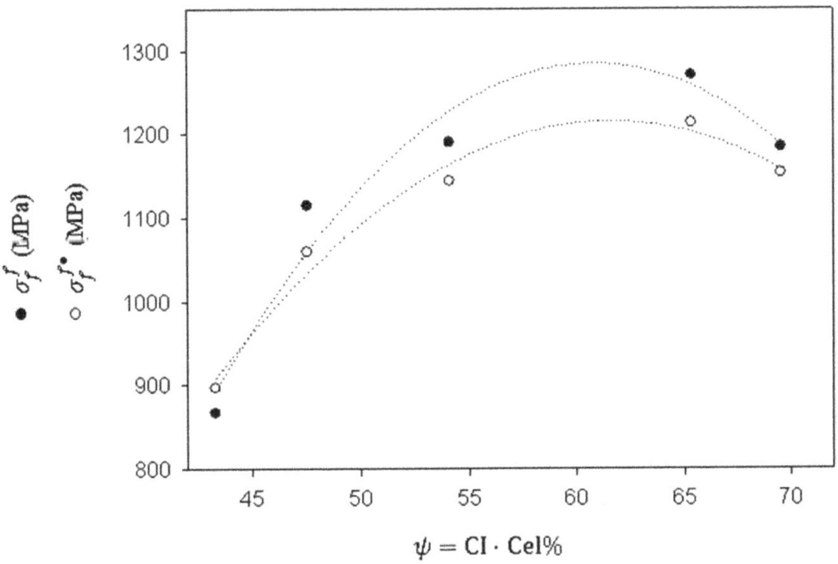


**Figure 6.** Intrinsic tensile strength of the jute strands as a function of crystalline cellulose contents and nonlinear regression quadratic curves obtained from the data (shown as dotted regression lines).

Considering that the literature stated the impact of MFA on the $\sigma_t^F$ of strands, the authors explored possible correlations to MFA. Mukherjee and Satyanarayana (1986) [51] studied the correlation of cos(MFA) to the intrinsic tensile strength of the strands. Thus, the authors explored the correlation of $\delta = \mathrm{Cel\%{\cdot}CI{\cdot}cos(MFA)}$ to the intrinsic flexural strengths, obtaining Equations (10) and (11):

$$\sigma_f^F = -1.3004(\delta)^2 + 156.55(\delta) - 3426.9,\ \text{with } 0.94\ \mathrm{R}^2 \tag{10}$$

$$\sigma_f^{F\bullet} = -0.9348(\delta)^2 + 114(\delta) - 2260.6,\ \text{with } 0.98\ \mathrm{R}^2 \tag{11}$$

The inclusion of cos(MFA) in the equations did not alter the coefficients of determination. The authors used an MFA $= 8°$, based on literature [45,48]. The low value of such an angle compared to the other factors and the fact that such cos(MFA) was constant are the main reasons for obtaining the same coefficients of correlation. The $\sigma_f^F$ is inversely correlated to MFA, thus, the authors explored the correlation of $\sigma_f^F$ to $\gamma = \mathrm{CI{\cdot}Cel\%/MFA}$, obtaining Equations (12) and (13):

$$\sigma_f^F = -81.612(\gamma)^2 + 1240.2(\gamma) - 3426.9,\ \text{with } 0.94\ \mathrm{R}^2 \tag{12}$$

$$\sigma_f^{F\bullet} = -58.666(\gamma)^2 + 903.16(\gamma) - 2260.6,\ \text{with } 0.98\ \mathrm{R}^2 \tag{13}$$

The results were like those obtained for cos(MFA), and for the same reasons. To evaluate the possible impact of MFA variations on the $\sigma_f^F$ of the jute strands, and taking into account, on the one hand, that MFA for jute varies from 7° to 12° [57], which is in line with the separation of nanofibrils across the cell wall [58], and on the other hand that MFA can increase due to the harshness of the treatments, the authors applied a range of MFA from 8° to 12° to J.0 to J.4, respectively (see Table 4). Using the equation that correlated the intrinsic flexural strengths with δ, the coefficients of correlations decreased to 0.91 and 0.95. These equations revealed that the intrinsic strength of the reinforcements is positively correlated with the crystalline cellulose contents of the strands and inversely correlated to the MFA. These equations can be used to explore the impact of these parameters on the intrinsic flexural strength, but not for evaluating the properties.

### *3.5. The Analysis of the Specific Flexural Properties of the Polylactic Acid, Its Composites, and Reinforcing Jute Strands*

The potential selection of PLA composites for light-mass applications, such as automotive, requires not only the corroboration of flexural properties suitable for products and parts subjected to bending loads but also requires the consideration of specific mechanical properties (e.g., the ratio between flexural strength and the material density). Dividing the mechanical properties of the material by its density normalizes them, allowing the objective cost comparison with other materials [59]. Specifically, the comparison of specific mechanical properties can be used to analyze weight savings, which is important in the automotive and aerospace industries [60]. As expected, the addition of jute strands to PLA led to specific $\sigma_f^C$ values higher than the PLA matrix by 15%, 39%, 38%, 44%, and 38%, respectively, for the composites reinforced with raw jute and subsequently delignified strands (see Table 5). These enhancement percentages are comparable with others reported for hemp fiber-reinforced PLA composites [61]. Moreover, the jute strands-reinforced composites respectively achieved 64%, 78%, 77%, 80%, and 77% of the specific $\sigma_f^C$ of the commercially available GF-reinforced PLA composite. The remarkable specific flexural performance of the PLA composites is proportional to the specific $\sigma_f^F$ of the reinforcing jute strands.

**Table 5.** Specific flexural mechanical properties of the PLA, its composites, and the delignified jute strands.

| **Stages** | **wt%** | $V^F$ | $\rho^M \vee \rho^C$ **(g cm$^{-3}$)** | $\frac{\sigma_f^M}{\rho^M} \vee \frac{\sigma_f^C}{\rho^C}$ **(MPa)** | $\frac{\varepsilon_f^M}{\rho^M} \vee \frac{\varepsilon_f^C}{\rho^C}$ **(%)** | $\rho^F$ **(g cm$^{-3}$)** | $\frac{\sigma_f^F}{\rho^F}$ **(MPa)** |
|---|---|---|---|---|---|---|---|
| NA | 0 | 0 | 1.24 | 55.1 | 2.64 | NA | NA |
| J.0 | 30 | 0.264 | 1.30 | 63.2 | 1.85 | 1.48 | 585.6 |
| J.1 | 30 | 0.262 | 1.31 | 76.4 | 2.00 | 1.50 | 743.3 |
| J.2 | 30 | 0.260 | 1.31 | 75.8 | 1.88 | 1.51 | 788.3 |
| J.3 | 30 | 0.258 | 1.32 | 79.2 | 1.93 | 1.53 | 829.8 |
| J.4 | 30 | 0.257 | 1.32 | 76.2 | 1.93 | 1.54 | 768.9 |
| GF [1] | 30 | 0.173 | 1.47 | 98.6 | NA | 2.55 | 941.2 |

[1] Recovered from [27].

Then, the potential practical implication of these findings is a positive effect on the future development trends in PLA biocomposites. For instance, the improvement of the PLA composites manufacturing based on different types, ratios, and shapes of natural fibers for specific applications. These results could contribute to the development of proper databases on natural fibers and biocomposites [1].

## 4. Conclusions

Herein, cellulose fiber-reinforced biocomposites have been manufactured and characterized. The delignification stages using sodium hypochlorite generate, apart from the decrease in lignin content, the progressive increase of cellulose in the delignified strands. The hemicelluloses and amorphous material content decreases, also progressively, given that the delignifying agent is not selective for lignin, unlike sodium chlorite. It is also evident that most of the extractives present in the primary wall, mainly, disappear, which notably favors the IFSS of the composite materials. Because of the change in the chemical composition of the strands, the $\rho^F$ increases slightly and progressively up to 4% concerning the raw jute strands.

The $\sigma_f^C$ increases significantly after the first delignification stage concerning the PLA matrix and composite manufactured with raw jute strands, reaching a stable behavior or "plateau" of around 100 MPa from the second delignification stage. The composite manufactured with strands subjected to the third delignification stage reaches its maximum value of 104.55 MPa. This value represents 66% and 61.3% of the $\sigma_f^C$ of two PLA composites reinforced with GF. The evolution of the $\sigma_f^C$ correlates reasonably well with the $\sigma_t^C$.

The intrinsic flexural strengths obtained from the relation between FTSM and FFSM and from the ratio between the flexural and tensile strengths of the composites show a good correlation. The intrinsic flexural strengths show a fair correlation to the intrinsic tensile strengths.

Concerning the intrinsic mechanical properties, nonlinear equations that correlate the intrinsic flexural strength of the jute strands ($\sigma_f^F$ and $\sigma_f^{F\bullet}$) with the CI, Cel%, and MFA were obtained for a constant MFA $= 8°$ with respect the delignification stages. To the best knowledge of the author, these correlations are a novelty for the prediction of the intrinsic mechanical strength of jute strands.

Finally, the specific $\sigma_f^C$ of the jute strands-reinforced PLA composites achieved more than 70% of commercially available PLA reinforced with GF. In addition, the lower and lesser scattered specific price of jute strands, and their environmental and processing advantages, indicate that these PLA composites could be considered for manufacturing products that need to be subjected to bending forces such as some GF-reinforced PLA products.

**Author Contributions:** Conceptualization, methodology and formal analysis, F.J.A.-M., P.M. and F.X.E.; Software, F.X.E.; Validation, F.X.E., Q.T., M.A., M.D.-A. and P.M.; Formal analysis, Writing—original draft preparation, F.J.A.-M. All authors have read and agreed to the published version of the manuscript.

**Funding:** Financial support of the Spanish Economy and Competitiveness Ministry to the project GREENCOMP, reference: MAT2017-83347-R.

**Institutional Review Board Statement:** Not applicable.

**Data Availability Statement:** Data are contained within the article.

**Acknowledgments:** Marc Delgado-Aguilar and Quim Tarrés are Serra Húnter Fellows.

**Conflicts of Interest:** The authors declare no conflict of interest.

## Abbreviations

| | |
|---|---|
| $\psi$ | The product of the crystallinity index and the cellulose content |
| $\delta$ | The product of the crystallinity index, the cellulose content, and the microfibrillar angle cosine |
| $\gamma$ | The product of the crystallinity index and the cellulose content, divided by the microfibrillar angle. |
| $l/d$ | length-diameter ratio |
| $V$ | Volume of the pycnometer |
| $D$ | Fiber diameter |
| S3 | Inner secondary layer of the cell wall |
| S2 | Middle secondary layer of the cell wall |
| S1 | Outermost secondary layer of the cell wall |
| PLA | Poly-(lactic acid) |
| MFI | Melt flow index |
| MFA | Microfibrillar angle |
| KN | Kappa number |
| J.4 | Jute strands after four subsequent delignification stages |
| J.3 | Jute strands after three subsequent delignification stages |
| J.2 | Jute strands after two subsequent delignification stages |
| J.1 | Jute strands after one delignification stage |
| J.0 | Raw jute strands |

| | |
|---|---|
| IFSS | Interfacial shear strength |
| GF | Glass fibers |
| FTSF | Fiber tensile strength factor |
| FFSF | Fiber flexural strength factor |
| CI | Crystallinity index |
| Cel% | Cellulose content |
| $\sigma_t^F$ | Fiber tensile strength |
| $\sigma_t^C$ | Composite tensile strength |
| $\sigma_f^{m*}$ | Contribution of the matrix to the composite flexural strength |
| $\sigma_f^M$ | Matrix flexural strength |
| $\sigma_f^{F\bullet}$ | Intrinsic flexural strength obtained from the ratio of flexural and tensile strengths. |
| $\sigma_f^F$ | Intrinsic flexural strength |
| $\sigma_f^C$ | Composite flexural strength |
| $\rho^F$ | Fiber density |
| $\rho^C$ | Composite density |
| $\rho^M$ | Matrix density |
| $\rho^{H_2O}$ | Distilled water density |
| $\varepsilon_f^M$ | Elongation at break of the matrix under flexural load |
| $\varepsilon_f^C$ | Elongation at break of the composite under flexural load |
| $w^M$ | Matrix weight |
| $w^F$ | Fiber weight |
| $w^C$ | Composite weight |
| $w^{H_2O}$ | Distilled water weight |
| $f_C^\bullet$ | Flexural coupling factor obtained from $\sigma_f^{F\bullet}$ |
| $f_C$ | Flexural coupling factor |
| $V^F$ | Fiber volume fraction |

## References


1. Siakeng, R.; Jawaid, M.; Ariffin, H.; Sapuan, S.M.; Asim, M.; Saba, N. Natural Fiber Reinforced Polylactic Acid Composites: A Review. *Polym. Compos.* **2019**, *40*, 446–463. [CrossRef]
2. Getme, A.S.; Patel, B. A Review: Bio-Fiber's as Reinforcement in Composites of Polylactic Acid (PLA). *Mater. Today Proc.* **2019**, *26*, 2116–2122. [CrossRef]
3. Mirmehdi, S.M.; Henrique, G.; Tonoli, D.; Dabbagh, F. Lignicellulose-Polyethylene Composite: Influence of Delignification, Filler Content and Filler Type. *Cellul. Chem. Technol.* **2017**, *51*, 341–346.
4. Joshi, S.V.; Drzal, L.T.; Mohanty, A.K.; Arora, S. Are Natural Fiber Composites Environmentally Superior to Glass Fiber Reinforced Composites? In Proceedings of the American Institute of Chemical Engineers (AIChE) 2002 Annual Meeting, Indianapolis, Indiana, 3–8 November 2002.
5. Jiang, N.; Yu, T.; Li, Y. Effect of Hydrothermal Aging on Injection Molded Short Jute Fiber Reinforced Poly(Lactic Acid) (PLA) Composites. *J. Polym. Environ.* **2018**, *26*, 3176–3186. [CrossRef]
6. Kian, L.K.; Saba, N.; Jawaid, M.; Sultan, M.T.H. A Review on Processing Techniques of Bast Fibers Nanocellulose and Its Polylactic Acid (PLA) Nanocomposites. *Int. J. Biol. Macromol.* **2019**, *121*, 1314–1328. [CrossRef] [PubMed]
7. Raju, A.; Shanmugaraja, M. Recent Researches in Polylactic Acid Reinforced with Natural Fiber Composites—A Literature Review. *IOP Conf. Ser. Mater. Sci. Eng.* **2020**, *988*, 012025. [CrossRef]
8. Laine, J.; Stenius, P.; Carlsson, G.; Strom, G. The Effect of ECF and TCF Bleaching on the Surface Chemical Composition of Kraft Pulp as Determined by ESCA. *Artic. Nord. Pulp Pap. Res. J.* **1996**, *11*, 201–210. [CrossRef]
9. Fang, C.C.; Zhang, Y.; Qi, S.Y.; Liao, Y.C.; Li, Y.Y.; Wang, P. Influence of Structural Design on Mechanical and Thermal Properties of Jute Reinforced Polylactic Acid (PLA) Laminated Composites. *Cellulose* **2020**, *27*, 9397–9407. [CrossRef]
10. Felix Sahayaraj, A.; Muthukrishnan, M.; Ramesh, M. Experimental Investigation on Physical, Mechanical, and Thermal Properties of Jute and Hemp Fibers Reinforced Hybrid Polylactic Acid Composites. *Polym. Compos.* **2022**, *43*, 2854–2863. [CrossRef]
11. Seculi, F.; Julián, F.; Llorens, J.; Espinach, F.X.; Mutjé, P.; Tarrés, Q. Methodologies to Evaluate the Micromechanics Flexural Strength Properties of Natural-Fiber-Reinforced Composites: The Case of Abaca-Fiber-Reinforced Bio Polyethylene Composites. *Polymers* **2023**, *15*, 3137. [CrossRef]

12. Hao, X.; Zhou, H.; Mu, B.; Chen, L.; Guo, Q.; Yi, X.; Sun, L.; Wang, Q.; Ou, R. Effects of Fiber Geometry and Orientation Distribution on the Anisotropy of Mechanical Properties, Creep Behavior, and Thermal Expansion of Natural Fiber/HDPE Composites. *Compos. Part B Eng.* **2020**, *185*, 107778. [CrossRef]
13. Oliver-Ortega, H.; Julian, F.; Espinach, F.X.; Tarrés, Q.; Ardanuy, M.; Mutjé, P. Research on the Use of Lignocellulosic Fibers Reinforced Bio-Polyamide 11 with Composites for Automotive Parts: Car Door Handle Case Study. *J. Clean. Prod.* **2019**, *226*, 64–73. [CrossRef]
14. Gassan, J.; Bledzki, A.K. Possibilities for Improving the Mechanical Properties of Jute/Epoxy Composites by Alkali Treatment of Fibres. *Compos. Sci. Technol.* **1999**, *59*, 1303–1309. [CrossRef]
15. Delgado-Aguilar, M.; Oliver-Ortega, H.; Alberto Méndez, J.; Camps, J.; Espinach, F.X.; Mutjé, P. The Role of Lignin on the Mechanical Performance of Polylactic Acid and Jute Composites. *Int. J. Biol. Macromol.* **2018**, *116*, 299–304. [CrossRef] [PubMed]
16. *TAPPI T204 cm-97*; Solvent Extractives of Wood and Pulp (Proposed Revision of T 204 Cm-97). Approved by Standard-Specific Interest Group for this Method. Technical Association of the Pulp and Paper Industry: Peachtree Corners, GA, USA, 2007.
17. *TAPPI T413 om-93*; Ash in Wood, Pulp, Paper and Paperboard: Combustion at 900 °C. Approved by Chemical Properties Committee of the Process and Product Quality Division. Technical Association of the Pulp and Paper Industry: Peachtree Corners, GA, USA, 1993.
18. *TAPPI T222*; Acid-Insoluble Lignin in Wood and Pulp (Reaffirmation of T 222 Om-02). Approved by Standard Specific Interest Group for this Test Method. Technical Association of the Pulp and Paper Industry: Peachtree Corners, GA, USA, 2006.
19. *TAPPI T 212 om-12 (2022)*; One Percent Sodium Hydroxide Solubility of Wood and Pulp. Technical Association of the Pulp and Paper Industry: Peachtree Corners, GA, USA, 2012.
20. *ISO 178:2001/Amd 1:2004*; Plastics—Determination of Flexural Properties—Amendment 1: Precision Statement. Approved by ISO/TC 61/SC 2—Mechanical Behavior. International Organization for Standarization—ISO Central Secretariat: Vernier, Switzerland, 2004.
21. *UNE-EN ISO 1183-1:2019*; Plastics—Methods for Determining the Density of Non-Cellular Plastics—Part 1: Immersion Method, Liquid Pycnometer Method and Titration Method (ISO 1183-1: 2019, Corrected Version 2019-05). Spanish Association for Standardization: Madrid, Spain, 2019.
22. Aruan Efendy, M.G.; Pickering, K.L. Comparison of Strength and Young Modulus of Aligned Discontinuous Fibre PLA Composites Obtained Experimentally and from Theoretical Prediction Models. *Compos. Struct.* **2019**, *208*, 566–573. [CrossRef]
23. Espinach, F.X.; Delgado-Aguilar, M.; Puig, J.; Julian, F.; Boufi, S.; Mutjé, P. Flexural Properties of Fully Biodegradable Alpha-Grass Fibers Reinforced Starch-Based Thermoplastics. *Compos. Part B Eng.* **2015**, *81*, 98–106. [CrossRef]
24. Gironès, J.; Lopez, J.P.; Vilaseca, F.; Bayer, R.; Herrera-Franco, P.J.; Mutjé, P. Biocomposites from Musa Textilis and Polypropylene: Evaluation of Flexural Properties and Impact Strength. *Compos. Sci. Technol.* **2011**, *71*, 122–128. [CrossRef]
25. Hashemi, S. Hybridisation Effect on Flexural Properties of Single- and Double-Gated Injection Moulded Acrylonitrile Butadiene Styrene (ABS) Filled with Short Glass Fibres and Glass Beads Particles. *J. Mater. Sci.* **2008**, *43*, 4811–4819. [CrossRef]
26. Wang, G.; Zhang, D.; Wan, G.; Li, B.; Zhao, G. Glass Fiber Reinforced PLA Composite with Enhanced Mechanical Properties, Thermal Behavior, and Foaming Ability. *Polymer* **2019**, *181*, 121803. [CrossRef]
27. RTP Company. *Glass Fiber Reinforced PLA Bioplastic: Sustainable Bio-Based Plastic Suitable for Durable and Semi-Durable Applications*; RTP Company: Winona, MN, USA, 2011.
28. Ma, H.; Joo, C.W. Structure and Mechanical Properties of Jute-Polylactic Acid Biodegradable Composites. *J. Compos. Mater.* **2011**, *45*, 1451–1460. [CrossRef]
29. Gibeop, N.; Lee, D.W.; Prasad, C.V.; Toru, F.; Kim, B.S.; Song, J. Il Effect of Plasma Treatment on Mechanical Properties of Jute Fiber/Poly (Lactic Acid) Biodegradable Composites. *Adv. Compos. Mater.* **2013**, *22*, 389–399. [CrossRef]
30. Bledzki, A.K.; Gassan, J. Composites Reinforced with Cellulose Based Fibres. *Prog. Polym. Sci.* **1999**, *24*, 221–274. [CrossRef]
31. Böras, L.; Gatenholm, P. Surface Composition and Morphology of CTMP Fibers. *Holzforschung* **1999**, *53*, 188–194. [CrossRef]
32. Hu, R.; Lim, J.K. Fabrication and Mechanical Properties of Completely Biodegradable Hemp Fiber Reinforced Polylactic Acid Composites. *J. Compos. Mater.* **2007**, *41*, 1655–1669. [CrossRef]
33. Alila, S.; Besbes, I.; Vilar, M.R.; Mutjé, P.; Boufi, S. Non-Woody Plants as Raw Materials for Production of Microfibrillated Cellulose (MFC): A Comparative Study. *Ind. Crops Prod.* **2013**, *41*, 250–259. [CrossRef]
34. Lau, K.T.; Hung, P.Y.; Zhu, M.H.; Hui, D. Properties of Natural Fibre Composites for Structural Engineering Applications. *Compos. Part B Eng.* **2018**, *136*, 222–233. [CrossRef]
35. Sedan, D.; Pagnoux, C.; Chotard, T.; Smith, A.; Lejolly, D.; Gloaguen, V.; Krausz, P. Effect of Calcium Rich and Alkaline Solutions on the Chemical Behaviour of Hemp Fibres. *J. Mater. Sci.* **2007**, *42*, 9336–9342. [CrossRef]
36. Huda, M.S.; Drzal, L.T.; Mohanty, A.K.; Misra, M. Chopped Glass and Recycled Newspaper as Reinforcement Fibers in Injection Molded Poly(Lactic Acid) (PLA) Composites: A Comparative Study. *Compos. Sci. Technol.* **2006**, *66*, 1813–1824. [CrossRef]
37. Tarrés, Q.; Oliver-Ortega, H.; Espinach, F.X.; Mutjé, P.; Delgado-Aguilar, M.; Méndez, J.A. Determination of Mean Intrinsic Flexural Strength and Coupling Factor of Natural Fiber Reinforcement in Polylactic Acid Biocomposites. *Polymers* **2019**, *11*, 1736. [CrossRef]
38. Placet, V.; Trivaudey, F.; Cisse, O.; Gucheret-Retel, V.; Boubakar, M.L. Diameter Dependence of the Apparent Tensile Modulus of Hemp Fibres: A Morphological, Structural or Ultrastructural Effect? *Compos. Part A Appl. Sci. Manuf.* **2012**, *43*, 275–287. [CrossRef]

39. Königsberger, M.; Lukacevic, M.; Füssl, J. Multiscale Micromechanics Modeling of Plant Fibers: Upscaling of Stiffness and Elastic Limits from Cellulose Nanofibrils to Technical Fibers. *Mater. Struct.* **2023**, *56*, 13. [CrossRef] [PubMed]
40. Naili, H.; Jelidi, A.; Limam, O.; Khiari, R. Extraction Process Optimization of Juncus Plant Fibers for Its Use in a Green Composite. *Ind. Crops Prod.* **2017**, *107*, 172–183. [CrossRef]
41. Meshram, J.H.; Palit, P. On the Role of Cell Wall Lignin in Determining the Fineness of Jute Fibre. *Acta Physiol. Plant.* **2013**, *35*, 1565–1578. [CrossRef]
42. Bourmaud, A.; Shah, D.U.; Beaugrand, J.; Dhakal, H.N. Property Changes in Plant Fibres during the Processing of Bio-Based Composites. *Ind. Crops Prod.* **2020**, *154*, 112705. [CrossRef]
43. Gassan, J.; Chate, A.; Bledzki, A.K. Calculation of Elastic Properties of Natural Fibers. *J. Mater. Sci.* **2001**, *34*, 3715–3720. [CrossRef]
44. Tenazoa, C.; Savastano, H.; Charca, S.; Quintana, M.; Flores, E. The Effect of Alkali Treatment on Chemical and Physical Properties of Ichu and Cabuya Fibers. *J. Nat. Fibers* **2021**, *18*, 923–936. [CrossRef]
45. Gerald Arul Selvan, M.T.; Binoj, J.S.; Mansingh, B.B.; Baby Sajin, J.A. Physico-Chemical Properties of Alkali Treated Cellulosic Fibers from Fragrant Screw Pine Prop Root. *J. Nat. Fibers* **2023**, *20*, 148–161. [CrossRef]
46. Bartos, A.; Anggono, J.; Farkas, Á.E.; Kun, D.; Soetaredjo, F.E.; Móczó, J.; Antoni; Purwaningsih, H.; Pukánszky, B. Alkali Treatment of Lignocellulosic Fibers Extracted from Sugarcane Bagasse: Composition, Structure, Properties. *Polym. Test.* **2020**, *88*, 106549. [CrossRef]
47. Gassan, J.; Mildner, I.; Bledzki, A.K. Influence of Fiber Structure Modification on the Mechanical Properties of Flax Fiber-Epoxy Composites. *Mech. Compos. Mater.* **1999**, *35*, 435–440. [CrossRef]
48. Mwaikambo, L.Y.; Ansell, M.P. Mechanical Properties of Alkali Treated Plant Fibres and Their Potential as Reinforcement Materials. I. Hemp Fibres. *J. Mater. Sci.* **2006**, *41*, 2483–2496. [CrossRef]
49. Rayirath, P.; Avramidis, S.; Mansfield, S.D. The Effect of Wood Drying on Crystallinity and Microfibril Angle in Black Spruce (Picea Mariana). *J. Wood Chem. Technol.* **2008**, *28*, 167–179. [CrossRef]
50. Wong, K.J.; Yousif, B.F.; Low, K.O. The Effects of Alkali Treatment on the Interfacial Adhesion of Bamboo Fibres. *Proc. Inst. Mech. Eng. Part L J. Mater. Des. Appl.* **2010**, *224*, 139–148. [CrossRef]
51. Mukherjee, P.S.; Satyanarayana, K.G. An Empirical Evaluation of Structure-Property Relationships in Natural Fibres and Their Fracture Behaviour. *J. Mater. Sci.* **1986**, *21*, 4162–4168. [CrossRef]
52. Bourmaud, A.; Morvan, C.; Bouali, A.; Placet, V.; Perré, P.; Baley, C. Relationships between Micro-Fibrillar Angle, Mechanical Properties and Biochemical Composition of Flax Fibers. *Ind. Crops Prod.* **2013**, *44*, 343–351. [CrossRef]
53. Heyn, A.N.J. The Microcrystalline Structure of Cellulose in Cell Walls of Cotton, Ramie, and Jute Fibers as Revealed by Negative Staining of Sections. *J. Cell Biol.* **1966**, *29*, 181–197. [CrossRef]
54. Karimah, A.; Ridho, M.R.; Munawar, S.S.; Adi, D.S.; Ismadi; Damayanti, R.; Subiyanto, B.; Fatriasari, W.; Fudholi, A. A Review on Natural Fibers for Development of Eco-Friendly Bio-Composite: Characteristics, and Utilizations. *J. Mater. Res. Technol.* **2021**, *13*, 2442–2458. [CrossRef]
55. Komuraiah, A.; Kumar, S.; Prasad, D. Chemical Composition of Natural Fibers and Its Influence on Their Mechanical Properties. *Mech. Compos. Mater.* **2014**, *50*, 359–376. [CrossRef]
56. Burgert, I. Exploring the Micromechanical Design of Plant Cell Walls. *Am. J. Bot.* **2006**, *93*, 1391–1401. [CrossRef]
57. Mohanty, S.; Nayak, S.K. Biobased Composites and Applications. In *Biopolymers: Biomedical and Environmental Applications*; Kalia, S., Avérous, L., Eds.; John Wiley & Sons: Hoboken, NJ, USA, 2011; pp. 225–266.
58. Bian, Z.X.; Miao, X.R.; Lin, J.Y.; Tian, F.; Bian, F.G.; Li, H. Extraction and Structural Investigation of Jute Cellulose Nanofibers. *Nucl. Sci. Tech.* **2018**, *29*, 106. [CrossRef]
59. Akonda, M.; Alimuzzaman, S.; Shah, D.U.; Rahman, A.N.M.M. Physico-Mechanical, Thermal and Biodegradation Performance of Random Flax/Polylactic Acid and Unidirectional Flax/Polylactic Acid Biocomposites. *Fibers* **2018**, *6*, 98. [CrossRef]
60. Öteyaka, M.Ö.; Aybar, K.; Öteyaka, H.C. Effect of Infill Ratio on the Tensile and Flexural Properties of Unreinforced and Carbon Fiber-Reinforced Polylactic Acid Manufactured by Fused Deposition Modeling. *J. Mater. Eng. Perform.* **2021**, *30*, 5203–5215. [CrossRef]
61. Pappu, A.; Pickering, K.L.; Thakur, V.K. Manufacturing and Characterization of Sustainable Hybrid Composites Using Sisal and Hemp Fibres as Reinforcement of Poly (Lactic Acid) via Injection Moulding. *Ind. Crops Prod.* **2019**, *137*, 260–269. [CrossRef]

polymers

*Article*

# Optical Limiting Properties of DNA Biopolymer Doped with Natural Dyes

**Petronela Gheorghe [1,*], Adrian Petris [1,*] and Adina Mirela Anton [1,2]**

1 National Institute for Laser, Plasma and Radiation Physics, 409 Atomistilor Street, 077125 Magurele, Romania; anastasiumirela@yahoo.com
2 Faculty of Chemical Engineering and Biotechnologies, National University of Science and Technology Politehnica Bucharest, 1-7 Polizu Street, 011061 Bucharest, Romania
* Correspondence: petronela.gheorghe@inflpr.ro (P.G.); adrian.petris@inflpr.ro (A.P.)

**Abstract:** The high-power lasers have important implications for present and future light-based technologies; therefore, the protection measures against their high-intensity radiation are extremely important. Currently, a great deal of interest is directed towards the development of new nonlinear optical materials for passive optical limiters, which are used to protect the human eye and sensitive optical and optoelectronic devices from laser-induced damage. Biopolymers doped with natural dyes are emerging as a new class of optical materials with interesting photosensitive properties. In this paper, the optical limiting capability of deoxyribonucleic acid bio-polymer functionalized with Turmeric natural dye has been demonstrated for the first time, to the best of our knowledge. The experimental investigation of the optical limit has been done by the Intensity-scan method in the NIR spectral domain at the important telecommunication wavelength of 1550 nm, using ultrashort laser pulses (~120 fs). Several optical properties of this natural dye are presented and discussed. The values of the optical transmittance in the linear regime, the saturation intensity of the nonlinear transmittance curves, and the coefficient of the nonlinear absorption have been determined. The influence of the DNA biopolymer and natural dye concentration on the optical limiting properties of the investigated biomaterials is reported and discussed. The photostability and thermal stability of the investigated solutions have also been evaluated by monitoring the temporal decay of the normalized absorption spectra under illumination with UVA light and heating, respectively. Our results evidence the positive influence of the DNA, which embeds Turmeric natural dye, on the optical limiting functionality itself and on the photostability and thermal stability of this novel material. The performed study reveals the potential of the investigated novel biomaterial for applications in nonlinear photonics, in particular in optical limiting.

**Keywords:** eco-friendly materials; DNA biopolymers; natural dyes; nonlinear optical response; optical limiting

## 1. Introduction

Lately, scientists around the world have been concerned about the protection of the environment and biodiversity by enhancing the sustainability and quality of eco-friendly materials. Natural dyes are emerging as an important class of biodegradable optical materials used for modern applications in optics and photonics, security, and safety and have been intensely studied in recent years due to their interesting photosensitive properties [1–3]. The deoxyribonucleic acid (DNA) functionalized with different synthetic dyes, e.g., Rhodamine B, Disperse Red 1 [4–6], represents novel materials with practical applications in photonics. Biopolymers doped with natural dyes are emerging as a new class of optical materials with interesting photosensitive properties.

One of the most promising practical applications of these biomaterials, based on the nonlinear absorption process is the optical limiting (OL) functionality. OL devices are of great practical importance for the protection of the human eye, light sensors, cameras, and other sensitive optical and optoelectronic devices against intense sources of laser radiation, which can irreversibly damage such elements when a safety damage threshold is exceeded [7–13]. The experimental dependence of the transmitted power/intensity on the same parameters of the incident beam is shown for an ideal OL (red line) and a real OL (green line), respectively (Figure 1). For the ideal OL, the transmitted power/intensity has a linear increase for incident power/intensity up to the OL threshold, remaining constant after this. For a real OL, the transmitted power/intensity depends nonlinearly on the incident one, being described by a saturation curve.

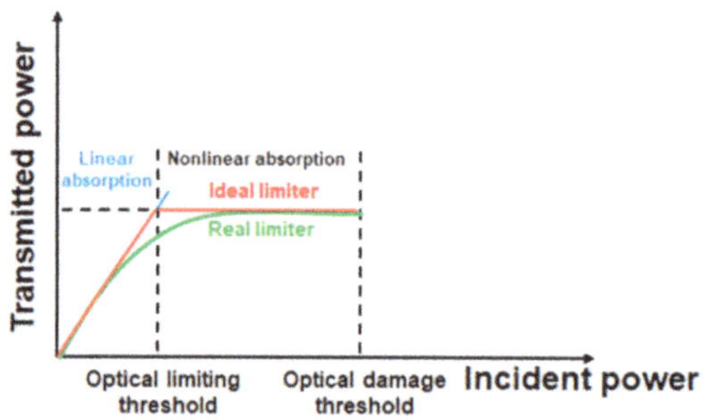


**Figure 1.** Optical limiting functionality.

The continuous development of this type of device is crucially dependent on the availability of suitable optical materials. A wide variety of materials are being studied to achieve effective OL [11,14–18]. For example, in many papers, the OL functionality for various groups of materials, including organic dyes, other common organic materials, graphene, and its derivatives, fullerenes, polymeric materials (organic and/or inorganic), inorganic semiconductors, and other materials, has been extensively investigated in recent years [11,19–22]. At the same time, it has been shown that the synthetic polymers used so far in optical limiters can be successfully replaced by the biopolymers that have some advantages over synthetic materials [12]. Among the studied materials with optical limiting properties are carbon-based materials, e.g., graphene and its derivates [15,19–21,23]. The OL functionality of these materials, in different forms as solutions, films, and bulk, has been mainly studied for visible and near-infrared nanosecond and picosecond laser pulses [24–28] and, in a smaller measure, for femtosecond laser pulses at 800 nm wavelength [29–33]. Very few papers have investigated the OL of femtosecond laser pulses in the IR band, which includes the wavelength of 1550 nm, an important wavelength for communications. B. N. S. Sooraj et al. reported in a recent study [34] the optical limiting of 35 ps laser pulses at 532 nm in silver, gold, and alloy clusters, investigated by Z-scan technique. The authors demonstrate that Ag and Au clusters exhibit almost the same efficiency for optical limiting, but the alloy clusters are less efficient in optical limiting. Other materials with optical limiting properties are glasses doped with different nanoparticles [35,36]. Measurements of OL response in synthetic polymers (Ethylene propylene-diene monomer (EPDM), Poly(methyl methacrylate) (PMMA)) doped with dye (Disperse Red 1, Disperse Yellow, respectively) are also reported using 532 nm wavelength (10 ns pulses and continuous wave) [37,38]. However, DNA-based biopolymers, which have been intensively studied recently, present several advantages over materials based on synthetic polymers. Biomaterials generally present unique properties, which are not easily replicated in conventional organic or inorganic materials. In addition, the DNA biopolymer comes from renewable natural resources.

To the best of our knowledge, the OL properties of the DNA-based materials functionalized with natural dyes, have not yet been investigated and represent an innovative and original approach. The nonlinear optical investigation of this new class of biopolymer-based materials is complex and represents a topic of high interest in photonics. One of the most important reasons why DNA appears to be exciting for OL functionality is the exceptional capability of DNA chains to interact with various dyes (nonbinding interactions, groove binding, and intercalation/semi-intercalation) [39,40] and the suitability of its properties. Its particular, double-stranded helical structure shows that it can be used as scaffolding for incorporating functional molecules, allowing the development of photonic, electronic, and opto-electronic devices [6,22,39–45]. Moreover, there is a lot of free space to functionalize it with different dyes to obtain and tailor desired properties for targeted practical applications [4,5,22,39–45]. DNA can be used in fast photoinduced responses, improved nonlinear optical effects in DNA compounds [46,47], as well as in other future applications in nanotechnology [48–50]. Another important property of DNA in photonics is its large transparency range, from 350 to about 1700 nm, with a cutoff wavelength around 320 nm [22].

DNA is an eco-friendly biodegradable material that can be obtained at a low cost from renewable resources such as the waste produced by the food processing industry. Pure DNA is known to be only soluble in water and exhibits poor thermal stability (denaturation at around 90 °C), making it difficult to process into good optical-quality thin films. On the other hand, it is well known that the complex of DNA with cetyltrimethyl-ammonium chloride (CTMA) surfactant is soluble in alcohols and can be processed into thin films with good optical quality [44]. Moreover, in the past few years, it was discovered that this solubility allows the functionalization of DNA-CTMA with synthetic/natural dyes, which contributes to the enhancement of the nonlinear optical properties of the resulting complexes and tuning their sensitivities in different spectral ranges. Recently, it was shown that the DNA–surfactant complex represents an interesting medium for photosensitive molecules, with significantly lower kinetic chemical and photo-thermal degradation constants [24,51,52], as observed for synthetic polymers. The DNA-CTMA complex has high thermal stability up to a temperature of 100 °C [45], maintaining its double-stranded helical structure [53]. The DNA and DNA–surfactant complexes decompose in the 220–230 °C temperature range [40,53–55]. Their high thermal conductivity ensures a lower laser heating of materials based on them compared to other synthetic polymers (e.g., PMMA) with similar optical absorption [46]. Another advantage of this new class of materials is the increased optical damage threshold after functionalization with natural dyes compared to synthetic polymers, [12,13,23,40,54–59], showing that the biopolymers are more resistant to high-energy laser pulses than some synthetic polymers like PC or PEG [40,55].

In the present work, we experimentally demonstrate the OL functionality of a new class of nonlinear optical materials (NLO), namely DNA biopolymer functionalized with Turmeric dye, by I-scan experiments with ultrashort laser pulses (~120 fs) in NIR, at the important telecommunication wavelength of 1550 nm. To the best of our knowledge, the OL properties of these new class of NLO materials have not been investigated yet and are reported for the first time in the literature, representing an innovative and original approach. The influence of the DNA presence and the variation of the dye concentrations in the prepared solutions on the linear transmittance and the optical limiting performance of these novel materials are discussed. The values of the optical transmittance in the linear regime, the saturation intensity of the nonlinear transmittance curves, and the coefficient of the nonlinear absorption have been determined. The photostability and the thermal stability of the investigated solutions have also been evaluated by monitoring the temporal decay of the normalized absorption spectra under the illumination with UVA light and heating, respectively.

## 2. Materials and Methods

The investigated materials consisted of DNA biopolymer functionalized with Turmeric natural dye. To demonstrate the influence of the biopolymer on the optical limiting properties we also investigated materials without DNA. The DNA was purchased from Ogata Research Laboratory, Ltd., Chitose, Japan. DNA is an eco-friendly biodegradable material that exhibits very interesting properties for its use as a matrix for dye molecules, with many interesting applications in photonics [6–9,13]. It can be used to replace synthetic polymers with several advantages, which are due to its particular helical structure [53]. This biopolymer is extracted from the waste produced by the food processing industry and its sources are practically unlimited, making it an inexpensive biopolymer. The DNA used in this study is extracted from salmon waste. It has a high molar mass of ~8 MDa. The processing steps involved in obtaining DNA are presented in detail in [60,61]. In the preparation of samples investigated in this study, the DNA was used in the form in which it was purchased. The DNA was not sonicated, thus keeping unaffected its double-stranded helical structure. It is well known that sonication can destroy the double-stranded structure of DNA (ds-DNA) by breaking nucleobase bonds resulting in single-stranded deoxyribonucleic acid (ss-DNA). This alteration process of DNA depends on sonication conditions.

The unprotected DNA is soluble in water and its degradation time is quite short. With the CTMA surfactant, the DNA-CTMA complex becomes insoluble in water, but soluble in organic solvents, such as alcohols, and can be functionalized with different photosensitive dyes. The choice of CTMA as a surfactant for DNA functionalization is based on several reasons described in detail in [62]. The DNA–CTMA complex becomes water-insoluble and more stable mechanically due to the alkyl chain of CTMA. Also, the DNA-CTMA complex is soluble in organic solvents, such as alcohols, which makes easier the doping of the complex with different photosensitive molecules.

The DNA functionalization with CTMA was performed in this study following the procedure described by Grote et al. [60]. This procedure is briefly described below. DNA is negatively charged, soluble in water, and reacts with positively charged surfactants forming a stable DNA-CTMA complex through ionic chemical interaction. The DNA-CTMA complex is stable and insoluble in water, but soluble in several organic solvents, which facilitates its doping with dyes. The DNA-CTMA complex was obtained by dropwise addition of the DNA solution to the CTMA solution. The final solution is stirred for approximately 4 h after the DNA solution has been added and allowed to settle. The obtained precipitate complex is filtered using a filter and then washed well until the wash water contains no more traces of surfactant. The water is removed with another filter paper by manual pressing. The DNA-CTMA complex is dried in a desiccator under vacuum at 60 °C. The dry precipitate obtained is grinded until the particle size is small enough for its use in solutions. Next, the DNA-CTMA complex is doped with Turmeric, the link between them being ensured by an electrostatic interaction with Turmeric.

The natural extract of Turmeric was obtained by the maceration technique, a simple method by which ground raw material is kept in contact with the solvent for a defined time. The resulting product was mixed at 400 rpm for 24 h and then filtered through filter paper. The obtained solutions were placed in the refrigerator to avoid degradation. For the functionalization of the DNA-CTMA complex with Turmeric, we used butanol as an organic solvent, because it has several advantages over other solvents. It has a low vapor pressure of 6.7 hPa at 20 °C [63], compared to other alcohols, such as, for example, ethanol which has a pressure of 59 hPa at the same temperature of 20 °C, which ensures a relatively slow evaporation. Also, butanol's moderate viscosity prevents evaporation [64]. The concentration of DNA-CTMA in butanol was 30 g/L. Solutions with different concentrations of Turmeric (3%, 5%, 7.5%, 10% and 15%) have been prepared. These concentrations represent the percentages of the Turmeric relative to the DNA-CTMA matrix.

In Table 1, we present the investigated solutions of DNA-CTMA-TURMERIC in butanol with different dye concentrations, and the solutions of Turmeric-natural dye in butanol, with similar dye concentrations.

**Table 1.** Denomination of the samples and Turmeric concentration in solutions.

| Sample | Turmeric Concentration in Solution g/L |
|---|---|
| T3-DNA/T3 | 3 |
| T5-DNA/T5 | 5 |
| T7.5-DNA/T7.5 | 7.5 |
| T10-DNA/T10 | 10 |
| T15-DNA/T15 | 15 |

## 2.1. Materials Characterization

The nonlinear optical properties are in strong correlation with the linear optical properties: the control of the linear optical properties can lead to a useful control of the nonlinear ones. The linear refractive index ($n_0$) of the prepared solutions was determined using the ABBE refractometer (Carl Zeiss, Jena, Germany). The values of the refractive index are slightly dependent on Turmeric concentration. The refractive index of the solutions with the lowest concentration of Turmeric is $n_0$ = 1.3969 for the T3 sample and $n_0$ = 1.4018 for the T3-DNA sample. The refractive index of the solutions with the highest concentration of Turmeric is $n_0$ = 1.3971 for the T15 sample and $n_0$ = 1.4021 for the T15-DNA sample. The refractive indices of solutions with intermediate concentrations are in between the values corresponding to solutions with the lowest and highest Turmeric concentrations, respectively (the difference is only at the third decimal).

In the use of this novel DNA-based optical material, it is important that the light does not produce irreversible changes in its optical properties (photodegradation), and also its thermal stability. To investigate these aspects, we performed several spectral analyses (UV-VIS-NIR spectroscopy, before and after its UV illumination/heating) to check its photodegradation and thermal stability. These analyses provide us with valuable information about the photochemical/thermal stability of the novel compounds.

### 2.1.1. UV-VIS-NIR Spectroscopy

The spectroscopic studies of the transmittance for the prepared solutions were done using a Perkin Elmer spectrophotometer (Perkin Elmer, Waltham, MA, USA), Lambda 950 model at a resolution of 1 nm. The UV-VIS-NIR transmission/absorption spectra of these innovative biomaterials, collected with air as reference were recorded between 300 and 1600 nm wavelengths. The recorded spectra are presented in Figures 2a and 2b, respectively. The inset from Figure 2a,b shows the absorption bands of prepared solutions for different concentrations, at the same interval of wavelengths. The used spectrophotometric cell has a 0.5 mm optical path to avoid obtaining a saturated signal in the spectral range of interest

From the UV-VIS-NIR spectra, shown above in Figure 2a, it can be seen that the values of the transmission curves of T3, 5, 7.5, 10, and 15 solutions are 60–80% in the VIS-NIR range, while for solutions based on DNA functionalized with Turmeric dye, it is slightly lower than in the case of solutions without DNA, but remaining sufficiently high for the eye or sensitive imaging devices [14]. Thus, we can say that these materials are transparent enough to see through them, a requirement for passive optical limiting materials in NIR. The spectra confirm, as expected, the presence of Turmeric and DNA, from the existence of absorption peaks at wavelength of 420 nm, specific to the Turmeric dye (Figure 2a,b) [65], and of DNA-CTMA at wavelengths in 200–300 nm range (Figure 2c), respectively. The DNA-CTMA compound does not have any absorption peak in the 300–800 nm wavelength range, as can be seen from Figure 2c. The spectra have a quasi-constant transmittance in the visible range (500–850 nm), followed by a higher transmittance in the near-infrared range (850–1300 nm), excepting the dip at 1200 nm and by a lower and decreasing transmittance in the wavelength range of 1350–1500 nm.

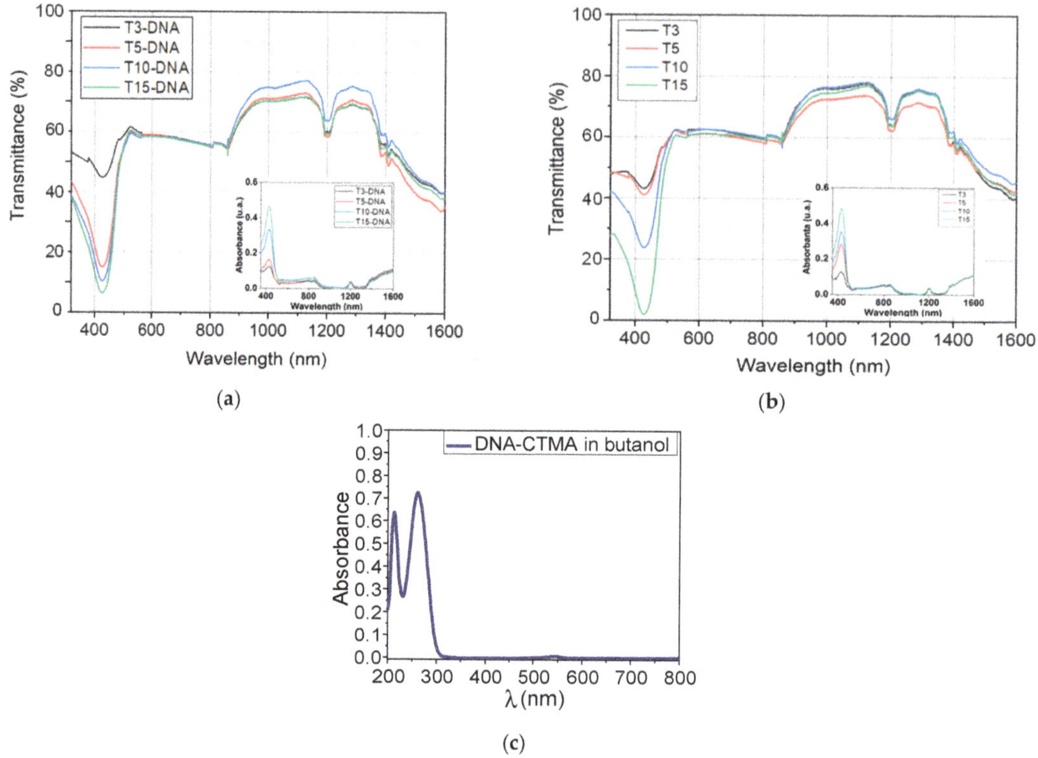


**Figure 2.** UV-VIS-NIR spectra of the DNA-CTMA-Turmeric solutions (**a**); Turmeric solutions (**b**), and DNA-CTMA solution (**c**) in butanol. (**a**) UV-VIS-NIR spectra of DNA-CTMA-Turmeric solutions in butanol. (**b**) UV-VIS-NIR spectra of Turmeric solutions in butanol. (**c**) UV-VIS-NIR spectra of DNA-CTMA solution in butanol.

Nevertheless, the transmittance does not drop below 40% in the considered VIS-NIR spectral range (500–1600 nm). As a general remark, a very slight decrease in the transmittance spectra can be observed in the case of solutions functionalized with DNA-CTMA compared to those without DNA; this fact does not impede the use of the DNA-CTMA-Turmeric compound in optical limiting. This difference may be due to the intercalation of the natural dye of Turmeric in the double-strand helical structure of DNA, which is an advantage that contributes to increasing the stability of the obtained DNA-CTMA complex. All these increase the resistance of solutions to photodegradation.

### 2.1.2. Photostability of Prepared Solutions

The chemical stability of turmeric and DNA-CTMA-Turmeric complex in butanol exposed to UVA light has been studied to investigate the resistance of synthesized solutions to the action of a light source. The photodegradation of the synthesized solutions in butanol was carried out using the Luzchem Photoreactor—4× (Luzchem Research Inc., Ottawa, ON, Canada), equipped with 8 lamps, each with a power of 12 W. The natural extract was exposed to UVA irradiation at a wavelength of 325 nm. The photoreactor is equipped with a fan to maintain a constant temperature throughout the experiment. To perform the photodegradation experiments, constant volumes of the synthesized solutions and covered quartz cuvettes were used to avoid evaporation. The absorption spectra of the biopolymers in butanol were recorded both at the initial moment and after each exposure to irradiation. The absorption spectra were recorded using the Thermo Scientific Spectrophotometer

(Fisher Scientific, Paisley, UK), Model 220. Figure 3 shows the overlays of the absorption spectra of investigated DNA-CTMA-Turmeric complexes with different concentrations and of the natural extracts of Turmeric without DNA (with the same concentrations) exposed to irradiation for different exposure times.

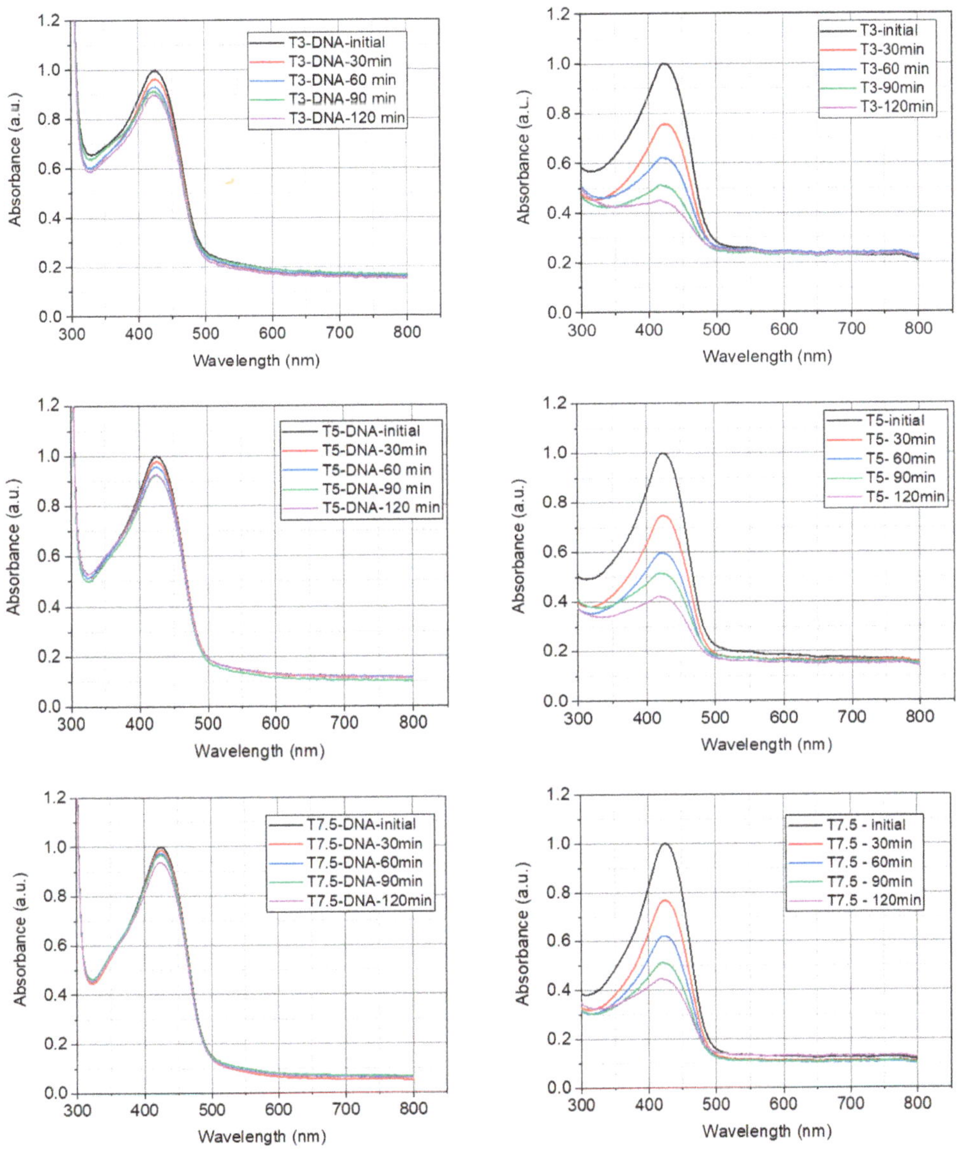


**Figure 3.** *Cont.*

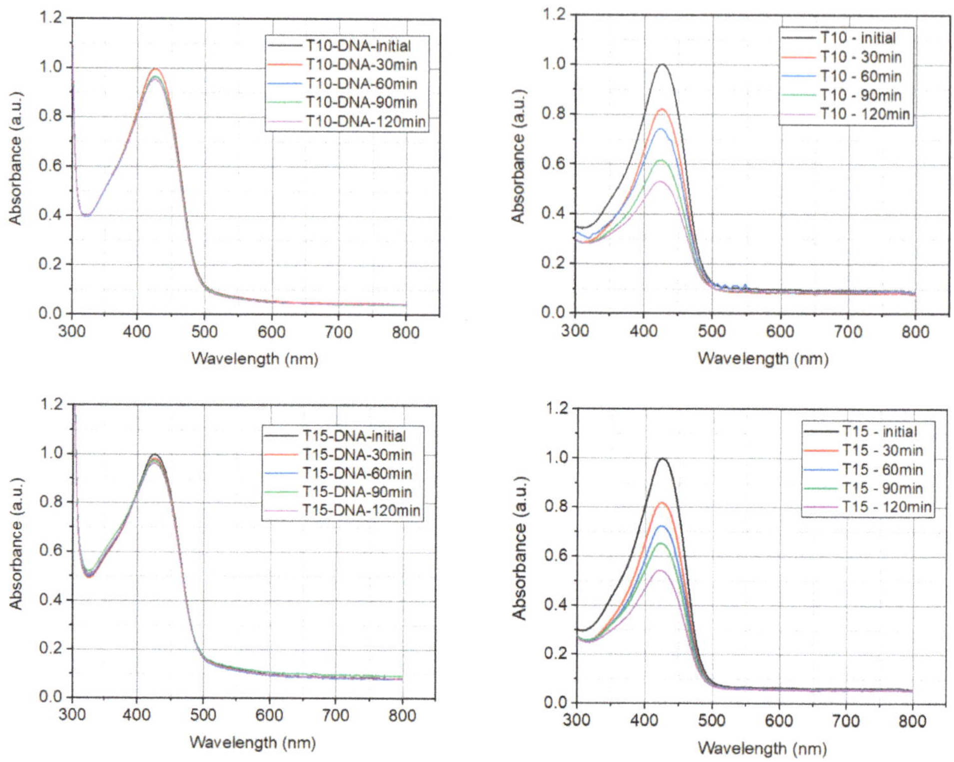


(a) (b)

**Figure 3.** The evolution of the UV-VIS absorption spectra of the prepared solutions in butanol, under the influence of UVA irradiation at several exposure times: (**a**) DNA-CTMA-Turmeric (T3-DNA ÷ T15-DNA) and (**b**) Turmeric solutions in butanol (T3 ÷ T15).

The absorption spectra acquired for each sample have been normalized to the maximum value of the amplitude which corresponds to the initial spectrum, recorded before the start of irradiation with UVA light. From the acquired spectrum, it can be seen that as the exposure time increases, the photodegradation of the Turmeric extract occurs much faster than the photodegradation of the DNA-CTMA-Turmeric, proved by a much higher decrease in absorbance. At the same time, increasing the concentration of Turmeric leads to a decrease in the photodegradation rate.

The dependence on time of the peaks of normalized absorbance under the influence of UVA irradiation of the investigated solutions is shown for both sets of samples, T3-DNA ÷ T15-DNA and T3 ÷ T15, in Figure 4a,b, respectively. In this figure the experimental points, for all investigated solutions, have been fitted with exponential decay functions, $y_{Pi}(t) = y_{0Pi} + A_{Pi} \cdot exp(-k_{Pi} \cdot t)$, $i = 1, \ldots, 5$, shown as continuous lines.

The parameters $y_{0Pi}$, $A_{Pi}$, and $k_{Pi}$ of the fitting functions for the experimental data from Figure 4a,b, corresponding to the investigated solutions, are shown in Table 2.

This kind of mathematical function, which describes the temporal decay of the normalized absorbance peak of investigated solutions in butanol, could be a consequence of the degradation of the dye molecules in the excited volume of the sample, namely its decrease over time, due to their photodegradation induced by UVA light.

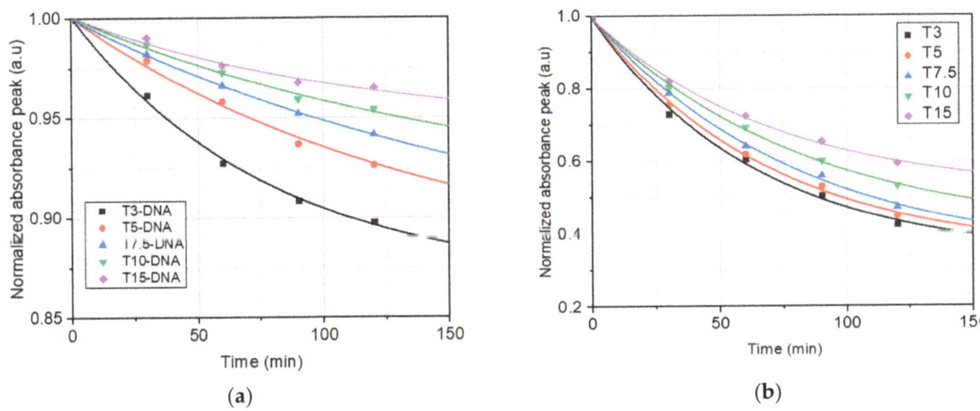


**Figure 4.** The temporal evolution of the normalized peak absorbance under the influence of UVA irradiation for: (**a**) DNA-CTMA-Turmeric solutions in butanol (T3-DNA ÷ T15-DNA) and (**b**) Turmeric solutions in butanol (T3 ÷ T15) (dots—experimental points; continuous lines—the fitting functions). The vertical scale is different in (**a**) with respect to (**b**).

**Table 2.** The parameters $y_{0Pi}$, $A_{Pi}$, and $k_{Pi}$, of the fitting functions for data points from Figure 4a,b.

| Fitting Parameter \ Film | T3-DNA/T3 | T5-DNA/T5 | T7.5-DNA/T7.5 | T10-DNA/T10 | T15-DNA/T15 |
|---|---|---|---|---|---|
| $y_{0Pi}$ | 0.8666/ 0.33452 | 0.87049/ 0.34271 | 0.87542/ 0.33081 | 0.90768/ 0.39744 | 0.94231/ 0.51004 |
| $A_{Pi}$ | 0.13429/ 0.66048 | 0.13029/ 0.65415 | 0.12578/ 0.66785 | 0.09363/ 0.59953 | 0.05685/ 0.48695 |
| $k_{Pi}$ (min$^{-1}$) | 0.01259/ 0.01569 | 0.00969/ 0.01453 | 0.00721/ 0.01405 | 0.00698/ 0.01246 | 0.00645/ 0.01117 |

The fit of the temporal decay of the normalized absorbance peak of investigated solutions in butanol under the influence of UVA irradiation (Figure 4) reveals that the photodegradation rates are higher in solutions with only Turmeric, compared to the solutions containing DNA. This is proof of the higher photostability of Turmeric dye in the DNA-CTMA-Turmeric solutions in butanol.

The photodegradation studies have shown that the combined use of DNA-CTMA and natural extract of Turmeric results in a slight improvement in the photostability of DNA-CTMA-Turmeric compared to Turmeric only. Moreover, these complexes show higher chemical stability under UVA illumination than synthetic polymers, such as PMMA, under similar irradiation conditions [66,67].

### 2.1.3. Thermal Degradation

Thermal analysis measurements were made to determine the thermal stability of Turmeric in butanol with/without DNA biopolymer.

To test the thermal degradation several cycles of heating the samples for 30 min at 60 °C have been performed. The absorption spectra have been recorded, using a Thermo Scientific Model Evolution 220 spectrophotometer (Fisher Scientific, Paisley, UK), both at the initial moment and after each exposure interval. The obtained absorption spectra are shown in Figure 5.

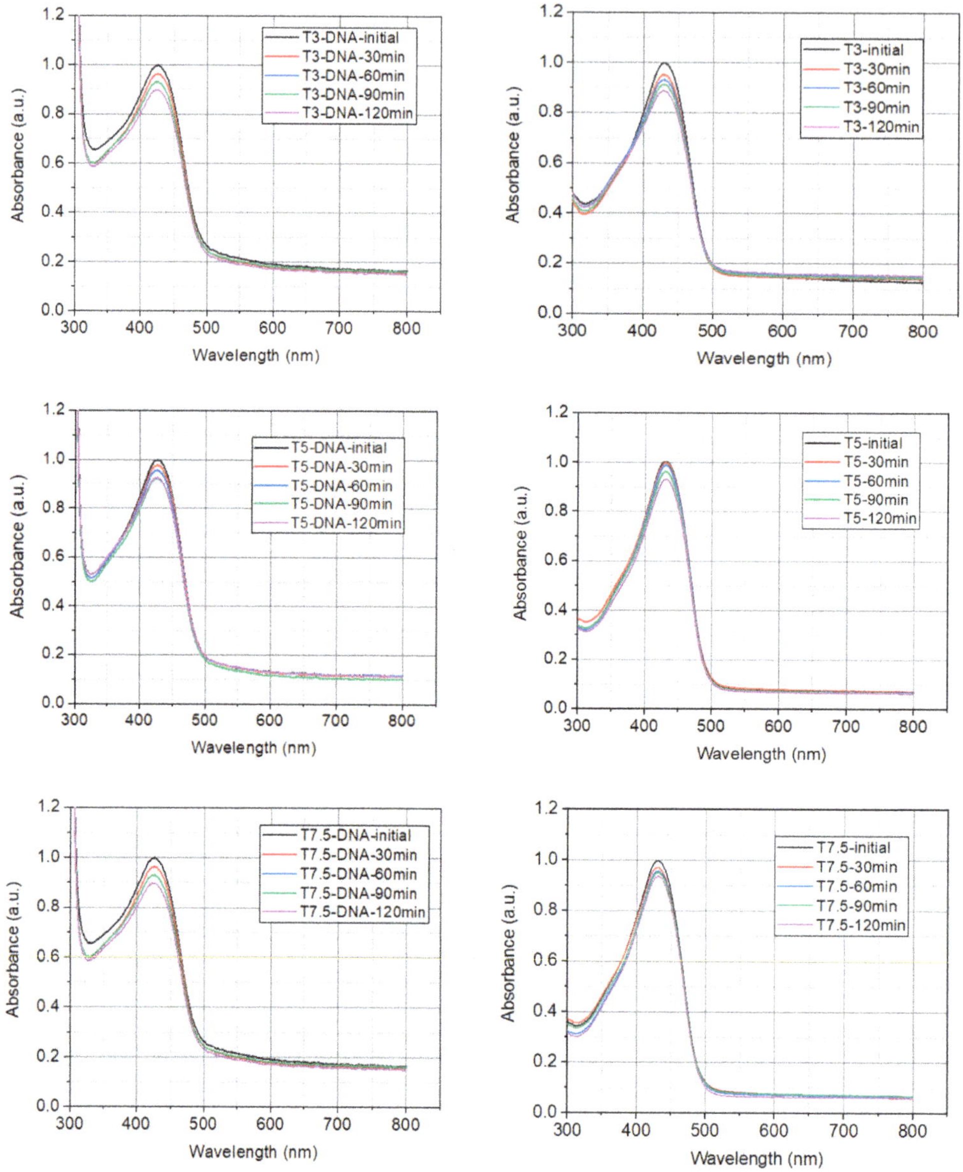


**Figure 5.** *Cont.*

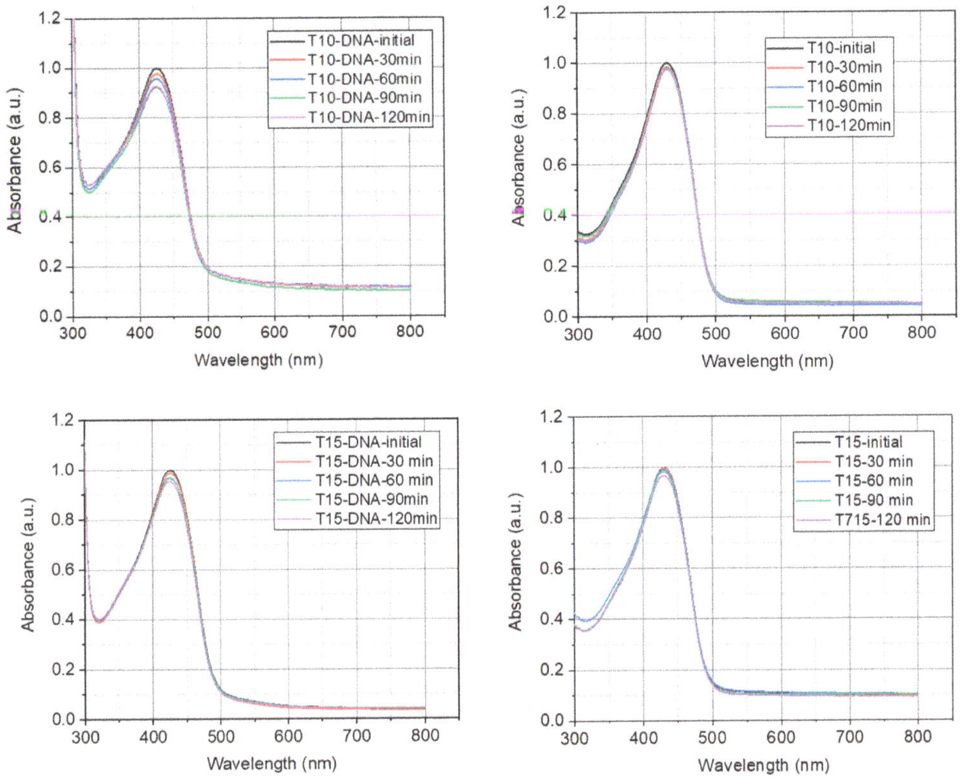


(**a**) (**b**)

**Figure 5.** The evolution of the absorption spectra of the prepared solutions in butanol under the influence of heating at 60 °C, for several exposure times, (**a**) DNA-CTMA-Turmeric (T3-DNA ÷ T15-DNA) and (**b**) Turmeric solutions in butanol (T3 ÷ T15).

The dependence on time of the peaks of normalized absorbance under the influence of heating at 60 °C of the investigated solutions is shown together for both sets of samples, T3-DNA ÷ T15-DNA and T3 ÷ T15, in Figure 6a,b, respectively. In this figure the experimental points for all investigated solutions have been fitted with exponential decay functions, $y_{Ti}(t) = y_{0Ti} + A_{Ti}{\cdot}exp(-k_{Ti}{\cdot}t)$, $i = 1, \ldots, 5$, shown as continuous lines.

The parameters $y_{0Ti}$, $A_{Ti}$, and $k_{Ti}$ of the fitting functions for the experimental data from Figure 6a,b, corresponding to the investigated solutions, are shown in Table 3.

**Table 3.** The parameters $y_{0Ti}$, $A_{Ti}$, and $k_{Ti}$, of the fitting functions for data points from Figure 6a,b.

| Film / Fitting Parameter | T3-DNA/T3 | T5-DNA/T5 | T7.5-DNA/T7.5 | T10-DNA/T10 | T15-DNA/T15 |
|---|---|---|---|---|---|
| $y_{0Ti}$ | 0.8839/ 0.87714 | 0.9000/ 0.90468 | 0.9158/ 0.91292 | 0.92052/ 0.9204 | 0.92547/ 0.9310 |
| $A_{Ti}$ | 0.11726/ 0.12183 | 0.10091/ 0.0953 | 0.08525/ 0.08672 | 0.08021/ 0.7964 | 0.07502/ 0.07001 |
| $k_{Ti}$ (min$^{-1}$) | 0.0141/ 0.01512 | 0.01159/ 0.01449 | 0.01151/ 0.01299 | 0.00882/ 0.01111 | 0.00703/ 0.00943 |

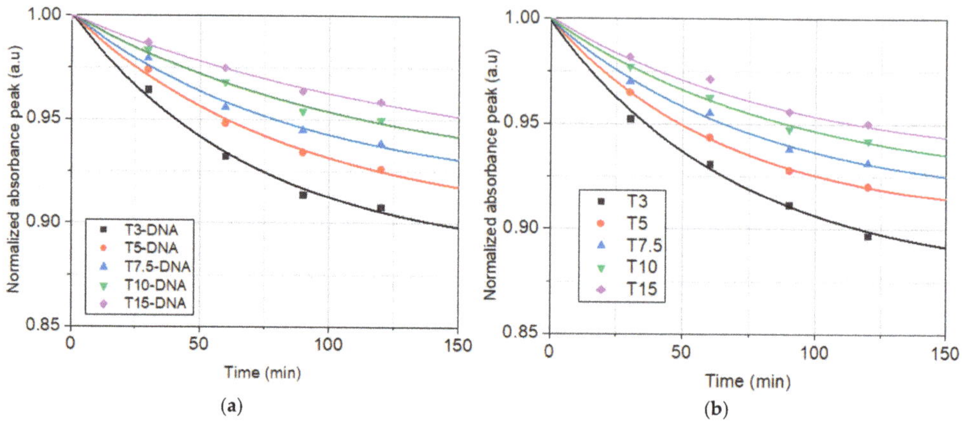


**Figure 6.** The temporal evolution of the normalized peak absorbance under the influence of heating at 60 °C for: (**a**) DNA-CTMA-Turmeric solutions in butanol (T3-DNA ÷ T15-DNA) and (**b**) Turmeric solutions in butanol (T3 ÷ T15) (dots—experimental points; continuous lines—the fitting functions).

The fit of the temporal decay of the normalized absorbance peak of investigated solutions heated at 60 °C (Figure 6) reveals that the differences between the thermal stability of the investigated solutions with and without DNA are much lower than the differences between their photostability (Figure 4). However, the thermal stability of DNA-CTMA-Turmeric solutions is slightly better than that of the Turmeric solutions. Thus, the analysis of the temporal decay of the normalized absorption spectra under illumination with UVA light and heating, respectively, revealed a better photo- and thermal stability of solutions containing DNA-CTMA.

## 3. Optical Limiting Capability: Measurements and Discussions

The OL capabilities, dependent on the nonlinear optical properties of these new biomaterials, were explored by the Intensity scan (I-scan) technique [65,66]. The I-scan is a sensitive and powerful optical experimental method for the investigation of nonlinear optical refraction and/or absorption. A schematic of the I-scan configuration is shown in Figure 7.

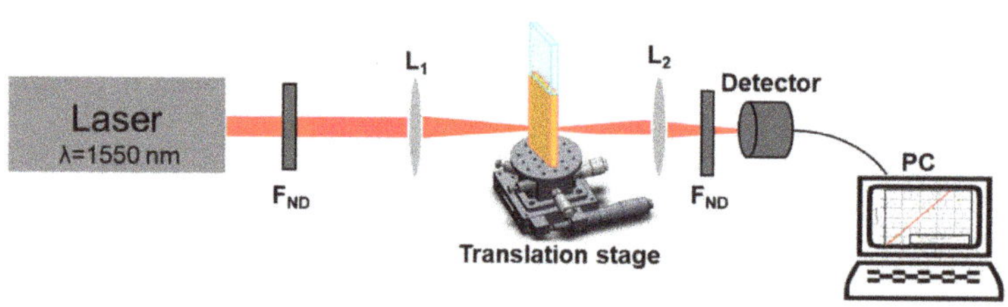


**Figure 7.** Schematic experimental configuration for optical limiting capability investigation by I-scan method.

In the I-scan measurements, we used as excitation source an Er-doped fiber laser (FemtoFiber Scientific FFS, TOPTICA Photonics AG, Munich, Germany) at the wavelength λ = 1550 nm, which generates ultrashort pulses about ~120 fs with a repetition rate of 76 MHz. The intensity incident on the sample is controlled with neutral density filters ($F_{ND}$) (Thorlabs, Munich, Germany), which changes the laser power. All the filters used in the I-scan measurements are calibrated by us at the wavelength of 1550 nm. The incident beam on the sample has a maximum average power of ~193 mW. The corresponding peak power and the pulse energy of the generated laser pulses are ~20 kW and ~2.5 nJ, respectively. The average incident laser powers are chosen to be below the values for which the laser-induced damage could occur.

The investigated biopolymers are placed in the focal plane of the lens $L_1$ which has a focal length of 5 cm and focuses the incident beam to a spot with a diameter of 26 μm on the sample. The investigated samples are in the form of solutions and they have been placed in special demountable cuvettes, with optical quality windows made from quartz (Hellma, Müllheim, Germany) of 0.5 mm internal thickness and 130 μL internal volume. Their spectral domain of transparency is 200–2500 nm. These cuvettes were fixed on the micrometric translation stages for the fine-tuning of their positions. The power of both incident and transmitted beams has been corrected to Fresnel reflections on external (air–quartz) and internal (quartz–solution) interfaces. The $L_2$ (focal length 6 cm) lens is used to adjust the spot size of the transmitted laser beam to the aperture of the detector which measures the beam's average power. For precise adjustment of the lenses' positions relative to the sample, we used micrometric translation stages (not shown in Figure 6). In front of the detector are placed different neutral density filters ($F_{ND}$) to keep the power of the light incident on the detector, below its maximum measurable power. The incident and the transmitted beams were measured using the OP-2IR sensor (Coherent, Portland, OR, USA) coupled to a power meter (FieldMax II-TOP, Coherent, Saxonburg, PA, USA). No optical damage was observed in the investigated samples in the range of the laser peak intensities available in our experiments.

The I-scan method [68] was developed as a variant of the Z-scan technique, presenting a series of advantages compared to the classic Z-scan method [69]. In the case of the Z-scan method, during the movement of the sample along the focused light path (Z axis), the size of the illuminated area of the sample is dependent on its position with respect to the focal plane of the focusing lens, and the collected signal is averaged on areas of different sizes. This could be a problem in the case of inhomogeneous samples. In the I-scan method [68,69], the sample position on the optical axis of the system is fixed. The laser intensity on the sample is modified by neutral density filters ($F_{ND}$).

The OL potential of the investigated DNA-based compounds was evaluated by measuring the transmitted peak intensities $I_{trans}$(peak) of the samples as a function of the incident peak intensities $I_{inc}$(peak). A deviation from the linear transmittance towards lower values is an indicator of optical limiting potential. This property of a nonlinear material is described by a linear transmission for low values of the incident light until a certain threshold value of the incident intensity is reached, after which the transmitted intensity remains constant when the incident intensity increases.

The experimental results obtained during the investigation of the optical limiting capability of the biopolymer samples are shown and discussed below. The experimental peak intensities $I_{trans}$(peak), of the beam transmitted through sample, in function of the peak intensities, $I_{inc}$(peak), of the incident beam, are shown in Figure 8.

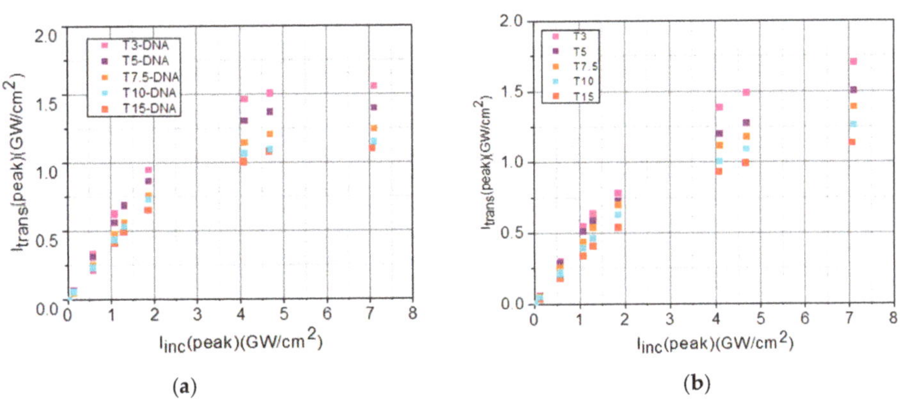


**Figure 8.** Transmitted peak intensities in function of the incident peak intensities for: (**a**) DNA-CTMA-Turmeric solutions in butanol (T3-DNA ÷ T15-DNA) and (**b**) Turmeric solutions in butanol (T3 ÷ T15).

The values of the experimentally determined transmitted peak intensities, $I_{trans}$(peak), for low incident peak intensities, $I_{inc}$(peak), shown in Figure 8, have been fitted with a linear dependence described by the Lambert–Beer law, Equation (1):

$$I_{trans,\ linear} = I_{inc} \cdot e^{-\alpha_0 \cdot L} \tag{1}$$

From the slope of this linear dependence (Figure 9), we determined the values of the linear absorption coefficient, $\alpha_0$, of each sample.

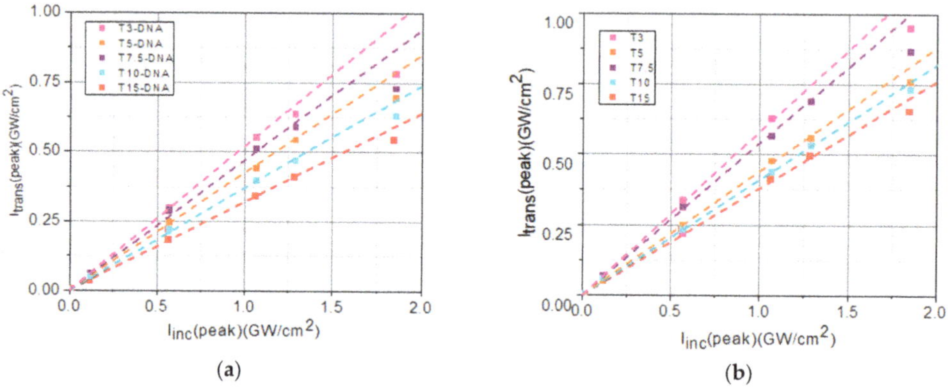


**Figure 9.** Transmitted peak intensities in function of the incident peak intensities, $I_{inc}$(peak), for low incident intensities, fitted (dash lines) with a linear dependence (Equation (1)) for: (**a**) DNA-CTMA-Turmeric solutions in butanol (T3-DNA ÷ T15-DNA) and (**b**) Turmeric solutions in butanol (T3 ÷ T15).

A nonlinear dependence of $I_{trans}$(peak) of the samples as a function of the incident peak intensities $I_{inc}$(peak) indicates that the absorption coefficient is no longer constant, being dependent on incident intensity. In this case, Equation (1) becomes:

$$I_{trans,\ NL} = I_{inc} \cdot e^{-\alpha(I_{inc})L} \tag{2}$$

In our experiments, the dependence $I_{trans}$(peak) of the samples as a function of the incident peak intensities $I_{inc}$(peak) shows a saturation trend. We considered for $\alpha(I_{inc})$ from Equation (2) the following equation, in which the saturation intensity, $I_{sat}$, and the nonlinear absorption coefficient, $\beta$, are considered [67]:

$$\alpha(I_{inc}) = \frac{\alpha_0}{1 + \frac{I_{inc}}{I_{sat}}} + \beta \cdot I_{inc} \tag{3}$$

The experimental data from Figure 8 have been fitted with de Equation (2), taking into account Equation (3), as shown in Figure 10. From the fit of the experimental data using Equations (2) and (3), the values of $I_{sat}$ and of $\beta$ have been determined.

The values of linear and nonlinear optical parameters (the linear transmittance, $T_L$, the linear absorption coefficient, $\alpha_0$, the saturation intensity, $I_{sat}$, and the nonlinear absorption coefficient, $\beta$) determined by fitting the experimental data from Figures 9 and 10 are summarized in Table 4.

The analysis of the data from Table 4 and of their graphical representation from Figure 11 reveals the influence of the DNA-CTMA complex on linear and nonlinear optical absorption properties of the investigated materials, which are involved in the passive OL functionality.

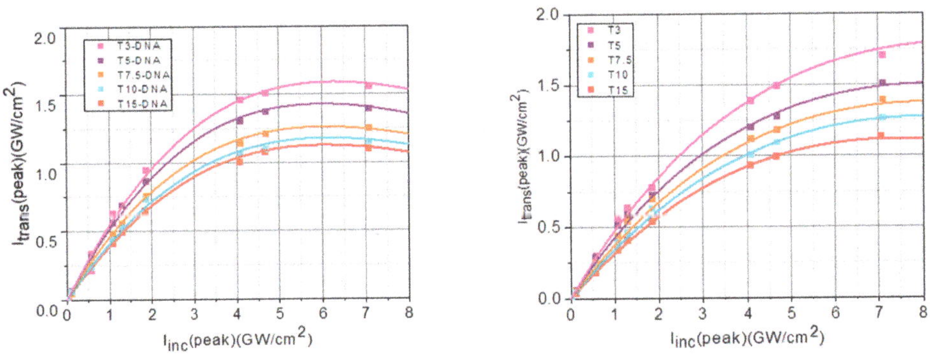


(a) (b)

**Figure 10.** Transmitted peak intensities in function of the incident peak intensities, $I_{inc}$(peak), fitted (continuous lines) with the nonlinear dependence given by Equation (2). for: (**a**) DNA-CTMA-Turmeric solutions in butanol (T3-DNA ÷ T15-DNA) and (**b**) Turmeric solutions in butanol (T3 ÷ T15).

**Table 4.** The linear transmittance, $T_L$, the linear absorption coefficient, $\alpha_0$, the saturation intensity, $I_{sat}$, and the nonlinear absorption coefficient, $\beta$, determined from the fit of the experimental data (Figures 9 and 10).

| Sample | Linear Transmittance | $\alpha_0$ (cm$^{-1}$) | $I_{sat}$ (GW/cm$^2$) | $\beta$ (cm/GW) |
|---|---|---|---|---|
| T3-DNA/T3 | 0.43/0.52 | 17/13 | 14/38 | 3.5/2.4 |
| T5-DNA/T5 | 0.38/0.47 | 19/15 | 12/35 | 3.6/2.7 |
| T7.5-DNA/T7.5 | 0.36/0.43 | 21/17 | 11/21 | 3.8/2.9 |
| T10-DNA/T10 | 0.31/0.37 | 23/20 | 7/18 | 3.9/3.0 |
| T15-DNA/T15 | 0.27/0.32 | 26/23 | 6/16 | 4.0/3.4 |

It can be observed that the set of solutions with DNA-CTMA-Turmeric has a lower linear transmittance than the samples that do not contain DNA-CTMA. At the same time, the increase in the dye concentration decreases the linear transmittance. Also, the solutions containing DNA-CTMA have lower linear transmittances compared to solutions with only Turmeric. The range of absorption coefficient values can thus be extended/restricted to higher/lower values in the investigated compounds, with a tunability given by the dye concentration. The results reveal the relationship between the dye concentrations and the magnitude of the absorption coefficient.

For samples with different concentrations, the OL is stronger for those with lower $I_{sat}$. Equation (2) provides information on the overall optical limiting capability of a sample, considering the nonlinear optical processes involved in this functionality. For samples with different linear transmittances, a larger deviation from the straight line (corresponding to linear transmission), at the same incident intensity, means better limitation.

The DNA-based matrix favorably influences the OL potential of the investigated DNA-CTMA-Turmeric samples, compared to Turmeric-only samples, in two different ways, one related to the OL process itself and the other one related to the photo- and thermal stability of the material. The performed OL experiments revealed that the OL process is more efficient in solutions containing DNA-CTMA, which is evidenced in Figures 9 and 10, and Table 4. Thus, the deviation of the saturation-type fit of the experimental data from the straight line corresponding to linear transmittance is larger in samples with DNA-CTMA, where the transition from the linear regime to the limiting regime starts at lower values of incident intensity (Figure 9) and, consequently, the intensity range of optical limiting is broader. Moreover, the saturation intensity, $I_{sat}$, is lower and the nonlinear absorption coefficient, $\beta$, is higher in these samples, as shown in Table 4 and Figure 11c,d. On the other hand, the results presented in Sections 2.1.2 and 2.1.3 revealed increased photo- and thermal stability of samples with DNA-CTMA, which is important for OL functionality. Thus, the DNA-CTMA matrix is beneficial for the OL potential of Turmeric.

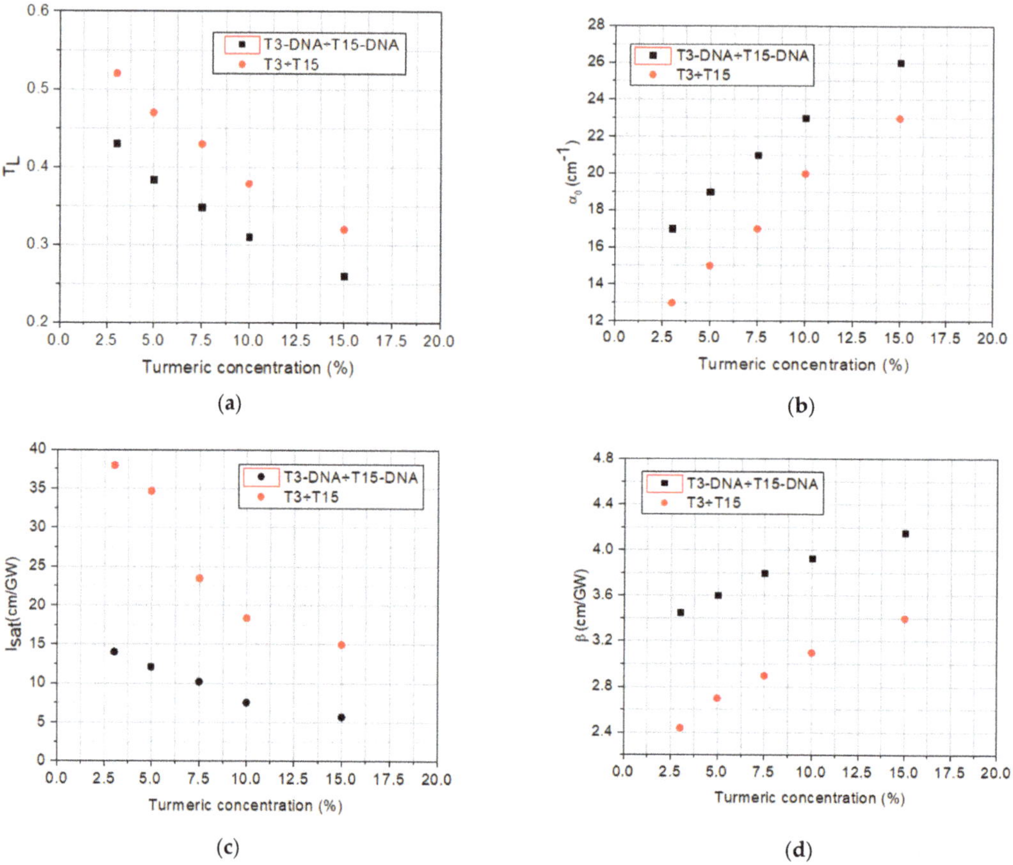


**Figure 11.** The dependences on Turmeric concentration of the linear transmittance, $T_L$ (**a**), linear absorption coefficient, $\alpha_0$ (**b**), saturation intensity, $I_{sat}$ (**c**), and the nonlinear absorption coefficient, $\beta$ (**d**).

## 4. Conclusions

The optical limiting in solutions in butanol of DNA-CTMA-Turmeric compound and of Turmeric with similar concentrations has been experimentally demonstrated, for the first time to the best of our knowledge, by Intensity scan experiments with ultrashort laser pulses (~120 fs) in NIR at 1550 nm wavelength. The photostability and thermal stability of these solutions have also been evaluated. The analysis of the temporal decay of the normalized absorption spectra under illumination with UVA light and heating, respectively, revealed a better photo- and thermal stability of Turmeric in solutions containing DNA-CTMA.

The efficiency of the passive optical limiting caused by the nonlinear optical absorption in the above-mentioned novel materials has been comparatively evaluated. The analysis of the experimental data obtained in the I-scan experiment allowed the determination of some important optical parameters. Thus, from the analysis of the initial part of the experimental curves, when the material's response is linear, the linear transmittance and the linear absorption coefficient have been determined. At higher incident intensities, the saturation trend of experimental curves and a nonlinear absorption have been considered. This analysis allowed us to determine the saturation intensity and the nonlinear absorption coefficient of the investigated materials, both of which are important for their OL capability. The study performed revealed a favorable effect of the DNA-CTMA matrix on OL in the investigated materials.

The obtained results are important in the applications of these novel DNA-based, eco-friendly materials for the protection of the human eye and sensitive devices against intense NIR laser radiation.

**Author Contributions:** Conceptualization, P.G. and A.P.; methodology, P.G. and A.P.; validation, P.G. and A.P.; investigation, P.G. and A.M.A.; resources, P.G.; data curation, P.G. and A.M.A.; writing—original draft preparation, P.G. and A.P.; writing—review and editing, P.G. and A.P.; visualization P.G. and A.P.; supervision, P.G. and A.P.; project administration, P.G.; funding acquisition, P.G. All authors have read and agreed to the published version of the manuscript.

**Funding:** This work was supported by two grants of the Romanian Ministry of Research, Innovation and Digitization, CNCS—UEFISCDI, project number PN-III-PI-1.1-TE-2021-1546 and PN-III-PI-1.1-TE-2021-0873, within PNCDI III and by Project 30N/2023 within program NUCLEU LAPLAS VII.

**Institutional Review Board Statement:** Not applicable.

**Data Availability Statement:** Data are contained within the article.

**Conflicts of Interest:** The authors declare no conflicts of interest.

## References


1. Das, B.C.; Reji, N.; Philip, R. Optical limiting behavior of the natural dye extract from Indigofera Tinctoria leaves. *Opt. Mater.* **2021**, *114*, 110925.
2. Marbello, O.; Valbuena, S.D.; Racedo, F.J.N. Optical limiting phenomenon study in oils of vegetable origin. *J. Appl. Res. Technol.* **2020**, *18*, 333–340.
3. Haripadmam, P.C.; Beryl, C.; Philip, R. Optical Limiting Properties of the Natural Dye Extract from *Alternanthera brasiliana* Leaves. *J. Electron. Mater.* **2022**, *51*, 3364–3371. [CrossRef]
4. Petris, A.; Gheorghe, P.; Rau, I. DNA—CTMA matrix influence on Rhodamine 610 light emission in thin films. *Polymers* **2023**, *15*, 3105. [CrossRef]
5. Petris, A.; Gheorghe, P.; Rau, I. Influence of continuous wave laser light at 532 nm on transmittance and on photoluminescence of DNA-CTMA-RhB solutions. *Heliyon* **2023**, *9*, e20410. [CrossRef]
6. Bazaru Rujoiu, T.; Petris, A.; Vlad, V.I.; Rau, I.; Manea, A.M.; Kajzar, F. Lasing in DNA–CTMA doped with Rhodamine 610 in butanol. *Phys. Chem. Chem. Phys.* **2015**, *17*, 13104–13111. [CrossRef]
7. Xia, T.; Hagan, D.J.; Dogariu, A.; Said, A.A.; Van Stryland, E.W. Optimization of optical limiting devices based on excited-state absorption. *Appl. Opt.* **1997**, *36*, 4110. [CrossRef]
8. Liu, Z.; Zhang, B.; Chen, Y. Recent Progress in Two-Dimensional Nanomaterials for Laser Protection. *Chemistry* **2019**, *1*, 17–43. [CrossRef]
9. Barbinta-Patrascu, M.-E.; Iordache, S.M. DNA—The fascinating biomacromolecule in optoelectronics and photonics applications. *J. Optoelectron. Adv. Mater.* **2022**, *24*, 563.
10. Lundén, H.; Glimsdal, E.; Lindgren, M.; Lopesa, C. How to assess good candidate molecules for self-activated optical power limiting. *Opt. Eng.* **2018**, *57*, 030802. [CrossRef]
11. Dini, D.; Calvete, M.J.F.; Hanack, M. Nonlinear Optical Materials for the Smart Filtering of Optical Radiation. *Chem. Rev.* **2016**, *116*, 13043–13233. [CrossRef]
12. Anton, A.M.; Rau, I.; Kajzar, F.; Simion, A.M.; Pirvu, C.; Radu, N.; Simion, C. Natural materials with enhanced optical damage threshold. *Opt. Mater.* **2018**, *86*, 1–6. [CrossRef]
13. Liang, L.; Fu, Y.; Wang, D.; Wei, Y.; Kobayashi, N.; Minari, T. DNA as Functional Material in Organic-Based Electronics. *Appl. Sci.* **2018**, *8*, 90. [CrossRef]
14. Aithal, S.; Aithal, P.S.; Bhat, G.K. Characteristics of ideal optical limiter and realization scenarios using nonlinear organic materials. *Int. J. Adv. Trends Eng. Technol.* **2016**, *1*, 2456.
15. Sun, Y.P.; Riggs, J.E. Organic and inorganic optical limiting materials. From fullerenes to nanoparticles. *Int. Rev. Phys. Chem.* **1999**, *18*, 43–90. [CrossRef]
16. Wang, J.; Werner, J.B. Inorganic and hybrid nanostructures for optical limiting. *J. Opt. Pure Appl. Opt.* **2009**, *11*, 024001. [CrossRef]
17. Parola, S.; Julián-López, B.; Carlos, L.D.; Sanchez, C. Optical Properties of Hybrid Organic–Inorganic Materials and their Applications—Part II: Nonlinear Optics and Plasmonics. In *Handbook of Solid State Chemistry*; Part 4 Nano and Hybrid Materials; Wiley: Hoboken, NJ, USA, 2017.

18. Jancy, J.; Vinoy, T.; Sujesh, B.; Ramakrishnan, J.; Sebastian, M.; Ibrahimkutty, R.; Abdulhassan, M. Open-aperture Z-scan and optical limiting of plasmonic silver-polymer system. *J. Optoelectron. Adv. Mater.* **2022**, *24*, 250.
19. Chen, Y.; Bai, T.; Dong, N.; Fan, F.; Zhang, S.; Zhuang, X.; Sun, J.; Zhang, B.; Zhang, X.; Wang, J.; et al. Graphene and its derivatives for laser protection. *Prog. Mater. Sci.* **2016**, *84*, 118–157. [CrossRef]
20. Wang, A.J.; Yu, W.; Fang, Y.; Song, Y.L.; Jia, D.; Long, L.L.; Cifuentes, M.P.; Humphrey, M.G.; Zhang, C. Facile Hydrothermal Synthesis and Optical Limiting Properties of $TiO_2$-Reduced Graphene Oxide Nanocomposites. *Carbon* **2015**, *89*, 130–141. [CrossRef]
21. Zhao, M.; Peng, R.; Zheng, Q.; Wang, Q.; Chang, M.J.; Liu, Y.; Song, Y.L.; Zhang, H.L. Broadband optical limiting response of a graphene–PbS nanohybrid. *Nanoscale* **2015**, *7*, 9268–9274. [CrossRef]
22. Rau, I.; Grote, J.G.; Kajzar, F.; Pawlicka, A. DNA–novel nanomaterial for applications in photonics and in electronics. *Comptes Rendus Phys.* **2012**, *13*, 853–864. [CrossRef]
23. Petris, A.; Vasiliu, I.C.; Gheorghe, P.; Iordache, A.M.; Ionel, L.; Rusen, L.; Iordache, S.; Elisa, M.; Trusca, R.; Ulieru, D.; et al. Graphene Oxide-Based Silico-Phosphate Composite Films for Optical Limiting of Ultrashort Near-Infrared Laser Pulses. *Nanomaterials* **2020**, *10*, 1638. [CrossRef]
24. Graphene Report 2020, Description. Available online: https://www.researchandmarkets.com/reports/4901148/the-graphene-report-2020 (accessed on 1 June 2020).
25. Liu, Z.; Wang, Y.; Zhang, X.; Xu, Y.; Chen, Y.; Tian, J. Nonlinear optical properties of graphene oxide in nanosecond and picosecond regimes. *Appl. Phys. Lett.* **2009**, *94*, 021902. [CrossRef]
26. Liaros, N.; Aloukos, P.; Kolokithas-Ntoukas, A.; Bakandritsos, A.; Szabo, T.; Zboril, R.; Couris, S. Nonlinear Optical Properties and Broadband Optical Power Limiting Action of Graphene Oxide Colloids. *J. Phys. Chem. C* **2013**, *117*, 6842–6850. [CrossRef]
27. Xu, Y.; Liu, Z.; Zhang, X.; Wang, Y.; Tian, J.; Huang, Y.; Ma, Y.; Zhang, X.; Chen, Y. A Graphene Hybrid Material Covalently Functionalized with Porphyrin: Synthesis and Optical Limiting Property. *Adv. Mater.* **2009**, *21*, 1275–1279. [CrossRef]
28. Liaros, N.; Orfanos, I.; Papadakis, I.; Couris, S. Nonlinear optical response of some Graphene oxide and Graphene fluoride derivatives. *Optofluid. Microfluid. Nanofluid.* **2016**, *3*, 53–58. [CrossRef]
29. Stathis, A.; Stavrou, M.; Papadakis, I.; Obratzov, I.; Couris, S. Enhancing and Tuning the Nonlinear Optical Response and Wavelength-Agile Strong Optical Limiting Action of N-octylamine Modified Fluorographenes. *Nanomaterials* **2020**, *10*, 2319. [CrossRef]
30. Jiang, X.F.; Polavarapu, L.; Neo, S.T.; Venkatesan, T.; Xu, Q. Graphene Oxides as Tunable Broadband Nonlinear Optical Materials for Femtosecond Laser Pulses. *J. Phys. Chem. Lett.* **2012**, *3*, 785–790. [CrossRef]
31. Zheng, Z.; Zhu, L.; Zhao, F. Nonlinear Optical and Optical Limiting Properties of Graphene Oxide Dispersion in Femtosecond Regime. *Proc. SPIE* **2014**, *9283*, 92830V-1.
32. Ren, J.; Zheng, X.; Tian, Z.; Li, D.; Wang, P.; Jia, B. Giant third-order nonlinearity from low-loss electrochemical graphene oxide film with a high power stability. *Appl. Phys. Lett.* **2016**, *109*, 221105. [CrossRef]
33. Oluwafemi, O.S.; Sreekanth, P.; Philip, R.; Thomas, S.; Kalarikkal, N. Improved nonlinear optical and optical limiting properties in non-covalent functionalized reduced graphene oxide/silver nanoparticle (NF-RGO/Ag-NPs) hybrid. *Opt. Mater.* **2016**, *58*, 476–483.
34. Sooraj, B.N.S.; Pradeep, T. Chapter 4—Optical properties of metal clusters. In *Atomically Precise Metal Nanoclusters*; Elsevier: Amsterdam, The Netherlands, 2023; pp. 83–101.
35. Xing, F.; Wang, Y.; Wang, J.; Zhou, S.; Zhao, J.; Xie, Z. Highly dispersed antimonene oxide quantum dots and their hybrid gel glasses for broadband nonlinear optical limiting. *J. Mater. Chem. C* **2021**, *9*, 10084–10088. [CrossRef]
36. Sun, X.; Hu, X.; Sun, J.; Xie, Z.; Zhou, S. Strong optical limiting properties of Ormosil gel glasses doped with silver nano-particles. *New J. Chem.* **2019**, *43*, 6274–6278. [CrossRef]
37. Zidan, M.D.; Ajji, Z. Optical limiting behavior of disperse red 1 dye doped polymer. *Opt. Laser Technol.* **2011**, *43*, 934–937. [CrossRef]
38. Aithal, S.; Aithal, P.S.; Bhat, G.K. CW Optical Limiting Study in Disperse Yellow Dye-Doped PMMA-MA Polymer Films. *IRA-Int. J. Appl. Sci.* **2016**, *5*, 129–146. [CrossRef]
39. Rau, I.; Kajzar, F. Biopolymers for application in photonics. *Sci. J. Volgograd State Univ.* **2014**, *4*, 29–41.
40. Moldoveanu, M.; Meghea, A.; Popescu, R.; Grote, J.; Kajzar, F.; Rau, I. On the stability and degradation of DNA based thin films. *Mol. Cryst. Liq. Cryst.* **2010**, *523*, 182–190. [CrossRef]
41. Steckl, A.J. DNA—A new material for photonics? *Nat. Photonics* **2007**, *1*, 3–5. [CrossRef]
42. Petris, A.; Gheorghe, P.; Vlad, V.I.; Rau, I.; Kajzar, F. Interferometric method for the study of spatial phase modulation induced by light in dye-doped DNA complexes. *Rom. Rep. Phys.* **2015**, *67*, 1373–1382.
43. Gheorghe, P.; Petris, A.; Vlad, V.I.; Rau, I.; Kajzar, F.; Manea, A.M. Temporal evolution of the laser recording of gratings in DNA-CTMA:Rh610 films. *Rom. Rep. Phys.* **2015**, *67*, 1412–1420.
44. Khazaeinezhad, R.; Kassani, S.H.; Paulson, B.; Jeong, H.; Gwak, J.; Rotermund, F.; Yeom, D.; Oh, K. Ultrafast nonlinear optical properties of thin-solid DNA film and their application as a saturable absorber in femtosecond mode-locked fiber laser. *Sci. Rep.* **2017**, *7*, 41480. [CrossRef]
45. Grote, J. Biotechnology in biopolymers: Developments, applications & challenging areas. *SPIE Newsroom* **2008**, *15*.
46. Petris, A.; Gheorghe, P.S.; Rau, I.; Manea-Saghin, A.M.; Kajzar, F. All-optical spatial phase modulation in films of dye-doped DNA biopolymer. *Eur. Polym. J.* **2019**, *110*, 130–137. [CrossRef]
47. Dancus, I.; Vlad, V.I.; Petris, A.; Bazaru Rujoiu, T.; Rau, I.; Kajzar, F.; Meghea, A.; Tane, A. Z-Scan and I-Scan methods for characterization of DNA Optical Nonlinearities. *Rom. Rep. Phys.* **2013**, *65*, 966.

48. Gonzalez, A.M.; Vilhena, J.G.; Perez, R.; Herrero, F.M. A molecular view of DNA flexibility. *Q. Rev. Biophys.* **2021**, *54*, e8.
49. Gonzalez, A.M.; Vilhena, J.G.; Herrero, F.M.; Perez, R. DNA Crookedness Regulates DNA Mechanical Properties at Short Length Scales. *Phys. Rev. Lett.* **2019**, *122*, 048102. [CrossRef]
50. Gonzalez, A.M.; Vilhena, J.G.; Herrero, F.M.; Perez, R. Sequence-dependent mechanical properties of double-stranded RNA. *Nanoscale* **2019**, *11*, 21471–21478. [CrossRef]
51. Moldoveanu, M.; Popescu, R.; Pîrvu, C.; Grote, J.G.; Kajzar, F.; Rau, I. Biopolymer thin films for optoelectronics applications. *Mol. Cryst. Liq. Cryst.* **2010**, *522*, 530–539.
52. Rau, I.; Tane, A.; Zgarian, R.; Meghea, A.; Grote, J.G.; Kajzar, F. Stability of Selected Chromophores in Biopolymer Matrix. *Mol. Cryst. Liq. Cryst.* **2012**, *554*, 43–55. [CrossRef]
53. Grote, J.G. Materials Science of DNA—Conclusions and Perspectives. In *Materials Science of DNA*; Jin, J.-I., Grote, J., Eds.; CRC Press: Boca Raton, FL, USA, 2011; pp. 311–317.
54. Mindroiu, M.; Manea, A.-M.; Rau, I.; Grote, J.G.; Oliveira, H.; Pawlicka, A.; Kajzar, F. DNA- and DNA-CTMA: Novel bio-nanomaterials for application in photonics and in electronics. *Proc. SPIE* **2013**, *8882*, 888202.
55. Anton, A.-M.; Rau, I.; Kajzar, F.; Simion, A.-M.; Simion, C. Third order nonlinear optical properties of DNA-based biopolymers thin films doped with selected natural chromophores. *Opt. Mater.* **2019**, *88*, 181. [CrossRef]
56. Nizioł, J.; Makyła-Juzak, K.; Marzec, M.M.; Ekiert, R.; Marzec, M.; Gondek, E. Thermal stability of the solid DNA as a novel optical material. *Opt. Mater.* **2017**, *66*, 344–350. [CrossRef]
57. Nizioł, J.; Fiedor, J.; Pagacz, J.; Hebda, E.; Marzec, M.; Gondek, E.; Kityk, I.V. DNA-hexadecyltrimethyl ammonium chloride complex with enhanced thermostability as promising electronic and optoelectronic material. *J. Mater. Sci Mater. Electron.* **2017**, *28*, 259–268. [CrossRef]
58. Manea-Saghin, A.M.; Paduret, C.C.; Kajzar, F. Spectroscopy and non-linear optical properties of DNA—Bilberry complex. *Opt. Mater.* **2020**, *100*, 109669. [CrossRef]
59. Manea-Saghin, A.M.; Paduret, C.C.; Kajzar, F. Optical properties of DNA doped with blackcurrant and bilberry extracts. *Opt. Mater.* **2020**, *101*, 109721. [CrossRef]
60. Grote, J.G.; Diggs, D.E.; Nelson, R.L.; Zetts, J.S.; Hopkins, F.K.; Ogata, N.; Hagen, J.A.; Heckman, E.; Yaney, P.P.; Stone, M.O.; et al. DNA photonics [deoxyribonucleic acid]. *Mol. Cryst. Liq.* **2005**, *426*, 3–17. [CrossRef]
61. Singh, T.B.; Sariciftci, N.S.; Grote, J.G. Bio-Organic Optoelectronic Devices Using DNA. In *Organic Electronics, Advances in Polymer Science*; Springer: Berlin/Heidelberg, Germany, 2009.
62. Wang, L.; Yoshida, I.; Ogata, N. Self-Assembled Supramolecular Films Derived from Marine Deoxyribonucleic Acid (DNA)−Cationic Surfactant Complexes: Large-Scale Preparation and Optical and Thermal Properties. *Chem. Mater.* **2001**, *13*, 1273–1281. [CrossRef]
63. Prenting, M.M.; Shilikin, M.; Dreier, T.; Schultz, C.; Endres, T. Characterization of tracers for two-color laser-induced fluorescence thermometry of liquid-phase temperature in ethanol, 2–ethylhexanoic-acid/ethanol mixtures, 1-butanol, and o-xylene. *Appl. Opt.* **2021**, *60*, C98–C113. [CrossRef]
64. You, H.; Spaeth, H.; Linhard, V.N.L.; Steckl, A.J. Role of Surfactants in the Interaction of Dye Molecules in Natural DNA Polymers. *Langmuir* **2009**, *25*, 11698–11702. [CrossRef]
65. Patra, D.; Barakat, C. Synchronous fluorescence spectroscopic study of solvatochromic curcumin dye. *Spectrochim. Acta Part A Mol. Biomol. Spectrosc.* **2011**, *79*, 1034–1041. [CrossRef]
66. Shanti, R.; Hadi, A.N.; Salim, Y.S.; Chee, S.Y.; Ramesh, S.; Ramesha, K. Degradation of ultra-high molecular weight poly(methyl methacrylate-co-butyl acrylate-co-acrylic acid) under ultra violet irradiation. *RSC Adv.* **2017**, *7*, 112–120. [CrossRef]
67. Anton, A.M.; Rau, I.; Kajzar, F.; Simion, A.M.; Simion, C. Stability studies of some DNA based materials doped with natural extracts. *U.P.B. Sci. Bull. B* **2019**, *81*, 121–129.
68. Taheri, A.; Liu, H.; Jassemnejad, B.; Appling, D.; Powell, R.C.; Song, J.J. Intensity scan and two photon absorption and nonlinear refraction of C60 in toluene. *Appl. Phys. Lett.* **1996**, *68*, 1317–1319. [CrossRef]
69. SheikBahae, M.; Said, A.A.; Wei, T.; Hagan, D.J.; Van Stryland, E.W. Sensitive measurement of optical nonlinearities using a single beam. *IEEE J. Quantum Electron.* **1990**, *26*, 760–769. [CrossRef]

polymers

MDPI

Article

# Effect of Starch Plasticization on Morphological, Mechanical, Crystalline, Thermal, and Optical Behavior of Poly(butylene adipate-co-terephthalate)/Thermoplastic Starch Composite Films

Xiaoyan He [1], Fuhong Zhang [2], Congcong Li [3], Weiwei Ding [2], Yuanyuan Jin [1], Lisheng Tang [1,*] and Ran Huang [1,4,*]

1 Department of Material Science and Engineering, Taizhou Institute of Zhejiang University, Taizhou 318000, China
2 Sanmen Megatron Tech. Co., Ltd., Taizhou 318000, China
3 Center for Biotechnology and Biomedical Engineering, Yiwu Research Institute of Fudan University, Yiwu 322000, China; licongcong@ywfudan.cn
4 Academy for Engineering and Applied Technology, Fudan University, Shanghai 200433, China
* Correspondence: tangls@rizt.ac.cn (L.T.); huangran@fudan.edu.cn (R.H.)

**Abstract:** Starches plasticized with glycerol/citric acid/stearic acid and tributyl 2-acetylcitrate (ATBC), respectively, were processed with poly (butylene adipate-Co-terephthalate (PBAT) via extrusion and a film-blown process. All the composite films were determined for morphology, mechanical, thermal stability, crystalline, and optical properties. Results show that the most improved morphology was in the 30% glycerol plasticized PBAT/thermoplastic starch (TPS) composite films, characterized by the smallest and narrowest distribution of TPS particle sizes and a more uniform dispersion of TPS particles. However, the water absorption of PBAT/TPS composite films plasticized with glycerol surpassed that observed with ATBC as a plasticizer. Mechanical properties indicated insufficient plasticization of the starch crystal structure when using 10% ATBC, 20% ATBC, and 20% glycerol as plasticizers, leading to poor compatibility between PBAT and TPS. This resulted in stress concentration points under external forces, adversely affecting the mechanical properties of the composites. All PBAT/TPS composite films exhibited a negative impact on the initial thermal decomposition temperature compared to PBAT. Additionally, the haze value of PBAT/TPS composite films exceeded 96%, while pure PBAT had a haze value of 47.42%. Films plasticized with 10% ATBC, 20% ATBC, and 20% glycerol displayed lower transmittance values in the visible light region. The increased transmittance of films plasticized with 30% glycerol further demonstrated their superior plasticizing effect compared to other PBAT/TPS composite films. This study provides a simple and feasible method for preparing low-cost PBAT composites, and their extensions are expected to further replace general-purpose plastics in daily applications.

**Keywords:** PBAT; plasticized starch; biodegradable films; optical property

## 1. Introduction

Poly (butylene adipate-co-terephthalate) (PBAT) has attracted great interest as a potential substitute for low-density polyethylene (LDPE), particularly in packaging applications such as lawn waste bags, food containers, or film wraps [1]. PBAT is particularly desirable due to its biodegradability, excellent flexibility, thermal stability, and outstanding water barrier properties [2,3]. Moreover, as an aliphatic-aromatic co-polyester, PBAT combines the advantageous mechanical properties of aromatic polyesters with the biodegradability expected of aliphatic polyesters [4,5]. Despite these merits, PBAT is seldom used alone in food packaging due to its inadequate oxygen barrier properties and high cost [6,7].

Among natural polymers, starch has emerged as a promising option for food packaging, given its biodegradability, film-forming properties, abundance, low cost, and good oxygen barrier properties [8]. However, starch alone lacks the mechanical strength and thermal stability required for effective polymer use. Native starch has to be modified in order to be melt-processed as a thermoplastic material because the melting temperature (Tm) of pure dry starch is close to 220–240 °C, which is higher than its degradation temperature (220 °C). Gelatinization of starch at elevated temperatures, typically in the presence of a plasticizer such as water or glycerol, is a common approach to producing thermoplastic starch (TPS). This process significantly enhances the flexibility and processability of the material [9]. Although products made from TPS exhibit good oxygen barrier properties, biodegradability, and compostability, some deficiencies, such as strong water absorption, relatively poor mechanical properties, and low thermal stability, limit their uses [10–13]. Furthermore, products made from TPS swell and deform upon exposure to high-moisture media. To alleviate these deficiencies, several strategies have been used, including melt blending of TPS with other conventional or biodegradable polymers. PBAT, with high flexibility and stability in the extrusion process, becomes an ideal complement. Therefore, blending TPS with PBAT facilitates film production and achieves the desired flexibility for wrapping and sealing processes.

However, with the usage of starch materials, small molecular plasticizers will progressively migrate out, which is starch retrogradation, resulting in a reduction in the quality of the materials [14,15]. In recent years, researchers have explored various plasticizers and coupling agents to modify PBAT/starch composites. Dammak et al. investigated the effect of maleated PBAT (PBATg-MA) and maleic anhydride (MA), citric acid (CA) coupling agents on the mechanical properties and morphology of PBAT/TPS composites, In the absence of the compatibilizer and in the presence of $PBAT_g$-MA, the PBAT was the continuous phase, while the TPS became the continuous one in the presence of MA and CA. These evolutions were explained by the change in the melt rheological properties of TPS in the presence of the compatibilizer and its aptitude to promote interfacial adhesion between TPS and PBAT phases [1]. Zhang et al. found that tartaric acid (TA) plays diverse roles in PBAT/TPS-TA composites [16]. TA primarily serves as an acid catalyst, reducing both the molecular weight of starch and the shear viscosity of TPS. As a result, TPS-TA particles can disperse uniformly in the PBAT matrix without agglomeration. When TA contents are below 1%, another crucial role of TA is the coupling effect, enhancing compatibility between TPS and PBAT. However, it also decreases the interface interaction between TPS and PBAT, leading to a phase morphology reversal in PBAT/TPS-TA-4 from a homogeneous phase to a "sea-island" structure. Li et al. combined CS particles (native starch and mechanically activated starch) with ATBC and PBS together through melt blending. The impact strength of composites was increased by 126%, and the elongation at break increased by more than five times compared to PBS, showing a ductile fracture trend. Moreover, the water absorption test and contact angle test demonstrate a significant improvement in water resistance [17].

To sum up, the synergistic plasticization of starch with small molecular compatibilizers, combined with glycerol-plasticized starch, provides a viable solution for the preparation of TPS. ATBC demonstrates the potential to plasticize starch, imparting improved ductile properties to PBS composites. No published studies have conducted comparisons between the effects of glycerol and ATBC as plasticizers on the properties of PBAT/TPS films. Recognizing the promising complementary properties between TPS and PBAT composites, this study focuses on the preparation of TPS using two different primary plasticizers. The physical properties of the PBAT/TPS mixtures are compared, in particular, TPS plasticized with glycerol/citric acid/stearic acid and with ATBC. The objective is to assess their effectiveness and examine how they influence the properties of the PBAT/TPS composite.

## 2. Materials and Methods

### 2.1. Materials

Poly (butylene adipate-co-terephthalate) (PBAT, MFR 4–6 g/10 min at 190 °C) was purchased from Xinjiang Blue Ridge Tunhe Chemical Industry Co., Ltd., Changji, China. Native corn starch containing 73% amylopectin and 27% amylose (purity 99%) was provided by Binzhou Jinhui Corn Development Co., Ltd., Binzhou, China. Stearic acid was obtained from Oledchemicals Industry, Kuala Lumpur, Malaysia. Citric acid (purity 99.5%), glycerol (purity 99.5%), tributyl 2-acetylcitrate (ATBC, purity 97%), and all other chemicals were supplied by Macklin Chemical Co., Ltd., Shanghai, China.

### 2.2. Gelatinisation of Starch

The corn starch was dried at 90 °C in the oven for 48 h to make the moisture content lower than 0.1% and then mixed with glycerol, citric acid, and stearic acid in a high-speed mixer at the mass ratio of 100:20:0.6:0.3 and 100:30:0.6:0.3 for 10 min, respectively. The temperature of the mixer was cooled and kept at 30 °C during the mixing process.

Two doses of corn starch plasticized by ATBC were prepared by adding ATBC at mass ratios of 100:10 and 100:20 and stirring thoroughly. Subsequently, all the mixtures were kept in a closed vessel to avoid absorption of moisture and were kept for 24 h before processing with an extruder.

### 2.3. Preparation of PBAT/TPS Composite Films

The melt mixing of PBAT/TPS biodegradable composites was carried out in a co-rotating twin screw extruder (KYmach SJ-30, Nanjing, China), with a variable temperature difference of 90–140 °C from the feed to the die zone, and a screw speed of 300 rpm. PBAT and TPS were compounded at a ratio of 70:30. The drawing rods were then cooled, pelletized, and stored in a sealed package until further use.

PBAT/TPS blended films were prepared by film blown using a single screw extruder with a screw diameter of 170 mm and an L/D ratio of 30 (Leibo, Hangzhou, China). The film-blown temperature was independently controlled at four zones along the extruder barrel, and a strand die was used to achieve a temperature profile in the range of 135–150 °C. The screw speed was set at 30 rpm, and the feed rate was 12 rpm. Films were named PBAT, PBAT/TPS-10% ATBC, PBAT/TPS-20% ATBC, PBAT/TPS-20% glycerol, and PBAT/TPS-30% glycerol. The thicknesses of PBAT/TPS composite films were 65 ± 5 μm.

### 2.4. Characterization of PBAT/TPS Composite Films

The surface morphology of PBAT/TPS composite films was examined using an optical microscope (OLYMPUS CX31, Church Hill, TN, USA). Images were taken at brightfield, transmission mode, with a magnification of 4–10×.

For each composition, five pieces of film were cut into 20 mm × 20 mm and stored at 55% RH for 7 days before testing and then dried in the oven at 105 °C for 24 h. The water absorption test was determined by immersing oven-dried samples into distilled water at room temperature. The weight of the samples was monitored and recorded at regular intervals (2 h, 4 h, 6 h, 8 h, 24 h, and 48 h). The water absorption ratio was calculated using Equation (1).

$$\text{Yield}\ (\%) = \frac{M_t - M_i}{M_i} \times 100 \tag{1}$$

where $M_i$ is the initial mass of the sample, and $M_t$ is the mass after water absorption.

The X-ray diffraction (XRD) was used to conduct the crystallinity of samples. The samples were analyzed in the X-ray diffractometer (Model: Rigaku Iniflex 600, Tokyo, Japan) using Cu-Kα radiation (λ = 0.154 nm) at 40 kV and 30 mA with a goniometer speed of 0.02 $s^{-1}$. The spectra were measured for 2θ in the range of 5–50°. The X-ray detector used was a scintillation counter with a detector angle of 40°, placed at a distance of 300 mm.

The mechanical properties of films were meticulously assessed through traction tests, according to the ASTM Standard method D882-12 (ASTM, 2012) [18]. Rectangular films with a width of 15 mm and a length of 150 mm were carefully prepared and positioned in an electronic tensile testing instrument with a charge cell of 500 N. The samples were then subjected to a controlled stretching rate of 300 mm/min, with an initial gauge distance of 50 mm. Samples were conditioned at 23 °C and 50% relative humidity for 72 h before testing. Six specimens were tested for each composition.

Thermal weight loss analysis was conducted using a thermogravimetric instrument (TGA 2, Mettler Toledo, Zurich, Switzerland). In this analysis, approximately 8–10 mg of the polymeric materials were placed in an aluminum dish under the nitrogen ($N_2$) atmosphere at room temperature. The sample was then subjected to heating at a rate of 10 °C/min using a high-resolution dynamic mode until it reached a temperature of 600 °C. Throughout the heating process, the weight loss of the sample in response to the temperature change was recorded.

The haze properties of the films were measured using an optical tester (YH1200, 3NH, Shenzhen, China), and different positions across the film were assessed for both transmittance (T) and haze (H).

The color of the different samples was determined using a color spectrophotometer (Leeuwarden, The Netherlands). A spectrophotometer was used to measure the color of the sample obtained. This apparatus measures the color in the CIE Lab (L*, a*, b*) scale. Each set of samples underwent three individual measurements, and the resulting average value was recorded.

The UV and visible light transmittance of the films was determined using a U-4100 UV–vis spectrophotometer (HITACHI, Japan), operating in the range of 300–800 $cm^{-1}$ and with a scan speed of 300 nm/min.

## 3. Results and Discussion

### *3.1. Surface Morphology of Films*

Figure 1 illustrates the surface morphology of neat PBAT film and PBAT/TPS composite films. The surface of neat PBAT film appeared smoother and more uniform than that of the 10% ATBC, 20% ATBC, and 20% glycerol plasticized films. When 10% ATBC and 20% ATBC served as the plasticizer, PBAT/TPS films (Figure 1b,c) exhibited rough and uneven features; the size of the phases and the lack of adhesion between the two phases evidenced immiscibility. Figure 1d shows the typical sea-island morphology of uncompatibilized composites; it can be seen that TPS particles were not uniformly dispersed in the PBAT matrix, and large particles were observed, which can be ascribed to the agglomeration of some TPS particles. Between Figure 1d,e, a gradual reduction in domain size and increase in distribution uniformity of TPS particles was observed with increasing glycerol content of the respective compatibilized composites. Among all the composites, the highest improved morphology had the smallest and narrowest distribution of TPS particle sizes in addition to a more uniform distribution of TPS particles, which can be seen in 30% glycerol plasticized PBAT/TPS composite films. It can be explained that 30% glycerol served as the optimum amount of plasticizer content can cause better interfacial bonding between two phases, resulting in finer and improved blend morphology.

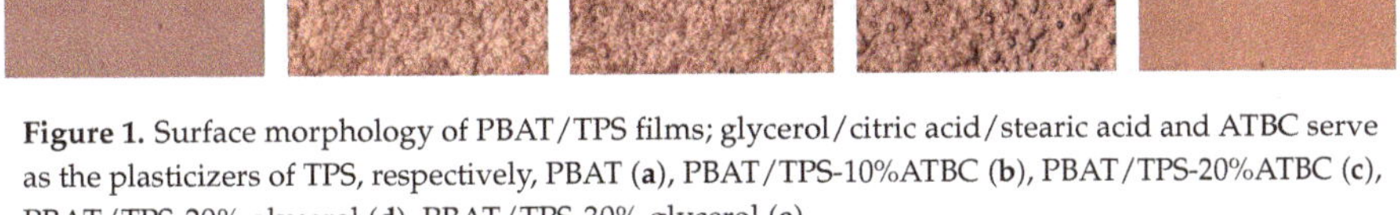


**Figure 1.** Surface morphology of PBAT/TPS films; glycerol/citric acid/stearic acid and ATBC serve as the plasticizers of TPS, respectively, PBAT (**a**), PBAT/TPS-10%ATBC (**b**), PBAT/TPS-20%ATBC (**c**), PBAT/TPS-20% glycerol (**d**), PBAT/TPS-30% glycerol (**e**).

### 3.2. Water Absorption

Water absorption is one of the main shortcomings that restrict the application of starch-based materials. One of the objectives of compounding TPS with hydrophobic polyester is to reduce the water absorption of the film to meet the requirements for maintaining acceptable levels of mechanical properties. The water absorption rates of PBAT/TPS composite films after immersion into distilled water for 48 h are presented in Figure 2, which is consistent with previously published results [19]. Given its hydrophobic nature, the water absorption of PBAT film presented a relatively low water absorption rate (0.44%). The PBAT/TPS composite films presented higher water absorption capacity (maximum 15.09%) after a 48 h incubation period, as compared to the PBAT film, with a significant difference between them [1]. This increase can be attributed to the numerous hydroxyl groups and hydrophilic character of starch. The water absorption of composite films increased rapidly during the first 2 h and then gradually approached the saturation plateau. With 20% ATBC as a plasticizer, the water absorption of the composite films was lower compared to when 10% ATBC was used as the plasticizer, suggesting that increasing the plasticizer content is advantageous for suppressing water absorption. This effect can be attributed to the interaction between the plasticizer and the hydroxyl groups of starch, leading to a weakening of intermolecular hydrogen bonding. A slightly higher water absorption rate was observed in the composite films when glycerol/citric acid/stearic acid were used as plasticizers compared to films with ATBC as the plasticizer. The full continuity in TPS allowed water to access the completely hydrophilic TPS phase through diffusion from the surface via network paths of the TPS phase [1]. Moreover, the incorporation of glycerol introduced more hydrophilic hydroxyl groups during interaction with starch, whereas ATBC introduced ester bonds with stronger hydrophobic characteristics.

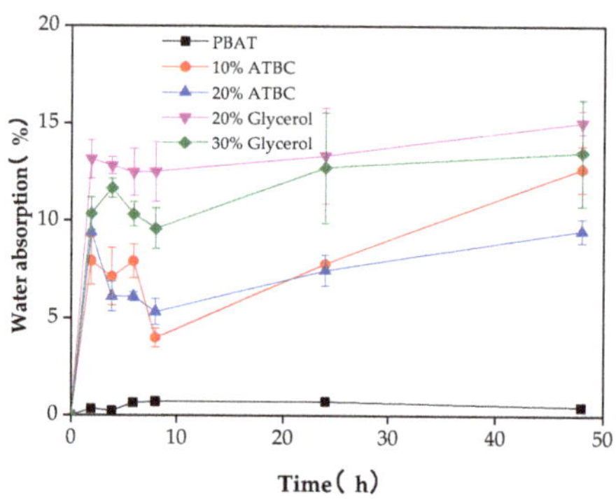


**Figure 2.** Water absorption of PBAT/TPS films in 48 h.

### 3.3. X-ray Diffraction Analysis of Films

X-ray diffraction patterns of neat PBAT and PBAT/TPS composite films are shown in Figure 3. Peaks at 2θ angles of 17.54°, 20.64°, and 23.14°are characteristic of the crystalline phase of PBAT [20,21]. The starch granule (which presents a certain degree of molecular organization) is partially crystalline and has a degree of crystallinity ranging from 20% to 45% [22]. The corn starch has A-type crystallinity and diffraction peaks in 2θ = 15° [23]. With the addition of glycerol-plasticized TPS, new characteristic peaks appeared at 2θ angles of 13.1° and 16.1° and were characteristic peaks of TPS, which correspond to the $V_H$-type amylose crystal [24,25]. The diffraction peaks at 13.38°, 16.24°, 17.54°, 20.64°, 23.14°, and 24.72° were seen in all PBAT/TPS composites, indicating that all composites had a similar crystalline structure. A minor diffraction peak at 20.08 emerged in the 30% glycerol-plasticized PBAT/TPS film. This occurrence may be attributed to the rapid recrystallization of starch during processing, resulting in the formation of a single helical structure complexed with glycerol [26]. It was reported that the transesterification reaction that occurs widely in PBAT/PBS composites not only improved the interfacial compatibility of the composites but also changed their original crystallization types [27]. PBAT/TPS composite films plasticized with 10% ATBC and 20% glycerol showed peaks at 2θ = 29.58°, attributed mainly to the starch retrodegradation during the preservation period [28].

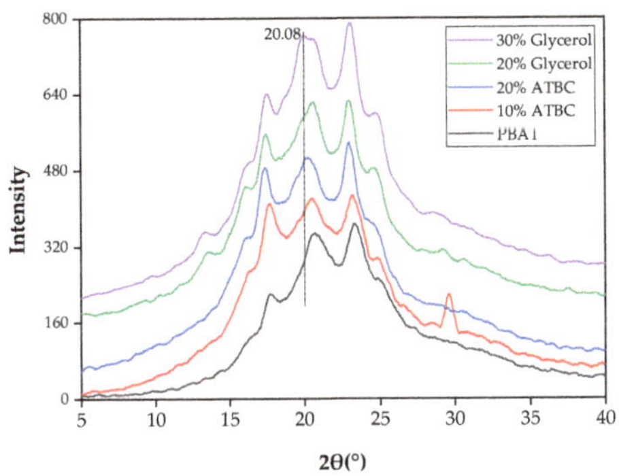


**Figure 3.** XRD patterns of PBAT/TPS films.

### *3.4. Tensile Property of Films*

As shown in Table 1, neat PBAT exhibited a tensile strength of 21.57 MPa and elongation at break of 561.08%, demonstrating high flexibility and ductility. In contrast to the mechanical properties of pure PBAT, PBAT composites with plasticized starch exhibited distinct and varied intrinsic results. The incorporation of ATBC did not enhance the toughness of PBAT when combined with TPS. Table 1 shows that all PBAT/TPS composites not only exhibited more brittle fracture but also demonstrated reduced strength (less than 9.34 MPa), even with an increase in ATBC concentration. The phenomena can be attributed to the compatibility and interface interaction between the polymer matrix and starch [17,29]. An additional factor was the insufficient plasticization process for the crystal structure of starch, which contributed to the poor compatibility between PBAT and TPS. TPS tended to agglomerate in the PBAT matrix, creating stress concentration points when subjected to external force, thereby comprising the mechanical properties of composites. After adding 30% glycerol as the plasticizer, the tensile strength of PBAT/TPS decreased to 17.41 MPa, and the elongation at break decreased to 493.25%, showing a negative impact on tensile properties. The smaller particle size and more regular surface morphology observed in the films plasticized by 30% glycerol aligned with the findings in morphological analyses. This allowed glycerol to penetrate more easily into the molecular structure of PBAT, enhancing compatibility with PBAT. However, the insertion of a small molecule plasticizer and starch between the molecular chains of PBAT weakened the attraction between the molecular chains. The weak intermolecular binding force resulted in the loose crystallization zone of PBAT, thus reducing the strength of PBAT films [17].

**Table 1.** Mechanical properties of films during 48 months of storage at ambient temperature.

| | 1 Week | | 48 Months | |
|---|---|---|---|---|
| | **Tensile Strength (MPa)** | **Elongation at Break (%)** | **Tensile Strength (MPa)** | **Elongation at Break (%)** |
| PBAT | 21.57 ± 1.47 | 561.08 ± 6.73 | 16.89 ± 2.11 | 499.72 ± 7.84 |
| 10% ATBC | 6.81 ± 0.56 | 451.92 ± 9.26 | 3.28 ± 0.81 | 12.87 ± 0.48 |
| 20% ATBC | 7.29 ± 1.62 | 465.26 ± 6.78 | 3.51 ± 0.39 | 43.23 ± 6.61 |
| 20% Glycerol | 9.34 ± 1.71 | 378.27 ± 8.43 | 2.92 ± 0.41 | 54.85 ± 7.04 |
| 30% Glycerol | 17.41 ± 2.69 | 493.25 ± 5.40 | 8.59 ± 1.26 | 433.18 ± 8.43 |

After incubating the films at ambient temperature for 48 months, the reassessment of the mechanical properties revealed general decline trends in both tensile strength and elongation at break for all films. The mechanical performance of PBAT composite films containing TPS exhibited a reduction exceeding 50%. However, the elongation at break of the PBAT/TPS composite film plasticized with 30% glycerol still exceeded 400%, possibly due to glycerol migration from the TPS to the PBAT continuous phase [30]. In addition, this indicates that PBAT/TPS composite films are prone to starch retrogradation and degradation during storage. It is crucial to balance the shelf life and biodegradability in practical application.

### 3.5. Thermal Analysis of Films

The weight loss curve and DTG curve of PBAT/TPS composite films are shown in Figure 4, and characteristic decomposition temperatures are summarized in Table 2. For all the composite films, minor mass losses were observed in the temperature range of 60–160 °C due to the evaporation of moisture or low molecular weight compounds (Figure 4b). The initial decomposition temperature ($T_{onset}$) was defined as the temperature at which a 5% weight loss of the initial mass occurs (Figure 4a and Table 2). $T_{onset}$ of plasticized PBAT/TPS composite films was lower than that of neat PBAT films. The initial decomposition temperature systematically shifted to a lower temperature with an increase in plasticizer content, which may be attributed to the volatilization of ATBC and glycerol [31]. The degradation stage occurred at a temperature range of 284–340 °C, representing the major decomposition of starch [32]. Further heating at a temperature range of 350–415 °C induced the highest thermal decomposition rate of PBAT, which was reflected by the drastic weight reduction. This stage was ascribed to the elimination of hydrogen groups, decomposition, and depolymerization of the starch and PBAT carbon chains. The DTG peak of 10% ATBC plasticized PBAT/TPS composite films in the temperature range of 430–500 °C shifted to a lower temperature range, indicating the decreased thermal stability of the benzene ring in PBAT compared to other composite films. The final thermal event at 492 °C suggests that all the films decomposed before 500 °C.

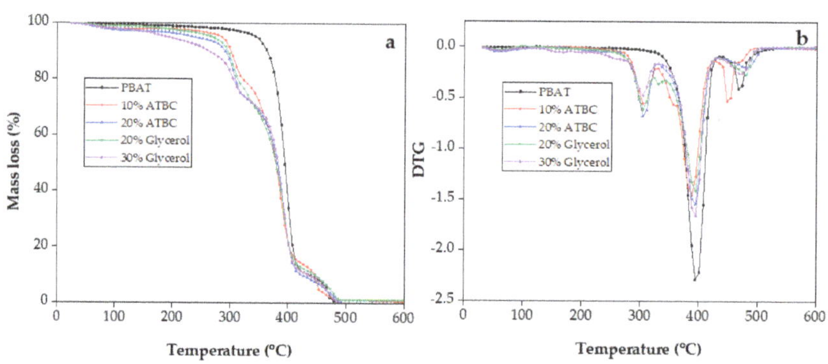


**Figure 4.** TGA (**a**) and DTG (**b**) curves of PBAT/TPS films during thermal decomposition.

**Table 2.** Characteristic decomposition temperature ($T_{onset}$ and $T_{max}$) of PBAT and PBAT/TPS composite films.

| | $T_{onset}$ | $T_{max1}$ | $T_{max2}$ | $T_{max3}$ |
|---|---|---|---|---|
| PBAT | 346 | 304 | 395 | 466.5 |
| 10% ATBC | 285 | 304/356 | 388.5 | 447/466.5 |
| 20% ATBC | 239 | 304 | 395 | 479.5 |
| 20% Glycerol | 268 | 304/330 | 395 | 479.5 |
| 30% Glycerol | 193.5 | 304 | 395 | 479.5 |

In summary, all PBAT/TPS composite films demonstrated a negative impact on the initial thermal decomposition temperature compared to PBAT. This can be attributed to the lower vaporization and decomposition temperatures of the plasticizers. It is noteworthy that in PBAT/TPS films with 10% ATBC as a plasticizer, the rapid decomposition temperature range of PBAT shifted towards lower temperature regions. This shift may be because ATBC weakened the intermolecular forces of PBAT, leading to less tightly packed stacking and, consequently, fewer crystalline regions in both starch and PBAT.

### 3.6. Effect of TPS on Haze and Transparency Property of Films

The optical properties of films are crucial factors that influence the aesthetics of the films and the purchasing experience for consumers [33]. Figure 5 shows the haze(H) and transparency (T) of PBAT and PBAT/TPS composite films. PBAT/TPS composite films exhibited a slightly lower transparency compared to the PBAT film, particularly evident in films plasticized with 10% ATBC, where T decreased by 7%. This aligns with the findings in haze properties, as all PBAT/TPS composite films exhibited a haze value exceeding 96%, while the haze value for pure PBAT was 47.42%. The inclusion of TPS led to higher haze and opacity and better absorption of visible light (the haze of PBAT/TPS films increased by approximately 50% ($p < 0.05$)) in comparison to the PBAT film. The pore structure of plasticized starch contributed to a larger specific surface area and an irregular microstructure, resulting in an increased refractive index of light and decreased transparency. If the packaged food contains a large amount of lipids, and opacity is desired to minimize light-catalyzed oxidation reactions, the PBAT/TPS composite films would be more suitable than the PBAT film.

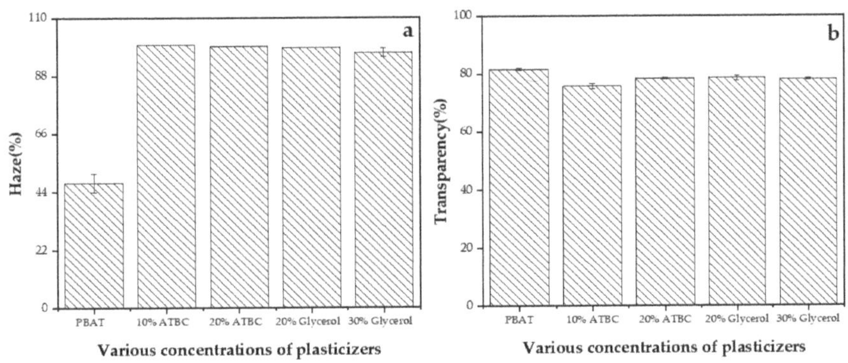


**Figure 5.** Haze (**a**) and transparency (**b**) of the PBAT/TPS composite films.

### 3.7. Color Variation in Films

Color parameters (L*, a*, b*) of blown films are described in Table 3. The incorporation of 10% ATBC, 20% ATBC, and 20% glycerol as plasticizers slightly increased the luminosity index of the PBAT film, except for the PBAT/TPS films plasticized with 30% glycerol. The brownish color of the starch contributed to heightened redness (a*) and yellowing (b*) in films containing ATBC [34]. Color shifts toward brown and yellow during the melting process may occur due to the loss of structure and crystallinity of starch granules. Samples with lower lightness (L* = 48.88) and comparable yellowing to the PBAT film were observed in the 30% glycerol plasticized films. Elevated amounts of glycerol resulted in films with a darker color, which can be explained by its action as a plasticizer, allowing higher molecular displacement and a less rigid structure. The color analysis complements the UV–vis results, indicating that the chemical constitution of the incorporated starch directly affected the color properties of the developed polymeric films.

**Table 3.** Color parameters of extruded films.

| | L* | a* | b* |
|---|---|---|---|
| PBAT | 50.91 ± 2.34 | −1.02 ± 0.03 | −8.34 ± 0.32 |
| 10% ATBC | 54.44 ± 0.37 | −0.76 ± 0.08 | −4.45 ± 0.26 |
| 20% ATBC | 54.99 ± 0.71 | −0.69 ± 0.09 | −4.44 ± 0.17 |
| 20% Glycerol | 55.27 ± 0.97 | −1.24 ± 0.16 | −6.62 ± 0.08 |
| 30% Glycerol | 48.88 ± 0.53 | −1.10 ± 0.09 | −8.30 ± 0.01 |

### *3.8. UV-vis Transmittance of Films*

To further investigate the impact of plasticizers on the transparency of PBAT/TPS composite films, the UV-visible light transmittance of PBAT and PBAT/TPS composite films was measured, as shown in Figure 6. Above 500 nm, transmittance increased with the plasticizer content in the films for both ATBC and glycerol, which was attributed to improved gelatinization. The transmittance of PBAT/TPS composite films, plasticized with 20% and 30% glycerol, exceeded that of films containing 10% and 20% ATBC. The transmittance of light in polymeric films is not only associated with their chemical structure and molar mass but also and, most importantly, dependent on film morphology [35,36]. The morphology of films with starch was affected by the presence of starches that were not completely gelatinized [34]. These particles, which are not fully molten during extrusion, are responsible for refracting most of the light. Films plasticized with 10% ATBC, 20% ATBC, and 20% glycerol exhibited lower transmittance values in the visible light region, likely due to a higher concentration of starch granules not completely gelatinized, as observed in the micrographs. The increased transmittance of films plasticized with 30% glycerol further proved their superior plasticizing effect compared to other PBAT/TPS composite films. The manufacture of biodegradable films with a higher optical barrier is important for most packaging applications [37].

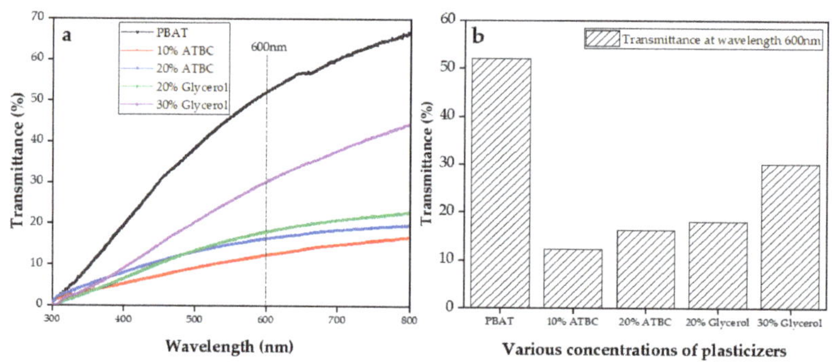


**Figure 6.** Transmittance of PBAT and PBAT/TPS composite films (**a**), transmittance of films at wavelength 600 nm (**b**).

## 4. Conclusions

This study investigated the properties of PBAT/TPS composite films, including starch plasticized using ATBC (10%, 20%) and glycerol (20%, 30%). The films exhibited improved finer morphology, particularly at 30% glycerol concentrations. However, the water absorption of the composite films plasticized with glycerol surpassed that observed with ATBC as a plasticizer; this result can be attributed to the abundant hydroxyl groups in glycerol and full continuity in TPS. The addition of 30% glycerol as the plasticizer led to a decrease in the tensile strength of PBAT/TPS to 17.41 MPa, and the elongation at break reduced to 493.25%, indicating a negative impact on the tensile properties. All PBAT/TPS composite films showed a negative impact on the initial thermal decomposition temperature compared to PBAT, which is attributed to the lower vaporization and decomposition temperatures of the plasticizers. The inclusion of TPS led to higher haze and opacity, demonstrating better absorption of visible light (the haze of PBAT/TPS films increased by approximately 50% ($p < 0.05$)) compared to the PBAT film. Color analysis aligned with UV–vis results, indicating that the chemical composition of the incorporated starch directly influenced the color properties of the developed polymeric films. Films plasticized with 10% ATBC, 20% ATBC, and 20% glycerol exhibited lower transmittance values in the visible light region, likely due to a higher concentration of starch granules not completely gelatinized, as observed in the micrographs. The increased transmittance of films plasticized with 30% glycerol further confirmed the superior plasticizing effect compared to other PBAT/TPS composite films. The production of biodegradable films with a higher optical barrier is crucial for most packaging applications.

As a result, PBAT/TPS composite films present an important option for the development of environmentally friendly and energy-saving packaging materials at a low cost.

**Supplementary Materials:** The following supporting information can be downloaded at: https://www.mdpi.com/article/10.3390/polym16030326/s1, Figure S1: Stress and strain curves of PBAT/TPS composites after 48 months of storage.

**Author Contributions:** Methodology, F.Z.; Investigation, C.L.; Resources, W.D.; Data curation, Y.J.; Writing—original draft preparation, X.H.; Writing—review and editing, L.T.; Supervision, R.H. All authors have read and agreed to the published version of the manuscript.

**Funding:** This work is financially supported by the Postdoctoral fellows of Taizhou (275104), Taizhou Municipal Science and Technology Program (21gya24, 22gya19), the National Key Research and Development Program China (2022YFC2009500), the Medical Engineering Fund of Fudan University (yg2021-005, yg2022-008), Guangdong Provincial Science and Technology Program (2023A0505050146), and the RIZT Industrial Program (2021ZSS04, 2022ZSS09, 2023ZM01, 2023CLG01, 2023CLG01PT).

**Institutional Review Board Statement:** Not applicable.

**Data Availability Statement:** Data are contained within the article and Supplementary Materials.

**Conflicts of Interest:** Authors Fuhong Zhang and Weiwei Ding were employed by the company Sanmen Megatron Tech. Co., Ltd. The remaining authors declare that the research was conducted in the absence of any commercial or financial relationships that could be construed as a potential conflict of interest.

## References


1. Dammak, M.; Fourati, Y.; Tarrés, Q.; Delgado-Aguilar, M.; Mutjé, P.; Boufi, S. Blends of PBAT with plasticized starch for packaging applications: Mechanical properties, rheological behaviour and biodegradability. *Ind. Crops Prod.* **2020**, *144*, 112061. [CrossRef]
2. Ferreira, F.V.; Cividanes, L.S.; Gouveia, R.F.; Lona, L.M. An overview on properties and applications of poly(butylene adipate-co-terephthalate)–PBAT based composites. *Polym. Eng. Sci.* **2019**, *59*, E7–E15. [CrossRef]
3. Pan, H.; Ju, D.; Zhao, Y.; Wang, Z.; Yang, H.; Zhang, H.; Dong, L. Mechanical properties, hydrophobic properties and thermal stability of the biodegradable poly(butylene adipate-co-terephthalate)/maleated thermoplastic starch blown films. *Fibers Polym.* **2016**, *17*, 1540–1549. [CrossRef]
4. Larrañaga, A.; Lizundia, E. A review on the thermomechanical properties and biodegradation behaviour of polyesters. *Eur. Polym. J.* **2019**, *121*, 109296. [CrossRef]
5. Madera-Santana, T.; Misra, M.; Drzal, L.; Robledo, D.; Freile-Pelegrin, Y. Preparation and characterization of biodegradable agar/poly(butylene adipate-co-terephatalate) composites. *Polym. Eng. Sci.* **2009**, *49*, 1117–1126. [CrossRef]
6. Jian, J.; Xiangbin, Z.; Xianbo, H. An overview on synthesis, properties and applications of poly(butylene-adipate-co-terephthalate)–PBAT. *Adv. Ind. Eng. Polym. Res.* **2020**, *3*, 19–26. [CrossRef]
7. Zhai, X.; Wang, W.; Zhang, H.; Dai, Y.; Dong, H.; Hou, H. Effects of high starch content on the physicochemical properties of starch/PBAT nanocomposite films prepared by extrusion blowing. *Carbohydr. Polym.* **2020**, *239*, 116231. [CrossRef]
8. Khan, B.; Bilal Khan Niazi, M.; Samin, G.; Jahan, Z. Thermoplastic starch: A possible biodegradable food packaging material—A review. *J. Food Process Eng.* **2017**, *40*, e12447. [CrossRef]
9. Chang, C.C.; Trinh, B.M.; Mekonnen, T.H. Robust multiphase and multilayer starch/polymer (TPS/PBAT) film with simultaneous oxygen/moisture barrier properties. *J. Colloid Interface Sci.* **2021**, *593*, 290–303. [CrossRef]

10. Patil, M.; Mathad, S.N.; Patil, A.Y.; Arshad, M.N.; Alorfi, H.S.; Puttegowda, M.; Asiri, A.M.; Khan, A.; Azum, N. Synthesis and Characterization of Microwave-Assisted Copolymer Membranes of Poly(vinyl alcohol)-g-starch-methacrylate and Their Evaluation for Gas Transport Properties. *Polymers* **2022**, *14*, 350. [CrossRef]
11. Aversa, C.; Barletta, M. Addition of thermoplastic starch (TPS) to binary blends of poly(lactic acid)(PLA) with poly(butylene adipate-co-terephthalate)(PBAT): Extrusion compounding, cast extrusion and thermoforming of home compostable materials. *Chin. J. Polym. Sci.* **2022**, *40*, 1269–1286. [CrossRef]
12. Li, M.; Jia, Y.; Shen, X.; Shen, T.; Tan, Z.; Zhuang, W.; Zhao, G.; Zhu, C.; Ying, H. Investigation into lignin modified PBAT/thermoplastic starch composites: Thermal, mechanical, rheological and water absorption properties. *Ind. Crops Prod.* **2021**, *171*, 113916. [CrossRef]
13. Bai, J.; Pei, H.; Zhou, X.; Xie, X. Reactive compatibilization and properties of low-cost and high-performance PBAT/thermoplastic starch blends. *Eur. Polym. J.* **2021**, *143*, 110198. [CrossRef]
14. Montilla-Buitrago, C.E.; Gómez-López, R.A.; Solanilla-Duque, J.F.; Serna-Cock, L.; Villada-Castillo, H.S. Effect of plasticizers on properties, retrogradation, and processing of extrusion-obtained thermoplastic starch: A review. *Starch-Stärke* **2021**, *73*, 2100060. [CrossRef]
15. Kahvand, F.; Fasihi, M. Plasticizing and anti-plasticizing effects of polyvinyl alcohol in blend with thermoplastic starch. *Int. J. Biol. Macromol.* **2019**, *140*, 775–781. [CrossRef] [PubMed]
16. Zhang, S.; He, Y.; Lin, Z.; Li, J.; Jiang, G. Effects of tartaric acid contents on phase homogeneity, morphology and properties of poly(butyleneadipate-co-terephthalate)/thermoplastic starch bio-composities. *Polym. Test.* **2019**, *76*, 385–395. [CrossRef]
17. Li, Y.; Li, Z.; Sheng, S.; Li, Y.; Zhong, J.-R.; Tan, J.; Zhang, Y.-F. Preparation and properties of rapidly plasticized poly(butylene succinate)/mechanically activated cassava starch biocomposite. *Polym. Bull.* **2023**, 1–17. [CrossRef]
18. *ASTM D882-12*; Standard Test Method for Tensile Properties of Thin Plastic Sheeting. ASTM International: West Conshohocken, PA, USA, 2012.
19. Šmídová, N.; Peidayesh, H.; Baran, A.; Fričová, O.; Kovaľaková, M.; Králiková, R.; Chodák, I. Influence of Air Humidity Level on the Structure and Mechanical Properties of Thermoplastic Starch-Montmorillonite Nanocomposite during Storage. *Materials* **2023**, *16*, 900. [CrossRef] [PubMed]
20. dos Santos Silva, A.; Rodrigues, B.V.M.; Oliveira, F.C.; Carvalho, J.O.; de Vasconcellos, L.M.R.; de Araújo, J.C.R.; Marciano, F.R.; Lobo, A.O. Characterization and in vitro and in vivo assessment of poly(butylene adipate-co-terephthalate)/nano-hydroxyapatite composites as scaffolds for bone tissue engineering. *J. Polym. Res.* **2019**, *26*, 53. [CrossRef]
21. Zhou, Y.; Lei, L.; Yang, B.; Li, J.; Ren, J. Preparation and characterization of polylactic acid (PLA) carbon nanotube nanocomposites. *Polym. Test.* **2018**, *68*, 34–38. [CrossRef]
22. da Silva, N.M.C.; Correia, P.R.C.; Druzian, J.I.; Fakhouri, F.M.; Fialho, R.L.L.; de Albuquerque, E.C.M.C. PBAT/TPS composite films reinforced with starch nanoparticles produced by ultrasound. *Int. J. Polym. Sci.* **2017**, *2017*, 4308261. [CrossRef]
23. Lopez-Silva, M.; Bello-Perez, L.A.; Agama-Acevedo, E.; Alvarez-Ramirez, J. Effect of amylose content in morphological, functional and emulsification properties of OSA modified corn starch. *Food Hydrocoll.* **2019**, *97*, 105212. [CrossRef]
24. Xiaofei, M.; Jiugao, Y.; Feng, J. A mixed plasticizer for the preparation of thermoplastic starch. *Chin Chem Lett* **2004**, *15*, 741–744.
25. Niu, Z.; Chen, F.; Zhang, H.; Liu, C. High Content of Thermoplastic Starch, Poly(butylenes adipate-co-terephthalate) and Poly(butylene succinate) Ternary Blends with a Good Balance in Strength and Toughness. *Polymers* **2023**, *15*, 2040. [CrossRef]
26. Raquez, J.-M.; Nabar, Y.; Srinivasan, M.; Shin, B.-Y.; Narayan, R.; Dubois, P. Maleated thermoplastic starch by reactive extrusion. *Carbohydr. Polym.* **2008**, *74*, 159–169. [CrossRef]
27. Muthuraj, R.; Misra, M.; Mohanty, A.K. Injection molded sustainable biocomposites from poly(butylene succinate) bioplastic and perennial grass. *ACS Sustain. Chem. Eng.* **2015**, *3*, 2767–2776. [CrossRef]
28. Wang, S.; Li, C.; Copeland, L.; Niu, Q.; Wang, S. Starch Retrogradation: A Comprehensive Review. *Compr. Rev. Food Sci. Food Saf.* **2015**, *14*, 568–585. [CrossRef]
29. Zhang, S.; He, Y.; Yin, Y.; Jiang, G. Fabrication of innovative thermoplastic starch bio-elastomer to achieve high toughness poly(butylene succinate) composites. *Carbohydr. Polym.* **2019**, *206*, 827–836. [CrossRef]
30. Garalde, R.A.; Thipmanee, R.; Jariyasakoolroj, P.; Sane, A. The effects of blend ratio and storage time on thermoplastic starch/poly(butylene adipate-co-terephthalate) films. *Heliyon* **2019**, *5*, e01251. [CrossRef]
31. Müller, P.S.; Carpiné, D.; Yamashita, F.; Waszczynskyj, N. Influence of pinhão starch and natural extracts on the performance of thermoplastic cassava starch/PBAT extruded blown films as a technological approach for bio-based packaging material. *J. Food Sci.* **2020**, *85*, 2832–2842. [CrossRef]
32. Nordin, N.; Othman, S.H.; Rashid, S.A.; Basha, R.K. Effects of glycerol and thymol on physical, mechanical, and thermal properties of corn starch films. *Food Hydrocoll.* **2020**, *106*, 105884. [CrossRef]
33. Mehta, M.J.; Kumar, A. Ionic liquid assisted gelatin films: Green, UV shielding, antioxidant, and antibacterial food packaging materials. *ACS Sustain. Chem. Eng.* **2019**, *7*, 8631–8636. [CrossRef]
34. Rodrigues, S.C.; De Mesquita, F.A.S.; De Carvalho, L.H.; Alves, T.S.; Folkersma, R.; dos RM Araújo, R.S.; Oliveira, A.D.; Barbosa, R. Preparation and characterization of polymeric films based on PLA, PBAT and corn starch and babassu mesocarp starch by flat extrusion. *Mater. Res. Express* **2021**, *8*, 035305. [CrossRef]
35. Orsuwan, A.; Sothornvit, R. Effect of banana and plasticizer types on mechanical, water barrier, and heat sealability of plasticized banana-based films. *J. Food Process. Preserv.* **2018**, *42*, e13380. [CrossRef]

36. Supthanyakul, R.; Kaabbuathong, N.; Chirachanchai, S. Random poly(butylene succinate-co-lactic acid) as a multi-functional additive for miscibility, toughness, and clarity of PLA/PBS blends. *Polymer* **2016**, *105*, 1–9. [CrossRef]
37. Bonilla, J.; Fortunati, E.; Vargas, M.; Chiralt, A.; Kenny, J.M. Effects of chitosan on the physicochemical and antimicrobial properties of PLA films. *J. Food Eng.* **2013**, *119*, 236–243. [CrossRef]

polymers

MDPI

*Article*

# Effect of Pulping Waste Liquid on the Physicochemical Properties and the Prediction Model of Wheat Straw Residue Granular Fuels

**Lanxin Xue, Guihua Yang *, Zhaoyun Lin *, Jinze Li, Bo He and Jiachuan Chen ***

State Key Laboratory of Biobased Material and Green Papermaking, Qilu University of Technology (Shandong Academy of Sciences), Jinan 250353, China; xuelanxin1024@163.com (L.X.); x1062310349@163.com (J.L.); hebo123@126.com (B.H.)

* Correspondence: ygh2626@126.com (G.Y.); linzhaoyun123@126.com (Z.L.); chenjc@qlu.edu.cn (J.C.)

**Abstract:** Herein, wheat straw residue and pulping waste liquid were collected from pulping mill and mixed to prepare bio-based granular fuels by using compression molding technology, and to explore the comprehensive utilization of the industrial waste of pulping and papermaking. The effects of pulping waste liquid on granular fuel properties were analyzed systemically. Further study of the function of pulping waste liquid, cellulose and hemicellulose was used to replace wheat straw residue and avoid the interference factors. Therefore, the prediction models of granular fuels were established with influencing factors that included cellulose, hemicellulose and pulping waste liquid. The granular fuels had the best performance with 18.30% solid content of pulping waste liquid. The highest transverse compressive strength of granular fuel was 102.61 MPa, and the activation energy was 81.71 $KJ \cdot mol^{-1}$. A series of curve fitting prediction models were established to clarify the forming process of granular fuel, and it turned out that the pulping waste liquid could improve the adhesion between solid particles and increase their compression resistance.

**Keywords:** pulping waste liquid; bio-based granular fuel; forming process

## 1. Introduction

Biomass granular fuel, as a renewable and environmentally friendly energy, is produced and expected to replace coal, oil, electricity and natural gas [1,2]. Therefore, the development of biomass granular fuel technology could compensate the shortage of conventional energy and provide significant environmental benefits. However, the traditional granular fuels made by molding technology has low quality, and various pollutants are released during the combustion process [3,4]. Consequently, selecting suitable additives is a crucial solution to improve the quality of granular fuels and reduce power consumption during granularizing processes [5].

Several industrial wastes as additives have been reported to improve the properties of biomass granular fuel and the physical properties and combustion characteristics of biomass granular fuel. It was discovered that using coal tar residue (CTR) as a binder for biomass/lignite granular fuel significantly enhances the granular fuel' mechanical strength and heating value [6]. With an increase in the proportion of CTR, the ignition temperature and activation energy of wheat straw granular fuel increased gradually. At the same time, the maximum burning rate and maximum weight loss percentage decreased. The bio-waste could be successfully densified into good-quality granular fuel by drum granulation using silicate-based binders, and the link between the properties of granules and the conditions of granulation process was established [7]. Meanwhile, the combustion behaviors of sludge–biomass mixed granular fuel with different sludge ratios in air atmosphere are investigated by thermogravimetric analysis [8]. Four main stages are dehydration, volatile oxidation, char burning, and burnout during the thermal decomposition of granular fuel. The reactivities of sludge–biomass granular fuels are improved by the increment of the mixing ratio of biomass. However, the above experiments have high emission pollution and ash slagging potential, so further research is needed. The co-granularization of agricultural and forestry waste using refuse-derived fuel lowered energy consumption during granularization slightly improved particle and bulk densities, which resulted in a significant increase in the breaking load and tensile strength of the granular fuel and reduced the ash content of the roasted biomass granular fuel. However, the durability, net calorific value, and energy density were impoverished as the refuse derived fuel in the blend increased [9,10]. The effect of paper sludge was studied as an additive to the quality of granular fuels [11]. They observed that it improved the characteristic ash melting temperature of straw granular fuel and the mechanical properties of wheat straw granular fuel. In our previous work, the effect of pulping waste liquid on forming effect and performance of wheat straw granular fuel were studied [12]. The proper amount of pulping waste liquid is added to improve the relaxed density and durability of the fuel. However, the compositional variables in the selected natural biomass materials may interfere with the experimental results to a certain extent, and the forming process of industrial waste to granular fuel has not been fully explained. Therefore, a granular fuel model was prepared for this experiment to explore this problem. The forming process of the solid content of pulping waste liquid on biomass granular fuel was explored by changing the amount of pulping waste liquid.

The pulping waste liquid is constituted with lignin, hemicellulose, sugars, organic acids, inorganic salts, etc. [13,14], which can be considered to be used as binder to be incorporated with wheat straw residues and to prepare granular fuels. In this study, cellulose, hemicellulose, and pulping waste liquid were used as raw materials to prepare biomass granular fuel models, so the function of pulping waste liquid on the properties of granular fuels was explored systemically. Thereafter, the granular fuels were characterized by the thermogravimetric analysis method and the Coats-Redfern method to analyze their combustion performance. Based on the data relationship between the granular fuel and prediction models, the function of pulping waste liquid was clarified. These research results provide good guidance to their practical utilization, and will provide a theoretical basis for the further development of biomass granular fuel.

## 2. Materials and Methods

### 2.1. Experimental Materials

Wheat straws residues with a particle size of >0.22 mm were attained from a pulp and paper mill in Weifang, China. The wheat straws were pre-treated with pectinase and lipase for 4 h at 55 °C, and then transferred in digester with 4% of NaOH, liquid ratio of 1:4 at 100 °C for 20 min; cellulose was filtered for papermaking, and the pulping waste liquid was attained. Cellulose (particle size:180 μm) was purchased from Shanghai Aladdin Co., Ltd. (Shanghai, China). Hemicellulose (xylan, 85%, Corncob) was purchased from Shanghai Yuanye Bio-Technology Co. Ltd. (Shanghai, China).

### 2.2. Detection of Pulping Waste Liquid

The lignin content in the pulping waste liquid was measured according to GB/T 2677.8-1994 and 72% sulfuric acid method. An ion chromatograph (ICS-5000+, Thermo Fisher Scientific, Waltham, MA, USA) was applied to determine the sugar content in the pulping waste liquid. A revolving rheometer (ARES-G2, TA Instruments, New Castle, DE, USA) was used to determine the viscosity of the pulping waste liquid.

### 2.3. Sample Preparation

Preparation of Wheat Straws Granular Fuel and Granular Fuel Model

The granular fuels with pulping waste liquid were prepared as Granular Fuel-1 in Table 1: (1) The oven dry wheat straws residue and pulping waste liquid were mixed, and then dried in an oven at 60 °C for 24 h; (2) 1.00 g of the mixed materials was molded by powder compressing machine (FW-4A, Tianjin Botian Shengda Technology Development Co., Ltd., Tianjin, China) at room temperature with a pressure of 8 MPa and a diameter of 13 mm. As a control group, the granular fuels with water were prepared as Granular Fuel-2 in Table 1: the oven dry wheat straws residue and water were mixed, and then molded by powder compressing machine and marked as A, B, C, D, E.

**Table 1.** The material balance of granular fuels.

| | Sample | 1 | 2 | 3 | 4 | 5 |
|---|---|---|---|---|---|---|
| Granular Fuel-1 | Wheat Straws Residue/g | 1 | 1 | 1 | 1 | 1 |
| | Pulping Waste Liquid/g | 1 | 5 | 10 | 15 | 20 |
| | Sample | 1# | 2# | 3# | 4# | 5# |
| Granular Fuel Model-1 | Mixture of Cellulose and Hemicellulose/g | 1 | 1 | 1 | 1 | 1 |
| | Pulping Waste Liquid/g | 1 | 5 | 10 | 15 | 20 |
| | Sample | A | B | C | D | E |
| Granular Fuel-2 | Wheat Straws Residue/g | 1 | 1 | 1 | 1 | 1 |
| | Moisture Content/% | 0 | 2 | 4 | 6 | 8 |
| | Sample | A# | B# | C# | D# | E# |
| Granular Fuel Model-2 | Mixture of Cellulose and Hemicellulose/g | 1 | 1 | 1 | 1 | 1 |
| | Moisture Content/% | 0 | 2 | 4 | 6 | 8 |

In this material balance, the preparing parameters before drying were listed, the applied pulping waste liquid has a solid content of pulping waste liquid of 4.45%. After drying, the mass fraction of pulping waste liquid were 4.26%, 18.30%, 30.94%, 40.19% and 47.26%, which were corresponded to 1, 2, 3, 4, 5.

The granular fuel models with pulping waste liquid were prepared as follows: (1) The mixture of cellulose and hemicellulose with a mass ratio of 30.15:27.06 was prepared and corresponded to the mass ratio in wheat straw residues [9]; (2) and then the mixtures were mixed with pulping waste liquid as Granular Fuel Model-1 in Table 1, and dried in an oven at 60 °C for 24 h; (3) 1.00 g of the mixed materials was molded by powder compressing machine. The samples were marked as 1#, 2#, 3#, 4#, 5#. As a control group, the granular fuels model with water were prepared with the mixture of cellulose and hemicellulose as Granular Fuel Model-2 in Table 1; the preparing process was the same as above mentioned, and the samples were marked as A#, B#, C#, D#, E#, respectively.

### 2.4. *Physicochemical Performance Testing*

#### 2.4.1. Density

The cylindrical granular fuels were measured after setting 0 and 12 h using a thickness gauge. The diameter of the cylindrical granular fuels was kept at 13 mm, but their height was changed and recorded to calculate the relaxed densities of d ($\rho = m/V$).

#### 2.4.2. Compressive Strength

An electronic universal testing machine (WDW-10E, Jinan Wenteng Testing Instrument Co., Ltd., Jinan, China) was used to measure the transverse and longitudinal compressive strengths of granular fuel with a compressing rate of 10 mm/min. When the first crack occurred, the corresponding pressure was rendered as the maximum compression resistance.

#### 2.4.3. X-ray Micro-Tomography

The microstr uctural properties of the granular fuel model were investigated using X-ray microCT (Sky Scan-2211, Bruker, Ettlingen, DE, USA) with a non-destructive method.

The scaled image pixel size of 1 μm was settled, which was sufficient to detect the large features within the granular fuel model.

### 2.5. Combustion Performance Test

#### 2.5.1. Calorific Value

Wheat straws residue granular fuel and granular fuel model were measured using an automatic calorimeter (ZDHW-5, Hebi Huifa Instrument Co., Ltd., Hebi, China) with the Oxygen bomb method, and the test unit was recorded with J/g.

#### 2.5.2. Thermogravimetric Experiments

The sample heating thermogravimetric curve (TG) and derivative thermogravimetric curve (DTG) of the samples were achieved using a TGA thermogravimetric analyzer (Rigaku TG-DTA8122 model, Japan TG-DTA Instruments, Kyoto, Japan). This test was performed under air atmosphere with a flow of 50 mL/min, and the temperature ranged from 25 °C to 800 °C with a heating rate of 10 °C/min. The TG–DTG standard definition method was used to analyze the igniting temperature $T_i$ (°C), the maximum combustion rate $v_{max}$ (%/min) and temperature $T_{max}$ (°C); the average combustion rate v (%/min) and the burn-out temperature $T_f$ (°C).

The index S of comprehensive combustion characteristic [15] can be calculated with the Equation (1):

$$S = \frac{v_{max} \cdot v}{T_i^2 \cdot T_f} \quad (1)$$

where, S—the comprehensive combustion characteristics, $\%^2 \cdot min^{-2} \cdot {}^\circ C^{-3}$; $v_{max}$—the maximum combustion rate, %/min; v—the average combustion rate, %/min; $T_i$—the ignition temperature, °C; $T_f$—the burn-out temperature, °C.

The index $R_v$ of volatile components precipitation characteristics can be calculated with the Equation (2):

$$R_v = \frac{v_{max}}{T_{max} \cdot (T_{max} - T_i)} \quad (2)$$

where, $R_v$—the volatile components precipitation characteristics, $\% \cdot min^{-1} \cdot K^{-2}$; $v_{max}$—the maximum combustion rate, %/min; $T_{max}$—the temperature of maximum combustion rate, K; $T_i$—the ignition temperature, K.

The fire index $D_i$ [16] can be calculated with the Equation (3):

$$D_i = \frac{v_{max}}{T_i \cdot T_{max}} \quad (3)$$

where, $D_i$—the fire index, $\% \cdot min^{-1} \cdot {}^\circ C^{-2}$; $v_{max}$—the maximum combustion rate, %/min; $T_{max}$—the temperature of maximum combustion rate, °C; $T_i$—the ignition temperature, °C.

In this experiment, the combustion thermogravimetric experiment was carried out under a constant heating rate. The Coats-Redfern method was used to fit the kinetic parameters of the first-stage reaction. Calculation was based on the Alenius formula (4), which reflected the relationship between the chemical reaction rate constant and temperature [17]:

$$\frac{d\alpha}{dt} = k \cdot f(\alpha) = Ae^{-\frac{E}{RT}} \cdot (1 - \alpha)^n \quad (4)$$

where, k—the reaction constant, $min^{-1}$; A—the pre-exponential factor, $min^{-1}$; E—the activation energy of reaction, J/mol; R—the universal gas constant, 8.314 J/(mol·K); T—the absolute temperature, K; α—the conversion rate of reaction process, %; n—the reaction order.

$$\alpha = \frac{m_0 - m_t}{m_0 - m_f} \cdot 100\% \quad (5)$$

where, $m_0$—the mass of sample at the initial moment of pyrolysis, mg; $m_t$—the mass of sample at the moment t of pyrolysis, mg; $m_f$—the quality of sample at the time when the solution is terminated, mg.

$$\ln\left[\frac{-\ln(1-\alpha)}{T^2}\right] = \ln\left[\frac{AR}{\beta E}\cdot\left(1-\frac{2RT}{E}\right)\right]-\frac{E}{RT} \quad (6)$$

For biomass fuels, $\left(1-\frac{2RT}{E}\right)\approx 1$, the formula can be simplified to:

$$\ln\left[\frac{-\ln(1-\alpha)}{T^2}\right] = \ln\frac{AR}{\beta E} - \frac{E}{RT} \quad (7)$$

Let $y = \ln\left[\frac{-\ln(1-\alpha)}{T^2}\right]$, $x = \frac{1}{T}$, build a model according to the first-level reaction y = ax + b, make $a = -\frac{E}{R}$, $b = \ln\frac{AR}{\beta E}$, according to slope a and intercept b of the fitted curve; the activation energy and the pre-exponential factor can then be obtained.

### 2.6. The Analysis of Performance Index Equation

The performance indexes of granular fuel like S, $R_v$, $D_i$ and initial density, are calculated to establish equations with the following steps:

Firstly, the solid content of pulp waste liquid was set as x, and the performance index of the granular fuel model was labelled with $y_m$. The x-$y_m$ curve was fitted to obtain Equation (1) according to experimental data of S, $R_v$, $D_i$ and initial density of granular fuel model, respectively. Therefore, the relationship between the x and $y_m$ can be expressed as following:

$$y_m = f(x)$$

The performance index of the granular fuel model was set as $y_m$, the performance index of the granular fuel was labelled with y. The $y_m$-y curve was fitted to obtain Equation (2) according to experimental data of S, $R_v$, $D_i$ and initial density of granular fuel. Therefore, the relationship between the $y_m$ and y can be expressed as following:

$$y = g(y_m)$$

Equation (1) was integrated with Equation (2) to obtain Equation (3). Therefore, the relationship between the x and y could be expressed as following:

$$y = g(f(x))$$

Finally, the prediction accuracy of the granular model was evaluated by the relative error value.

## 3. Results and Discussion

### 3.1. The Composition and Adhesion of Pulping Waste Liquid

As is shown in Figure 1a, the pulping waste liquid contains a variety of chemical components, in which lignin has the highest concentration of 13,300 mg/L followed by, polysaccharide, which has a concentration of 370.98 mg/L, and then some other sugars like arabinose, galactose, glucose and xylose, which have a concentration below 10 mg/L. In the cooking process, lignin, cellulose and hemicellulose can be hydrolyzed as small molecular carbohydrates, which have a relatively small proportion. It is known to all that lignin and sugars are natural adhesive polymers in plants, and can act as adhesive binder in molding granule. Furthermore, lignin has a high heating value of 27,000 J/g due to its rich content of carbon and hydrogen [18], so lignin can also be used to provide the quality of heat for granular fuels. So pulping waste liquid can be used to improve the performance of granular fuels.

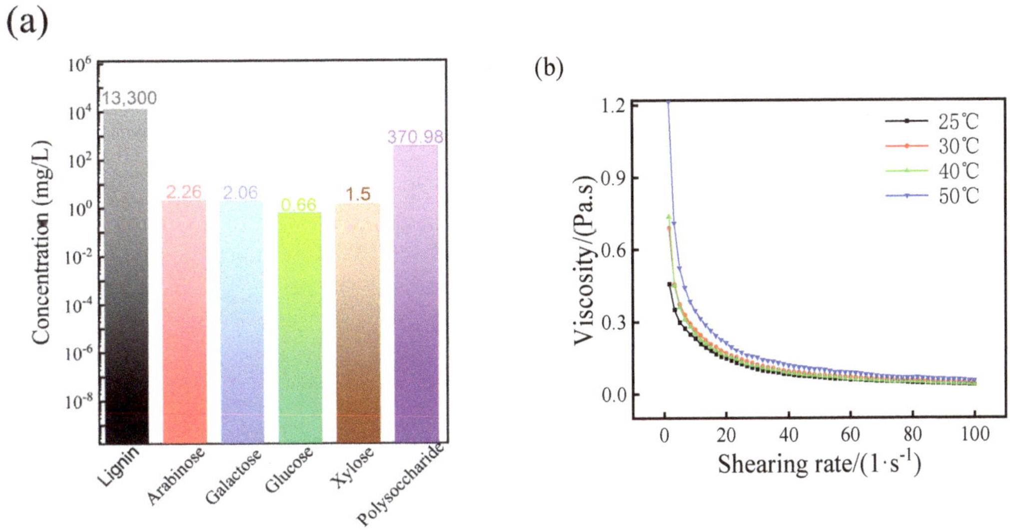


**Figure 1.** (**a**) Composition and (**b**) viscosity of pulping waste liquid.

The influence of shearing rate and temperature on pulping waste liquid are shown in Figure 1b. With the increasing shear rate, the viscosity of pulping waste liquor has a decreasing tendency, which shows a shearing–thinning behavior. Meanwhile, the viscosity of pulping waste liquid increases with the increasing temperature, the pulping waste liquid has the highest initial viscosity at 50 °C of about 5.32 Pa·s, and lowest initial viscosity at 25 °C of about 0.51 Pa·s. Lignin could dissolve in solvents at low temperature, but lignin would become soft with increasing temperature, along with the increasing viscosity. Actually, the natural adhesives like lignin, hydrolysates and other resins, waxes, are sensitive to temperature and pressure, and so they would also play binding role in the pressing process. With the increasing shearing rate, all the sample with various temperatures showed a similar viscosity of about 0.15 Pa·s, in which all the chains of the molecules have a similar directional arrangement under the shear force.

### *3.2. Effect of Pulping Waste Liquid on the Physical Properties of Wheat Straw Residue Granular Fuel*

The manufactured granular fuels are influenced by many factors, like additives, moisture content, dimensions, densities, compressive strength, heating value, etc. As the important constituents and main adhesives of granular fuels, the effect of pulping waste liquid would be studied systemically.

The relationship between the pulping waste liquid and the physical properties of granular fuel is displayed in Figure 2a. When the solid content of the pulping waste liquid increases from 4.26% to 30.94%, the density of granular fuel increases significantly, and the initial density reaches 1.37 g/cm$^3$. It turns out that the solids in the pulping waste liquid can fill the gap between the wheat straw residue particles. When the solid content of waste liquid increases by 47.26%, the initial density of granular fuels decreases by 1.38 g/cm$^3$, so the addition of pulping waste liquid would lead to more complicated mechanical engagement and intermolecular binding force, and finally result in a dense structure. However, the internal structure undergoes stress deformation over time, so the density of all granular fuels shows decreasing tendency with ongoing, time and the relaxation density tends to a constant after 4 h. In addition, the results show that the granular fuels with high addition of pulping waste liquid have a low reduction amplitude, and further demonstrate that the addition of pulping waste liquid improves the binding force of the granular fuels.

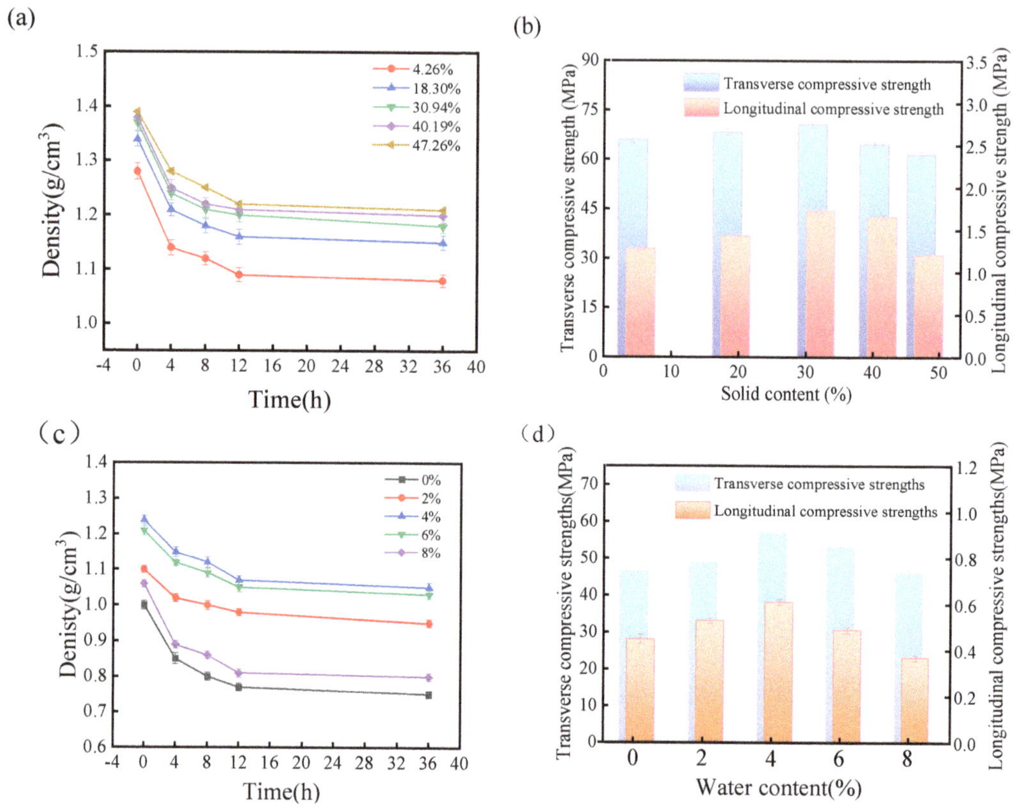


**Figure 2.** The physical properties of granular fuel with different pulping waste liquid: Density (**a**) and Compressive strength (**b**); and with water (**c**,**d**).

As shown in Figure 2b, the compressive strength of wheat straw residue granular fuel initially increases and then decreases with the increasing content of pulping waste liquid. Compared with the granular fuel with 4% water content (56.82 MPa), the best transverse compressive strength of the granular fuel increases by 9.27% with the adding of pulping waste liquid. It can be seen that pulping waste liquid has a more substantial enhancement effect on the compressive strength of straw residue granular fuel. When the solid content of pulping waste liquid is 30.94%, the transverse compressive strength reaches 62.09 MPa, which is 25.79% higher than the transverse compressive strength, with 4.26% pulping waste liquid. With the further increase of pulping waste liquid by 47.26%, the transverse compressive strength decreases to 58.40 MPa. The excessive solid content of pulping waste liquid has increased the lubricity of the internal structure [19], and thus wheat straw residue granular fuel cannot effectively support the higher external force.

Compared with the granular fuels with pulping waste liquid, the control group has low density and low physical strength and the results are listed in Figure 2c,d. Water is important in the molding process,. with increasing water proportion, water works as free radical and promotes the binding among particles, even with less external pressure. As can be seen in Figure 2c, the granular fuels have increasing transverse and longitudinal compressive strengths with increasing water content that then decreases after 4%.

### 3.3. *Effect of Pulping Waste Liquid on the Physicochemical Properties of Granular Fuel Model*

The granular fuels with wheat straw residue have complicate constituents, and it is hard to tell the function of pulping waste liquid clearly. Thereafter, the granular fuel models are prepared, and their physicochemical properties are analyzed and compared to clarify the forming process.

It can be seen from Figure 3a that the initial density and relaxation density show a downward tendency with the increasing moisture content. Meanwhile, the initial density and relaxation density (after 24 h) show an increasing tendency with the increasing pulping waste liquid solid content. The particle size of cellulose and hemicellulose powder is small, and their internal gap reduces without any additives. Adding water to the mixture to make the granules, the hydrogen bonding forms among cellulose, hemicellulose and water, which strengths the binding force of granular fuels. But the density decreases with increasing moisture content, and water becomes attached to the surface of particles and forms a layer of water film, which hinders the tight binding between the particles [20]. Adding the pulping waste liquid with the solid content of 4.29%, the initial density is 1.02 g/cm$^3$. The initial density then increases by 1.05 g/cm$^3$ with the solid content of 47.26%. Compared with the granules with water, the granules with pulping waster liquid have a smaller difference between initial and relaxed density. Lignin and sugars in the pulping waste liquid has good adhesive function [21,22] and elastic performance; they can produce plastic deformation after compression molding process, and have a fast stress transmission speed [23,24].

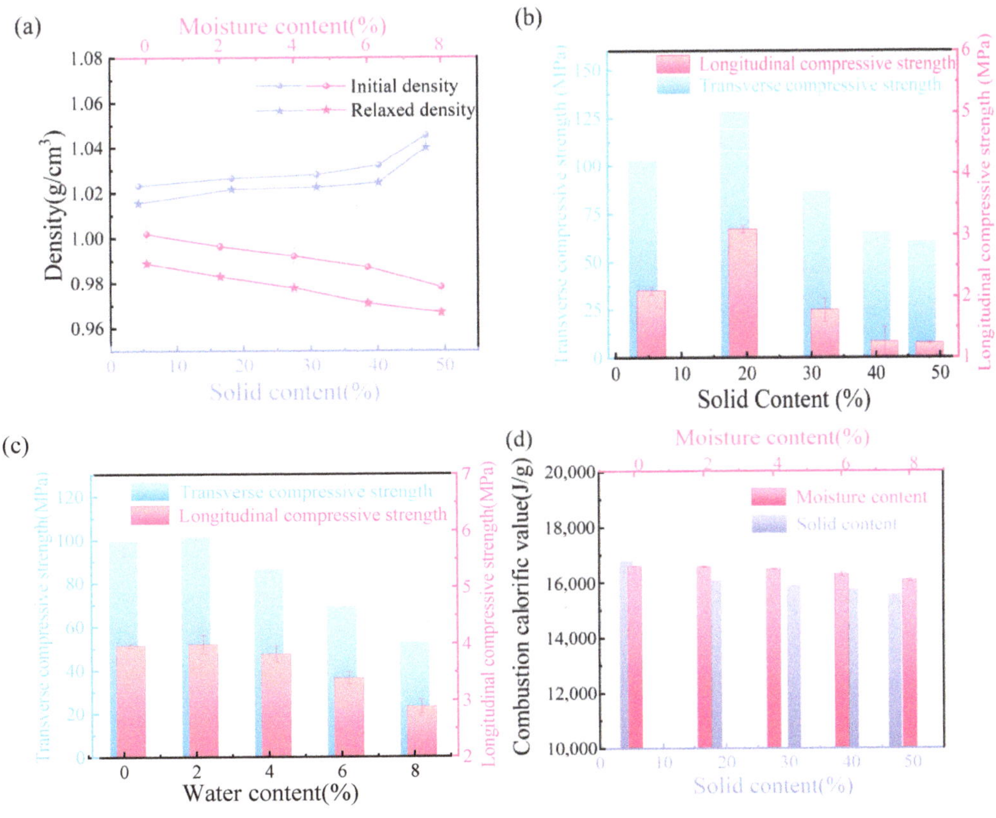


**Figure 3.** Effect of pulping waste liquor on physicochemical properties of granular fuel. (**a**) Density, (**b**,**c**) compressive strength and (**d**) combustion calorific value.

Figure 3b,c show that the transverse compressive strength is obviously higher than the longitudinal compressive strength of granular fuels, and they all show a descending tendency with increasing moisture. Excessive water (>2%) has a weakening effect on the compressive strength of granular fuel. However, the transverse and longitudinal compressive strength show the same tendency, which increases with increasing solid content and then decreases, and shows a maximum at 20%. The lignin and sugars in the pulping waste liquid show good bonding effect and contribute to an improved compressive strength. When the pulping waste liquid content exceeds 20%, too many spaces are filled with lignin and sugars, which creates great connectivity with cellulose and hemicellulose, and results in bad skeleton stability of the granule fuels. Therefore, the amount of pulping waste liquid should be kept at a suitable dosage.

The combustion calorific value of granular fuel model shows a downward trend with the increasing water content and pulping waste liquid solid content (Figure 3d). The pulping waste liquid containing lignin, polysaccharide etc. has a value of calorific value but is lower than cellulose/hemicellulose, which explains the reason for the decreasing tendency. Furthermore, the granular fuels with pulping waste liquid are higher than water, due to the existence of organic compounds.

### 3.4. *Effect of Pulping Waste Liquid on the Combustion Characteristics of Granular Fuel Model*

As shown in Figure 4, the evaporation temperature range of water is between 25 and 150 °C, the degradation temperature range of cellulose and hemicellulose is between 175 and 400 °C, while lignin degrades in the range from 350 to 600 °C. The higher the content of pulping waste liquid, the lower the combustion and decomposition rate of granular fuel. It is observed that the residual weight of wheat straws residue granular fuel with the most solid content (27.65%) is higher than that of wheat straws residue granular fuel without pulping waste liquor (18.96%). This is because of the presence of higher inorganic mineral content and ash in the pulping waste liquor.

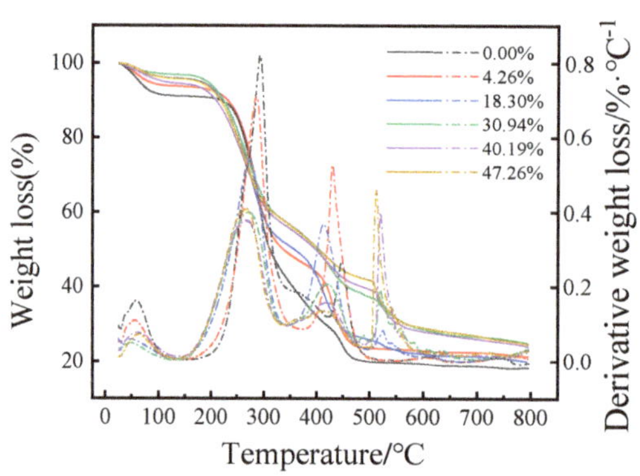


**Figure 4.** TG and DTG curves of wheat straws granular fuel with different pulping waste liquid content.

It can be seen from Figure 5a that the TG curve of granular fuel mainly have four stages. They can be concluded as follows: the water evaporation stage (I), the volatile component combustion stage (II), the fixed carbon combustion stage (III) and the burn-out stage (IV); the specific data are listed in Table 2. The evaporation temperature range of water is between 25 and 150 °C, the degradation temperature of volatile components ranges from 220 to 360 °C, the fixed carbon combustion temperature ranges from 360 to 570 °C, and granular fuels burn out at 600 °C. It is known that hemicellulose pyrolysis occurs from 220 to 315 °C, and cellulose pyrolysis from 315 to 360 °C [25]. Meanwhile, $CO_2$, CO and some small organic matter [26–28] are generated and a high weight loss of 75.45% occurs from 220 to 360 °C. At the third stage, the pyrolysis of the organic matter in the granular fuel is completed. At the last stage, the combustion is completed, and the residues are kept at 1.23%.

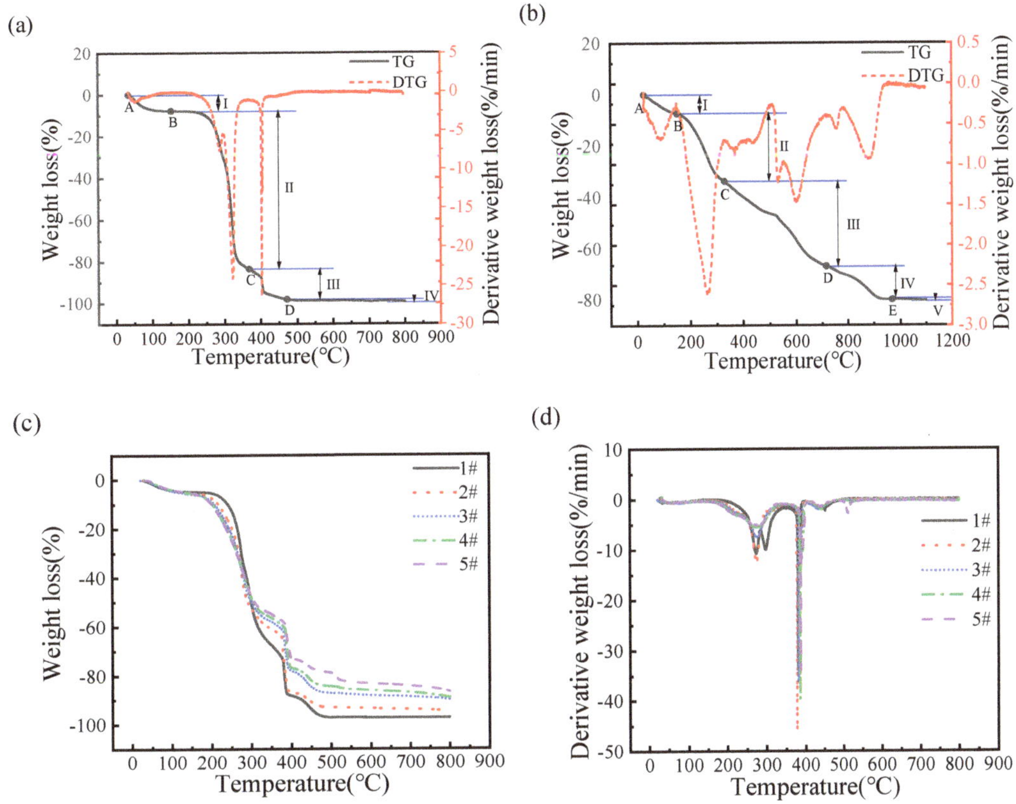


**Figure 5.** TG and DTG curves of granular fuel models with different pulping waste liquid content. (**a**) TG–DTG curves of the granular fuel model with water. (**b**) TG–DTG curves of the pulping waste liquid solids. (**c**) TG curves of granular fuel models with different pulping waste liquid content. (**d**) DTG curves of granular fuel models with different pulping waste liquid content.

**Table 2.** The combustion process analysis of the model without adding waste liquid fuel or solid content of waste liquid; granular fuel model with different solid content of the waste liquid.

| Sample | Water Evaporation Stage/°C | Volatile Component Combustion Stage/°C | Fixed Carbon Oxidation Stage/°C | Fourth Stage/°C | Burnout Period/°C | Burning Remaining Amount/% |
|---|---|---|---|---|---|---|
| A# | –204 | 204–367 | 367–511 | | 511–800 | 1.23 |
| Waste liquid solids | –149 | 149–327 | 327–717 | 717–955 | 955–1200 | 19.99 |
| 1# | –177 | 177–347 | 347–520 | | 520–800 | 3.15 |
| 2# | –158 | 158–350 | 350–558 | | 558–800 | 6.80 |
| 3# | –150 | 150–343 | 343–555 | | 555–800 | 6.82 |
| 4# | –150 | 150–342 | 342–494 | 494–552 | 552–800 | 14.38 |
| 5# | –138 | 138–340 | 340–483 | 483–557 | 557–800 | 17.30 |

In referring to Figure 5b, it can be seen that the pyrolysis of pulping waste liquid mainly includes five stages. They can be concluded as follows: the water evaporation stage (I), the volatile component combustion stage (II), the fixed carbon combustion stage (III), the fourth stage (IV), and the burn-out stage (V); the detailed temperature range is listed in Table 3. At the fourth stage (480–560 °C), the oxidation and decomposition of inorganic salt in the pulping waste liquid occurs, and alkali metal salts melt, evaporate, decompose, and burn to produce volatile components such as CO and $CO_2$ [29,30]. As can be seen from Table 3, with the increasing solid content of pulping waste liquid, the combustion residues of the granular fuel model also increase. The sample of A# is prepared without pulping waste liquid, and the burning residues amount of the granular fuel model is 1.23%. When the solid content of pulping waste liquid increases to 47.26%, the burning residues amount of granular fuel model increases to the maximum of 17.30%. Cellulose/hemicellulose granular fuel model with water can be completely burned, and the inorganic substances within pulping waste liquid result in residue. Therefore, when the more solid content of the pulping waste liquid is retained in the granular fuel model, more combustion residues are produced.

**Table 3.** Analysis table of the thermodynamic parameters of the volatile stage combustion.

| Sample | $T_i$/°C | $v_{max1}$/(%/min) | $T_{max1}$/°C | $v_{max2}$/(%/min) | $T_{max2}$/°C | $T_f$/°C | $\bar{v}$/(%/min) |
|---|---|---|---|---|---|---|---|
| A# | 298 | 24.04 | 321 | 26.43 | 403 | 511 | 4.64 |
| Waste liquid solids | 204 | 2.63 | 262 | 1.54 | 601 | 955 | 1.53 |
| 1# | 251 | 11.20 | 274 | 33.22 | 380 | 520 | 3.36 |
| 2# | 249 | 12.44 | 275 | 46.17 | 377 | 558 | 2.91 |
| 3# | 227 | 7.45 | 277 | 36.31 | 382 | 555 | 2.75 |
| 4# | 221 | 5.53 | 278 | 39.71 | 386 | 552 | 2.64 |
| 5# | 216 | 5.12 | 281 | 33.82 | 387 | 557 | 2.53 |

$T_i$: Ignition temperature; $T_f$: Burnout temperature; $T_{max1}$: Maximum burning temperature in the volatile combustion stage; $T_{max2}$: Maximum burning temperature in the fixed carbon stage; $v_{max1}$: Maximum burning rate in the volatile combustion stage; $v_{max2}$: Maximum burning rate in the fixed carbon stage; v: Average burning rate.

Figure 5c,d show the pyrolysis of granular fuel models with different pulping waste liquid content. The volatile combustion stage of granular fuel models range from 150 to 350 °C, in which the granular fuels form the initial carbon layer. With the increasing pulping waste liquid solid content, the inorganic substances proportion in the fuel model increases, followed by the melt reduction reaction recording a high initial temperature. The more inorganic substances lower the combustion temperature of the fuel model, indicating that the pulping waste liquid solid content has a synergistic effect on the combustion of the granular fuel model.

Table 3 shows that the ignition temperature of granular fuel model decreases with the increasing waste liquid solid contents. As inorganic substances are incombustible, they will decrease the ignition temperature of granular fuel model. For the granular fuel model, $v_{max1}$ and $v_{max2}$ show a downward trend with increasing pulping waste liquid solid content; the v of the A#, 1#, and 5# models are 4.64%·min$^{-1}$, 3.36%·min$^{-1}$ and 2.53%·min$^{-1}$, respectively. The inorganic substances reduce the calorific value of the granular fuel, which results in the decrease of combustion releasing heat. Furthermore, the inorganic substances cover the unburned organic matter during the combustion process, which increases the resistance to the flame spread and decreases the combustion rate [30,31].

Table 4 shows that increasing the pulping waste liquid solid content decreases the ignition index of the granular fuel model, and weakens the ignition performance. The $D_i$ of 1# fuel model is 3.91 × $10^{-5}$%·$min^{-1}$·°$C^{-2}$, and it has relatively good ignition performance; the $D_i$ of 5# fuel model is 1.89 × $10^{-5}$%·$min^{-1}$·°$C^{-2}$ with the worst ignition performance. The comprehensive combustion characteristic index of the 1# fuel model is 1.69 × $10^{-7}$$\%^2$·$min^{-2}$·°$C^{-3}$. The S of granular fuel decreases as the pulping waste liquid solid content increases. The $R_v$ of the 1# and 5# fuel models are 2.12 × $10^{-4}$ and 0.65 × $10^{-4}$%·$min^{-1}$·$K^{-2}$, respectively. Therefore, the increasing waste liquid solid contents result in decreasing $R_v$ of granular fuel, and a decreasing rate of the volatile components precipitation and combustion.

**Table 4.** Analysis table of the combustion characteristics of the granular fuel model.

| Sample | S/$\%^2$·$min^{-2}$·°$C^{-3}$ | $R_v$/%·$min^{-1}$·$K^{-2}$ | $D_i$ /%·$min^{-1}$·°$C^{-2}$ |
|---|---|---|---|
| A# | 4.36 × $10^{-7}$ | 3.45 × $10^{-4}$ | 7.07 × $10^{-5}$ |
| Waste liquid solids | 0.14 × $10^{-7}$ | 0.43 × $10^{-4}$ | 1.02 × $10^{-5}$ |
| 1# | 1.69 × $10^{-7}$ | 2.12 × $10^{-4}$ | 3.91 × $10^{-5}$ |
| 2# | 1.57 × $10^{-7}$ | 1.93 × $10^{-4}$ | 4.33 × $10^{-5}$ |
| 3# | 0.95 × $10^{-7}$ | 1.06 × $10^{-4}$ | 2.68 × $10^{-5}$ |
| 4# | 0.70 × $10^{-7}$ | 0.79 × $10^{-4}$ | 2.03 × $10^{-5}$ |
| 5# | 0.66 × $10^{-7}$ | 0.65 × $10^{-4}$ | 1.89 × $10^{-5}$ |

S: The comprehensive combustion characteristic index; $R_v$: The volatile component precipitation characteristic index; $D_i$: The ignition index.

The kinetic parameters of granular fuel model are obtained by the first-order reaction equation in each combustion stage, and the correlation coefficients are all above 0.95, indicating that the fitting results are feasible. The activation energy E and frequency factor A of granular fuel model are calculated by fitting the slope and intercept of linear equation in each combustion stage. Herein, E is the difference between the activated molecules' average energy and all molecules' energy in the combustion process. The combustion process of the granular fuel is easier with lower value of E. A is the frequency factor, which is a constant determined only by the chemical substance and used to measure the intensity of the combustion process of granular fuel.

Table 5 shows that the E of the A# granular fuel model in the volatile and fixed carbon combustion stages are 124.70 and 283.99 KJ·$mol^{-1}$, which are 136.35% and 561.67% higher than that of the waste liquid solid. The A of the A# granular fuel model in the volatile and fixed carbon combustion stages are 4.25 × $10^{10}$ and 3.69 × $10^{21}$ $min^{-1}$, which are 1.41 × $10^6$ and 7.62 × $10^{19}$ times that of the waste liquid solid. Thus, under the same conditions, the volatile and fixed carbon in the waste liquid solid are more flammable than those in the A# granular fuel model, and the combustion process is milder and less prone to deflagration. The E of the 5# granular fuel model in the volatile and fixed carbon combustion stages are 55.42 and 216.86 KJ·$mol^{-1}$, which are 55.56% and 23.64% lower than those of the A# model. This is because of the catalytic role of metal oxides in the pyrolysis process [32,33]. As the waste liquid solid content increases, metal oxides' content in the granular fuel model increase, the catalytic pyrolysis is strengthened, the energy required for the reaction decreases, and the activation energy decreases. The A of A# granular fuel model in the volatile and fixed carbon combustion stages are 9.79 × $10^5$ times and 1.24 × $10^5$ times that of the 5# granular fuel model. This indicates that the volatile and fixed carbon of the granular fuel combined with waste liquid is more flammable, more stable, and less prone to deflagration than those without the addition of waste liquid.

**Table 5.** Combustion kinetic parameters of the granular fuel model.

| Sample | Temperature Interval/°C | Fitting Equation | Correlation Coefficient $R^2$ | Activation Energy E/KJ·mol$^{-1}$ | Frequency Factor A/min$^{-1}$ |
|---|---|---|---|---|---|
| A# | 220–310 | y = −14999.15x + 12.55 | 0.9969 | 124.70 | $4.25 \times 10^{10}$ |
| | 375–420 | y = −34157.40x + 36.92 | 0.9613 | 283.99 | $3.69 \times 10^{21}$ |
| Waste liquid solids | 210–300 | y = −6345.99x − 0.75 | 0.9998 | 52.76 | $3.01 \times 10^{4}$ |
| | 380–425 | y = −5161.84x − 6.97 | 0.9931 | 42.92 | $4.84 \times 10^{1}$ |
| | 790–880 | y = −17846.32 + 1.55 | 0.9855 | 148.37 | $8.43 \times 10^{5}$ |
| 1# | 210–300 | y = −12898.62x + 10.15 | 0.9978 | 107.24 | $3.30 \times 10^{9}$ |
| | 360–405 | y = −28286.50x + 29.02 | 0.9681 | 235.17 | $1.13 \times 10^{18}$ |
| 2# | 210–300 | y = −9827.87x + 5.13 | 0.9915 | 81.71 | $1.66 \times 10^{7}$ |
| | 355–400 | y = −47576.01x + 58.70 | 0.9549 | 395.55 | $1.48 \times 10^{31}$ |
| 3# | 210–300 | y = −8237.74x + 2.31 | 0.9965 | 68.49 | $8.30 \times 10^{5}$ |
| | 365–410 | y = −42527.43x + 51.22 | 0.9594 | 353.57 | $7.49 \times 10^{27}$ |
| 4# | 210–300 | y = −6957.58x + 0.03 | 0.9979 | 57.85 | $7.20 \times 10^{4}$ |
| | 365–410 | y = −30575.91x + 32.41 | 0.9544 | 254.21 | $3.64 \times 10^{19}$ |
| | 500–545 | y = −54352.30x + 54.38 | 0.9647 | 451.89 | $2.25 \times 10^{29}$ |
| 5# | 210–300 | y = −6666.24x − 0.43 | 0.9976 | 55.42 | $4.34 \times 10^{4}$ |
| | 355–410 | y = −26084.06x + 25.46 | 0.95367 | 216.86 | $2.97 \times 10^{16}$ |
| | 500–525 | y = −77360.42x + 84.46 | 0.9641 | 643.17 | $3.72 \times 10^{42}$ |

### *3.5. Analysis of the Molding Process of Granular Fuel*

Figure 6 shows the 3D images of the internal and external microstructure of granular fuel. The granular fuel with 2% moisture shows a regular section with small pores and cracks, and the distribution of water is shown in Figure 6b. In comparison, the granular fuel with 18.30% pulping waste liquid has a denser structure, and the liquid is closely distributed with the solid particles. It is concluded that the pulping waste liquid makes a good fusion between particles, and a strengthened binding force is formed. This can be used to directly explain the strengthened physical properties and deduced density obtained with pulping waste liquid.

**Figure 6.** 3D images of micro-CT reconstructed granular fuel model with (**a**) 2% moisture content and (**c**) 18.30% pulping waste liquid solid content; Distribution of water (**b**) and pulping waste liquid solids (**d**) in the granular fuel model.

The granular fuel and granular fuel model with pulping waste liquid solid content of 4.26%, 18.30%, 30.94, 40.19%, 47.26% were calculated to obtain prediction models by curve fitting and equation simplifying, with their S, $R_v$, $D_i$ and density listed in the following:

$$y = 4.06 \times 10^{-7} - 1.84 \times 10^{-7} \cdot (0.03)^{e^{-\frac{x}{2.8}}} \quad (8)$$

In Equation (8), x refers to solid content of pulping waste liquid, and y refers to S. As is shown in Table 6, the relative error between measured data and predicated data are lower than 0.20, which means a high fitting degree exists. And then the derivation was calculated:

$$\frac{dy}{dx} = -1.84 \times 10^{-7} \cdot (0.03)^{e^{-\frac{x}{2.8}}} \cdot \ln(0.03) \cdot e^{-\frac{x}{2.8}} \cdot \left(-\frac{1}{2.8}\right) \quad (9)$$

$$y' = -\frac{1.84 \times 10^{-7} \cdot \ln(0.03) \cdot e^{\frac{x}{2.8}}}{2.8} \quad (10)$$

in which, $x \in [0, 50]$, and $y' < 0$. So, the value of S is negatively associated with the solid content of pulping waste liquid.

$$y = 0.28 \times \left(3.41 \times 10^{-4} - 6.00 \times 10^{-5} \cdot x^{0.39}\right)^{0.81} \quad (11)$$

**Table 6.** Prediction models of the comprehensive combustion characteristic index.

| | Solid Content/% | Measured Data | Predicted Data | Relatived Error |
|---|---|---|---|---|
| S | 0 | $4.01 \times 10^{-7}$ | $4.01 \times 10^{-7}$ | 0 |
| | 4.26 | $3.21 \times 10^{-7}$ | $3.22 \times 10^{-7}$ | 0.0031 |
| | 18.3 | $2.72 \times 10^{-7}$ | $2.23 \times 10^{-7}$ | 0.1801 |
| | 30.94 | $2.12 \times 10^{-7}$ | $2.22 \times 10^{-7}$ | 0.0472 |
| | 40.19 | $1.98 \times 10^{-7}$ | $2.22 \times 10^{-7}$ | 0.1212 |
| | 47.26 | $2.06 \times 10^{-7}$ | $2.22 \times 10^{-7}$ | 0.0777 |

In Equation (11), x refers to solid content of pulping waste liquid and y refers to $R_v$. As is shown in Table 7, the relative error between measured data and predicated data are also lower than 0.20, which means a high fitting degree exists. And then the derivation was calculated:

$$\frac{dy}{dx} = 0.81 \cdot 0.28 \cdot \left(3.41 \times 10^{-4} - 6.00 \times 10^{-5} \cdot x^{0.39}\right)^{-0.19} \cdot \left(-6.00 \times 10^{-5}\right) \cdot 0.39 \cdot x^{-0.61} \quad (12)$$

$$\frac{dy}{dx} = -[1.88 \times 10^{5} \cdot x^{0.61} \cdot (3.41 \times 10^{-4} - 6 \times 10^{-5} \cdot x^{0.39})^{0.19}]^{-1} \quad (13)$$

$$y' = -\frac{5.31 \times 10^{-6}}{\left(3.41 \times 10^{-4} - 6.00 \times 10^{-5} \cdot x^{0.39}\right)^{0.19} \cdot x^{0.61}} \quad (14)$$

in which, $x \in [0, 50]$ and $(3.41 \times 10^{-4} - 6.00 \times 10 - 5 \cdot x^{0.39}) > 0$, so $y' < 0$. So, the value of $R_v$ is negatively associated with the solid content of pulping waste liquid.

$$y = 0.0042 \cdot \left(7.03 \times 10^{-5} - 1.47 \times 10^{-5} \cdot x^{0.32}\right)^{0.38} \quad (15)$$

**Table 7.** Prediction models of the volatile component precipitation characteristic index.

| | Solid Content/% | Measured Data | Predicted Data | Relatived Error |
|---|---|---|---|---|
| $R_v$ | 0 | $4.13 \times 10^{-4}$ | $4.17 \times 10^{-4}$ | 0.0097 |
| | 4.26 | $3.48 \times 10^{-4}$ | $2.87 \times 10^{-4}$ | 0.1753 |
| | 18.3 | $2.16 \times 10^{-4}$ | $2.19 \times 10^{-4}$ | 0.0138 |
| | 30.94 | $1.49 \times 10^{-4}$ | $1.68 \times 10^{-4}$ | 0.1275 |
| | 40.19 | $1.21 \times 10^{-4}$ | $1.37 \times 10^{-4}$ | 0.1322 |
| | 47.26 | $1.38 \times 10^{-4}$ | $1.15 \times 10^{-4}$ | 0.1667 |

In Equation (15), x refers to solid content of pulping waste liquid and y refers to $D_i$. As is shown in Table 8, the relative error between measured data and predicated data are also lower than 0.20, which means a high fitting degree exists. And then the derivation was calculated:

$$\frac{dy}{dx} = 0.0042 \cdot 0.38 \cdot \left(7.03 \times 10^{-5} - 1.47 \times 10^{-5} \cdot x^{0.32}\right)^{-0.62} \cdot \left(-1.47 \times 10^{-5}\right) \cdot 0.32 \cdot x^{-0.68} \quad (16)$$

$$y' = -\frac{7.51 \times 10^{-9}}{\left(7.03 \times 10^{-5} - 1.47 \times 10^{-5} \cdot x^{0.32}\right)^{0.62} \cdot x^{0.68}} \quad (17)$$

in which, $x \in [0, 50]$ and $\left(7.03 \times 10^{-5} - 1.47 \times 10^{-5} \cdot x^{0.32}\right)^{0.62} > 0$, and $y' < 0$. So, the value of $D_i$ is negatively associated with the solid content of pulping waste liquid.

$$y = \frac{7.34 \cdot e^{\frac{x}{7.92}} + 552.62}{5.22 \cdot e^{\frac{x}{7.92}} + 417.17} \quad (18)$$

**Table 8.** Prediction models of the ignition index.

| | Solid Content/% | Measured Data | Predicted Data | Relatived Error |
|---|---|---|---|---|
| $D_i$ | 0 | $10.99 \times 10^{-5}$ | $11.11 \times 10^{-5}$ | 0.0109 |
| | 4.26 | $9.99 \times 10^{-5}$ | $9.53 \times 10^{-5}$ | 0.0460 |
| | 18.3 | $8.69 \times 10^{-5}$ | $8.35 \times 10^{-5}$ | 0.0391 |
| | 30.94 | $6.91 \times 10^{-5}$ | $7.65 \times 10^{-5}$ | 0.1071 |
| | 40.19 | $6.83 \times 10^{-5}$ | $7.21 \times 10^{-5}$ | 0.0556 |
| | 47.26 | $7.26 \times 10^{-5}$ | $6.89 \times 10^{-5}$ | 0.0610 |

In Equation (18), x refers to solid content of pulping waste liquid, and y refers to initial density. As is shown in Table 9, the relative error between measured data and predicated data are also lower than 0.20 with the range of solid content at 4.26–47.26%. With the high fitting degree of the equation, its derivation was calculated:

$$\frac{dy}{dx} = \frac{\left(\left(5.22 \cdot e^{\frac{x}{7.92}} + 417.17\right) \times 0.927 \cdot e^{\frac{x}{7.92}} - \left(7.34 \cdot e^{\frac{x}{7.92}} + 552.62\right) \times 0.659 \cdot e^{\frac{x}{7.92}}\right)}{\left(5.52 \cdot e^{\frac{x}{7.92}} + 417.17\right)^2} \quad (19)$$

$$y' = \frac{(177.35) \cdot e^{\frac{x}{7.92}}}{215.81 \cdot e^{\frac{2x}{7.92}} + 34493.62 \cdot e^{\frac{x}{7.92}} + 1378324.01} \quad (20)$$

in which, $x \in [0, 50]$ and $y' > 0$. So, the value of initial density is positively associated with the solid content of pulping waste liquid.

$$y = 1.19 - 0.24 \cdot e^{-\frac{x}{6.90}} \quad (21)$$

**Table 9.** Prediction models of the initial density.

| | Solid Content/% | Measured Data | Predicted Data | Relativized Error |
|---|---|---|---|---|
| Initial Density | 0 | 1.10 | 1.33 | 0.2091 |
| | 4.26 | 1.28 | 1.33 | 0.0391 |
| | 18.3 | 1.34 | 1.33 | 0.0075 |
| | 30.94 | 1.37 | 1.36 | 0.0071 |
| | 40.19 | 1.38 | 1.38 | 0 |
| | 47.26 | 1.39 | 1.39 | 0 |

In Equation (21), x refers to solid content of pulping waste liquid and y refers to relaxed density. As is shown in Table 10, the relative error between measured data and predicated data are also lower than 0.20, and the equation has high fitting degree. And then the derivation was calculated:

$$\frac{dy}{dx} = (-0.24)\cdot\left(-\frac{1}{6.90}\right)\cdot e^{-\frac{x}{6.90}} \tag{22}$$

$$y' = 3.48\times10^{-2}\cdot e^{-\frac{x}{6.90}} \tag{23}$$

in which, $x \in [0, 50]$, and $y' > 0$. So, the value of relaxed density is positively associated with the solid content of pulping waste liquid.

**Table 10.** Prediction models of the relaxed density.

| | Solid Content/% | Measured Data | Predicted Data | Relatived Error |
|---|---|---|---|---|
| Relaxed Density | 0 | 0.95 | 0.95 | 0 |
| | 4.26 | 1.08 | 1.06 | 0.0185 |
| | 18.3 | 1.15 | 1.17 | 0.0174 |
| | 30.94 | 1.18 | 1.19 | 0.0085 |
| | 40.19 | 1.2 | 1.19 | 0.0083 |
| | 47.26 | 1.21 | 1.20 | 0.0083 |

According to the above data, the initial density and relaxed density are positively associated with the solid content of pulping waste liquid, whereas, the S, $R_v$, $D_i$ are all negatively associated with the solid content of pulping waste liquid. Therefore, the addition of pulping waste liquid increased the tightness and physical performance of the granular fuels, but decreased the combustion. Combined with the characteristic of pulping waste liquid and the structure of granular fuels, this confirmed the binding function of pulping waste liquid; Meanwhile, within the same quantity, pulping waste liquid has inferior combustion performance than wheat straw resides and reduced the whole combustion performance.

## 4. Conclusions

In this study, the function of pulping waste liquid on the forming process and properties of granular fuel were investigated systemically.

(1) Lignin and sugars in the pulping waste liquid were natural adhesives, and the addition of pulping waste liquid could improve the mechanical strength of granular fuels. The granular fuels had the best performance with 18.30% solid content of pulping waste liquid, the highest transverse compressive strength was up to 102.61 MPa, and the activation energy was 81.71 $KJ\cdot mol^{-1}$ at the volatile carbon combustion stage.

(2) The predicated models were established to study the forming process, and the pulping waste liquid were evenly distributed in granular fuels. The addition of pulping waste liquid improved their density and density combustion indexes.

**Author Contributions:** L.X.: Conceptualization, software, validation, investigation, writing—original draft preparation. G.Y.: conceptualization, methodology, validation, resources, writing—review and editing, project administration, funding acquisition. Z.L.: methodology, writing—review and editing, visualization. J.C.: validation, supervision. J.L.: formal analysis. B.H.: data curation. All authors have read and agreed to the published version of the manuscript.

**Funding:** This research was funded by the National Key Research and Development Program of China (grant number 2019YFC1905900); Jinan Science and Technology Bureau Project (grant number 20233046); Pilot Project for Integrating Science, Education and Industry (grant number 2022JBZ01-05, 2022PY057).

**Institutional Review Board Statement:** The study was conducted in accordance with the Declaration of Helsinki, and approved by the Ethics Committee.

**Data Availability Statement:** All data are contained within the article.

**Conflicts of Interest:** The authors declare no conflicts of interest.

## References

1. Arulprakasajothi, M.; Beemkumar, N.; Parthipan, J.; Battu, N. Investigating the Physio-chemical Properties of Densified Biomass Pellet Fuels from Fruit and Vegetable Market Waste. *Arab. J. Sci. Eng.* **2020**, *45*, 563–574. [CrossRef]
2. Costa, S.; Barcelos, R.; Magnago, R. Solid biofuel from glycerol and agricultural waste as a source of energy. *Cellul. Chem. Technol.* **2017**, *51*, 765–774.
3. Nesterovic, A.; Djatkov, D.; Viskovic, M.; Martinov, M.; Adamovic, D. Air pollutants emissions from biomass combustion in the City of Novi Sad, Serbia. *Biomass Convers. Biorefinery* **2023**, *13*, 10935–10946. [CrossRef]
4. Woolley, K.E.; Dickinson-Craig, E.; Bartington, S.E.; Oludotun, T.; Kirenga, B.; Mariga, S.T.; Kabera, T.; Coombe, A.; Pope, F.D.; Singh, A.; et al. Effectiveness of interventions to reduce household air pollution from solid biomass fuels and improve maternal and child health outcomes in low- and middle-income countries: A systematic review protocol. *Syst. Rev.* **2021**, *10*, 33. [CrossRef]
5. Mohamed, O.; Zuaiter, H. Fresh Properties, Strength, and Durability of Fiber-Reinforced Geopolymer and Conventional Concrete: A Review. *Polymers* **2024**, *16*, 141. [CrossRef]
6. Cheng, J.; Zhou, F.; Si, T. Mechanical strength and combustion properties of biomass pellets prepared with coal tar residue as a binder. *Fuel Process. Technol.* **2018**, *179*, 229–237. [CrossRef]
7. Chen, H.; Mangwandi, C.; Rooney, D. Production of solid biofuel granules from drum granulation of bio-waste with silicate-based binders. *Powder Technol.* **2019**, *354*, 231–239. [CrossRef]
8. Jiang, L.; Yuan, X.; Li, H. Co-pelletization of sewage sludge and biomass: Thermogravimetric analysis and ash deposits. *Fuel Process. Technol.* **2016**, *145*, 109–115. [CrossRef]
9. García, R.; González, M.P.; Rubiera, F. Co-pelletization of pine sawdust and refused derived fuel (RDF) to high-quality waste-derived pellets. *J. Clean. Prod.* **2021**, *328*, 129635. [CrossRef]
10. Gnoffo, C.; Frache, A. Identification of plastics in mixtures and blends through pyrolysis-gas chromatography/mass spectrometry. *Polymers* **2024**, *16*, 71. [CrossRef]
11. Matúš, M.; Križan, P.; Šooš, Ľ. The effect of papermaking sludge as an additive to biomass pellets on the final quality of the fuel. *Fuel* **2018**, *219*, 196–204. [CrossRef]
12. Yang, G.; Hou, H.; Lin, Z. Effects of waste liquid on the formation and properties of wheat slag pellet fuel. *Trans. Chin. Soc. Agric. Eng.* **2022**, *38*, 197–203.
13. Zhao, L.; Wang, Y.; Wang, Q.; Liu, S.; Ji, X. Lignin Nanoparticles Produced from Wheat Straw Black Liquor Using γ-Valerolactone. *Polymers* **2024**, *16*, 49. [CrossRef] [PubMed]
14. Liu, X.; Zhang, J.; Zhang, B. Study on the Composition and Evaporation Characteristics of Pulping Black Liquor. *Trans. China Pulp Pap.* **2018**, *33*, 35–39.
15. Jia, G. Combustion Characteristics and Kinetic Analysis of Biomass Pellet Fuel Using Thermogravimetric Analysis. *Clean Combust. Emiss. Veh. Power Syst.* **2021**, *9*, 868. [CrossRef]
16. Li, B.; Liu, G.; Bi, M.; Li, Z.; Han, B.; Shu, C. Self-ignition risk classification for coal dust layers of three coal types on a hot surface. *Energy* **2021**, *216*, 119197. [CrossRef]
17. Raza, M.; Abu-Jdayil, B.; Al-Marzouqi, A.H.; Inayat, A. Kinetic and thermodynamic analyses of date palm surface fibers pyrolysis using Coats-Redfern method. *Renew. Energy* **2022**, *183*, 67–77. [CrossRef]
18. Maksimuk, Y.; Antonava, Z.; Krouk, V. Prediction of higher heating value based on elemental composition for lignin and other fuels. *Fuel* **2020**, *263*, 116727. [CrossRef]
19. Priyabrata, P.; Sanjay, M.M.; Amit, A. Production and utilization of fuel pellets from biomass: A review. *Fuel Process. Technol.* **2018**, *181*, 215–232.
20. Xu, Y.; Li, J.; Liu, Y.; Wu, W. Construction and Curing Behavior of Underwater In Situ Repairing Coatings for Offshore Structures. *Polymers* **2024**, *16*, 306. [CrossRef] [PubMed]

21. Anthony, A.; Jonas, B.; Gunnar, H.; Stefan, F.; Magnus, S. A review of the mechanism of bonding in densified biomass pellets. *Renew. Sustain. Energy Rev.* **2021**, *148*, 111249.
22. Rossi, L.; Wechsler, L.; Peltzer, M.A.; Ciannamea, E.M.; Ruseckaite, R.A.; Stefani, P.M. Sustainable Particleboards Based on Brewer's Spent Grains. *Polymers* **2024**, *16*, 59. [CrossRef] [PubMed]
23. Sadeq, A.; Heinrich, D.; Pietsch-Braune, S.; Heinrich, S. Influence of oscillating water content on the structure of biomass pellets. *Powder Technol.* **2023**, *426*, 118631. [CrossRef]
24. Akkenzheyeva, A.; Haritonovs, V.; Bussurmanova, A.; Merijs-Meri, R.; Imanbayev, Y.; Riekstins, A.; Serikbayeva, A.; Sydykov, S.; Aimova, M.; Mustapayeva, G. Study of the Viscoelastic and Rheological Properties of Rubber-Bitumen Binders Obtained from Rubber Waste. *Polymers* **2024**, *16*, 114. [CrossRef] [PubMed]
25. Wang, J.; Shen, B.; Kang, D.; Yuan, P.; Wu, C. Investigate the interactions between biomass components during pyrolysis using in-situ DRIFTS and TGA. *Chem. Eng. Sci.* **2019**, *195*, 767–776. [CrossRef]
26. Guo, F.; He, Y.; Hassanpour, A.; Gardy, J.; Zhong, Z. Thermogravimetric analysis on the co-combustion of biomass pellets with lignite and bituminous coal. *Energy* **2020**, *197*, 117147. [CrossRef]
27. Li, C.; Zou, Y.; Li, G.; Lu, Y.; Wang, W.; Duan, J. Hydrodynamic characteristics of pyrolyzing biomass particles in a multi-chamber fluidized bed. *Powder Technol.* **2023**, *421*, 118403. [CrossRef]
28. Sobek, S.; Tran, Q.K.; Junga, R.; Werle, S. Hydrothermal carbonization of the waste straw: A study of the biomass transient heating behavior and solid products combustion kinetics. *Fuel* **2022**, *314*, 122725. [CrossRef]
29. Ozgen, S.; Cernuschi, S.; Caserini, S. An overview of nitrogen oxides emissions from biomass combustion for domestic heat production. *Renew. Sustain. Energy Rev.* **2021**, *135*, 110113. [CrossRef]
30. Yang, W.; Pudasainee, D.; Gupta, R.; Li, W.; Wang, B.; Sun, L. An overview of inorganic particulate matter emission from coal/biomass/MSW combustion: Sampling and measurement, formation, distribution, inorganic composition and influencing factors. *Fuel Process. Technol.* **2021**, *213*, 106657. [CrossRef]
31. Kleinhans, U.; Wieland, C.; Frandsen, J.F.; Spliethoff, H. Ash formation and deposition in coal and biomass fired combustion systems: Progress and challenges in the field of ash particle sticking and rebound behavior. *Prog. Energy Combust. Sci.* **2018**, *68*, 65–168. [CrossRef]
32. Téllez-Salazar, W.N.; Ovalle-Encinia, O.; Ramírez-Rosales, D.; Ma, X.; Dorantes-Rosales, H.J.; Lara-García, H.A.; Ortiz-Landeros, J. Chemical synthesis and evaluation of $Co_3O_4/Ce_{0.9}Zr_{0.05}Y_{0.05}O_{2-\delta}$ mixed oxides for the catalytic-assisted combustion of soot. *Chem. Eng. Sci.* **2021**, *234*, 116443. [CrossRef]
33. Sheth, M.M.; Sigdel, S.; Harichandan, A.B.; Bhoraniya, R. Performance of fuel reactor in Chemical Looping Combustion system with various metal oxide particle size and operating temperature. *Int. J. Thermofluids* **2023**, *17*, 100295. [CrossRef]

MDPI AG
Grosspeteranlage 5
4052 Basel
Switzerland
Tel.: +41 61 683 77 34

*Polymers* Editorial Office
E-mail: polymers@mdpi.com
www.mdpi.com/journal/polymers

www.ingramcontent.com/pod-product-compliance
Lightning Source LLC
Chambersburg PA
CBHW040825250926
51850CB00007BA/1263
* 9 7 8 3 7 2 5 8 7 3 9 2 0 *